Cooling Techniques for Electronic Equipment

Cooling Techniques for Electronic Equipment

DAVE S. STEINBERG

Manager, Mechanical Engineering
Design Analysis Section
Litton Guidance and Control Systems
Woodland Hills, California

A WILEY-INTERSCIENCE PUBLICATION

JOHN WILEY & SONS, New York • Chichester • Brisbane • Toronto • Singapore

Copyright © 1980 by John Wiley & Sons, Inc.

All rights reserved. Published simultaneously in Canada.

Reproduction or translation of any part of this work beyond that permitted by Sections 107 or 108 of the 1976 United States Copyright Act without the permission of the copyright owner is unlawful. Requests for permission or further information should be addressed to the Permissions Department, John Wiley & Sons, Inc.

Library of Congress Cataloging in Publication Data:

Steinberg, Dave S   1923-
   Cooling techniques for electronic equipment.

  "A Wiley-Interscience publication."
  Includes index.
  1. Electronic apparatus and appliances—Cooling.

I. Title.
TK7870.25.S73   621.381   80-14141
ISBN 0-471-04403-2

Printed in the United States of America

10 9 8 7 6

To My Wife Annette
And To My Two Daughters Cori and Stacie

Preface

Electronic equipment is slowly reaching into almost every phase of modern living, from sewing machines and washing machines to mass transportation and atomic energy control systems. The reliability of these systems is of great importance to our comfort and safety. If a transistor fails in a television set, it may only cause a minor inconvenience. However, if an electronic control system should malfunction because it has overheated, it may result in substantial property damage and possible injury.

Electronic systems are rapidly shrinking in size, while their complexity and capability continue to grow at an amazing rate. As the power has been increasing, the volume has been decreasing, resulting in a dramatic increase in the heat density. As a result, the temperatures in many electronic systems have been rising rapidly, producing a large increase in the number of failures.

High failure rates in electronic boxes may also be caused by high thermal stresses in the solder joints of electronic components mounted on circuit boards. This is usually due to a high thermal expansion coefficient with insufficient strain relief in the component lead wire. This area is discussed in detail, together with recommendations for mounting components to prevent this type of failure.

The purpose of this book is to show designers and engineers quick methods for designing electronic hardware to withstand severe thermal environments without failing. Techniques are presented that will permit the development of many different types of reliable electronic systems *without* the aid of the powerful new high speed digital computers.

This book was developed from a series of seminars, lectures, and short courses for the cooling of electronic equipment, which I have presented at the University of Wisconsin–Extension every year since 1975. The book was influenced by my industrial experience in the mechanical design, packaging, and testing of many different types of sophisticated electronic components and systems during the past 25 years.

Mathematical modeling techniques using analog resistor networks are also included. For those who wish to use high speed digital computers to solve thermal problems, these techniques can be used to break up a complex system into many individual thermal resistors and nodes.

In an attempt to keep the book as simple as possible, the derivations of equations were minimized. The emphasis is always on the real electronic hardware that is used in today's sophisticated electronic systems. Many sample problems are presented, to demonstrate practical and cost effective methods for designing efficient, reliable cooling systems.

Although the metric system for weights and measures is widely used in Europe, at the present time it is used by only a few American companies. However, there appears to be a strong movement toward conversion from the present English system to the metric system. This book was therefore written with dual units, English and metric, to permit electronic equipment designers and engineers to work with either set of units effectively.

I thank Mr. Joel Newberger for his contributions in the section on induced draft cooling, and Mr. Joel Sloan for proofreading several sections of the book.

DAVE S. STEINBERG

Westlake Village, California
August 1980

Contents

Preface   vii

Symbols   xvii

1   Evaluating the Cooling Requirements   1

- **1.1**   Heat Sources,   1
- **1.2**   Heat Transmission,   2
- **1.3**   Steady State Heat Transfer,   4
- **1.4**   Transient Heat Transfer,   5
- **1.5**   Electronic Equipment for Airplanes, Missiles, Satellites, and Spacecraft,   6
- **1.6**   Electronic Equipment for Ships and Submarines,   8
- **1.7**   Electronic Equipment for Communication Systems and Ground Support Systems,   9
- **1.8**   Minicomputers, Microcomputers, and Microprocessors,   10
- **1.9**   Cooling Specifications for Electronics,   11
- **1.10**   Specifying the Power Dissipation,   12
- **1.11**   Dimensional Units and Conversion Factors,   14

2   Designing the Electronic Chassis   21

- **2.1**   Formed Sheet Metal Electronic Assemblies,   21
- **2.2**   Dip Brazed Boxes with Integral Cold Plates,   22
- **2.3**   Plaster Mold and Investment Castings with Cooling Fins,   24
- **2.4**   Die Cast Housings,   25

- 2.5 Large Sand Castings, 26
- 2.6 Extruded Sections for Large Cabinets, 26
- 2.7 Humidity Considerations in Electronic Boxes, 26
- 2.8 Conformal Coatings, 28
- 2.9 Sealed Electronic Boxes, 29
- 2.10 Standard Electronic Box Sizes, 33

3 Conduction Cooling for Chassis and Circuit Boards 35

- 3.1 Concentrated Heat Sources, Steady State Conduction, 35
- 3.2 Mounting Electronic Components on Brackets, 36
- 3.3 Sample Problem—Transistor Mounted on a Bracket, 39
- 3.4 Uniformly Distributed Heat Sources, Steady State Conduction, 41
- 3.5 Sample Problem—Cooling Integrated Circuits on a PCB, 44
- 3.6 Circuit Board with an Aluminum Heat Sink Core, 46
- 3.7 Sample Problem—Temperature Rise along a PCB Heat Sink Plate, 46
- 3.8 How to Avoid Warping on PCBs with Metal Heat Sinks, 47
- 3.9 Chassis with Nonuniform Wall Sections, 48
- 3.10 Sample Problem—Heat Flow along Nonuniform Bulkhead, 50
- 3.11 Two Dimensional Analog Resistor Networks, 53
- 3.12 Sample Problem—Two Dimensional Conduction on a Power Supply Heat Sink, 54
- 3.13 Heat Conduction across Interfaces in Air, 60
- 3.14 Sample Problem—Temperature Rise across a Bolted Interface, 64
- 3.15 Sample Problem—Temperature Rise across a Small Air Gap, 65
- 3.16 Heat Conduction across Interfaces at High Altitudes, 66
- 3.17 Outgassing at High Altitudes, 69
- 3.18 Circuit Board Edge Guides, 70

Contents xi

    3.19   Sample Problem—Temperature Rise across a PCB Edge Guide, 72

    3.20   Heat Conduction through Sheet Metal Covers, 72

    3.21   Radial Heat Flow, 73

    3.22   Sample Problem—Temperature Rise through a Cylindrical Shell, 74

4 Mounting and Cooling Techniques for Electronic Components   77

    4.1   Various Types of Components, 77

    4.2   Mounting Components on PCBs, 78

    4.3   Sample Problem—Hot Spot Temperature of an Integrated Circuit on a Plug-in PCB, 81

    4.4   How to Mount High Power Components, 87

    4.5   Sample Problem—Mounting High Power Transistors on a Heat Sink Plate, 89

    4.6   Electrically Isolating High Power Components, 91

    4.7   Sample Problem—Mounting a Transistor on a Heat Sink Bracket, 92

    4.8   Potted Modules, 94

    4.9   Sample Problem—Temperature Rise in a Potted Module, 95

    4.10   Component Lead Wire Strain Relief, 98

5 Practical Guides for Natural Convection and Radiation Cooling   106

    5.1   How Natural Convection Is Developed, 106

    5.2   Natural Convection for Flat Vertical Plates, 109

    5.3   Natural Convection for Flat Horizontal Plates, 109

    5.4   Heat Transferred by Natural Convection, 110

    5.5   Sample Problem—Vertical Plate Natural Convection, 111

    5.6   Turbulent Flow with Natural Convection, 113

    5.7   Sample Problem—Heat Lost from an Electronic Box, 114

    5.8   Finned Surfaces for Natural Convection Cooling, 117

    5.9   Sample Problem—Cooling Fins on an Electronic Box, 119

    5.10   Natural Convection Analog Resistor Networks, 121

    5.11   Natural Convection Cooling for PCBs, 123

- 5.12 Natural Convection Coefficient for Enclosed Air Space, 125
- 5.13 Sample Problem—PCB Adjacent to a Chassis Wall, 126
- 5.14 High Altitude Effects on Natural Convection, 129
- 5.15 Sample Problem—PCB Cooling at High Altitudes, 130
- 5.16 Radiation Cooling of Electronics, 132
- 5.17 Radiation View Factor, 136
- 5.18 Sample Problem—Radiation Heat Transfer from a Hybrid, 142
- 5.19 Sample Problem—Junction Temperature of a Dual FET Switch, 145
- 5.20 Radiation Heat Transfer in Space, 147
- 5.21 Effects of α/e on Temperatures in Space, 148
- 5.22 Sample Problem—Temperatures of an Electronic Box in Space, 150
- 5.23 Simplified Radiation Heat Transfer Equation, 151
- 5.24 Sample Problem—Radiation Heat Loss from an Electronic Box, 152
- 5.25 Combining Convection and Radiation Heat Transfer, 154
- 5.26 Sample Problem—Electronic Box in an Airplane Cockpit Area, 155
- 5.27 Equivalent Ambient Temperature for Reliability Predictions, 157
- 5.28 Sample Problem—Equivalent Ambient Temperature of an RC07 Resistor, 159
- 5.29 Increase in Effective Emittance on Extended Surfaces, 160

6 Forced Air Cooling for Electronics 164

- 6.1 Forced Cooling Methods, 164
- 6.2 Cooling Air Flow Direction for Fans, 165
- 6.3 Static Pressure and Velocity Pressure, 166
- 6.4 Losses Expressed in Terms of Velocity Heads, 170
- 6.5 Sample Problem—Air Flow Loss at a Fan Entrance, 171
- 6.6 Establishing the Flow Impedance Curve for an Electronic Box, 172

Contents   xiii

- 6.7 Sample Problem—Fan Cooled Electronic Box, 173
- 6.8 Hollow Core PCBs, 189
- 6.9 Cooling Air Fans for Electronic Equipment, 192
- 6.10 Air Filters, 194
- 6.11 Cutoff Switches, 195
- 6.12 Static Pressure Loss Tables and Charts, 195
- 6.13 High Altitude Conditions, 196
- 6.14 Sample Problem—Fan Cooled Box at 30,000 Feet, 199
- 6.15 Other Convection Coefficients, 203
- 6.16 Sample Problem—Cooling A TO-5 Transistor, 205
- 6.17 Conditioned Cooling Air from an External Source, 207
- 6.18 Sample Problem—Generating a Cooling Air Flow Curve, 208
- 6.19 Static Pressure Losses for Various Altitude Conditions, 209
- 6.20 Sample Problem—Static Pressure Drop at 65,000 Feet, 212
- 6.21 Total Pressure Drop for Various Altitude Conditions, 218
- 6.22 Sample Problem—Total Pressure Loss through an Electronic Box, 219
- 6.23 Finned Cold Plates and Heat Exchangers, 219
- 6.24 Pressure Losses in Multiple Fin Heat Exchangers, 221
- 6.25 Fin Efficiency Factor, 223
- 6.26 Sample Problem—Hollow Core PCB with a Finned Heat Exchanger, 225
- 6.27 Undesirable Air Flow Reversals, 239

7 Cooling Minicomputers, Microcomputers, and Microprocessors   242

- 7.1 Introduction, 242
- 7.2 Minicomputer Systems, 242
- 7.3 Sample Problem—PCB Mounted above a Minicomputer Floppy Disk, 244
- 7.4 Sample Problem—Fan Cooled Minicomputer, 249
- 7.5 Microcomputer Systems, 252
- 7.6 Sample Problem—Cooling a Microcomputer, 253

- 7.7 Microprocessor Development, 255
- 7.8 Mounting Microprocessors to Resist Vibration Fatigue, 256
- 7.9 Microprocessors in Severe Thermal Environments, 257
- 7.10 Sample Problem—Cooling a Microprocessor Mounted on a PCB, 257

8 Effective Cooling for Large Racks and Cabinets — 260

- 8.1 Induced Draft Cooling for Large Consoles, 260
- 8.2 Air Flow Losses for Large Cabinets, 261
- 8.3 Flotation Pressure and Pressure Loss, 262
- 8.4 Sample Problem—Induced Draft Cooling for a Large Cabinet, 262
- 8.5 Natural Cooling for Large Cabinets with Many Flow Restrictions, 267
- 8.6 Sample Problem—Temperature Rise of Cooling Air in a Cabinet with an Induced Draft, 268
- 8.7 Warning Note for Induced Draft Systems, 272
- 8.8 Tall Cabinets with Stacked Card Buckets, 273
- 8.9 Sample Problem—Induced Draft Cooling of a Console with Seven Stacked Card Buckets, 274
- 8.10 Electronics Packaged within Sealed Enclosures, 278
- 8.11 Small Enclosed Modules within Large Consoles, 281
- 8.12 Sample Problem—Small PCB Sealed within an RFI Enclosure, 283
- 8.13 Test Data for Small Enclosed Modules, 290
- 8.14 Pressure Losses in Series and Parallel Air Flow Ducts, 293
- 8.15 Sample Problem—Series and Parallel Air Flow Network, 294

9 Transient Cooling for Electronic Systems — 300

- 9.1 Simple Insulated Systems, 300
- 9.2 Sample Problem—Transient Temperature Rise of a Transformer, 301
- 9.3 Thermal Capacitance, 302

Contents

- **9.4** Time Constant, 303
- **9.5** Heating Cycle Transient Temperature Rise, 304
- **9.6** Sample Problem—Transistor on a Heat Sink, 304
- **9.7** Temperature Rise for Different Time Constants, 308
- **9.8** Sample Problem—Time for Transistor to Reach 95% of its Stabilized Temperature, 310
- **9.9** Cooling Cycle Transient Temperature Change, 310
- **9.10** Sample Problem—Transistor and Heat Sink Cooling, 310
- **9.11** Transient Analysis for Temperature Cycling Tests, 312
- **9.12** Sample Problem—Electronic Chassis in a Temperature Cycling Test, 317
- **9.13** Sample Problem—Methods for Decreasing Hot Spot Temperatures, 322
- **9.14** Sample Problem—Transient Analysis of an Amplifier on a PCB, 324

10 Special Applications for Tough Cooling Jobs   331

- **10.1** New Technology—Approach with Caution, 331
- **10.2** Heat Pipes, 331
- **10.3** Degraded Performance in Heat Pipes, 333
- **10.4** Typical Heat Pipe Performance, 334
- **10.5** Heat Pipe Applications, 336
- **10.6** Direct and Indirect Liquid Cooling, 340
- **10.7** Forced Liquid Cooling Systems, 341
- **10.8** Pumps for Liquid Cooled Systems, 342
- **10.9** Storage and Expansion Tank, 343
- **10.10** Liquid Coolants, 343
- **10.11** Simple Liquid Cooling System, 344
- **10.12** Mounting Components for Indirect Liquid Cooling, 344
- **10.13** Basic Forced Liquid Flow Relations, 347
- **10.14** Sample Problem—Transistors on a Water Cooled Cold Plate, 351

References   361

Index   365

Symbols

A	Area, ft², in², cm²
a	Length, ft, in, cm
b	Length, ft, in, cm
b	Temperature intercept, °F, °C
Btu	British thermal units, heat rate
C	Thermal capacitance, Btu/°F, cal/°C
°C	Temperature, degrees Celsius
cal	Calorie, heat rate
cfm	Cubic feet per minute, flow rate
cm	Centimeter
C_p	Specific heat, Btu/lb °F, cal/g °C
D	Diameter, ft, in, cm
d	Length, ft, in, cm
DIP	Dual inline package, electronic component
e	Emissivity (dimensionless)
e	Natural logarithm, 2.71828
F	Flow
f	View factor or friction factor (both dimensionless)
°F	Temperature, degrees Fahrenheit
ft	Foot
G	Weight velocity flow, lb/hr ft², g/sec cm²
g	Acceleration of gravity, 32.2 ft/sec², 386 in/sec², 980 cm/sec²
g	Gram weight
H	Head of water (H_2O), in, cm
H	Height, ft, in, cm
h	Convection coefficient, Btu/hr ft² °F, cal/sec cm² °C
Hg	Mercury
hr	Hour

H_2O	Water	
IC	Integrated circuit	
in	Inch	
J	Colburn factor (dimensionless)	
K	Thermal conductivity, Btu/hr ft °F, cal/sec cm °C	
K	Absolute temperature, degrees Kelvin	
k'	Thermal conductance, Btu/hr °F, cal/sec °C	
kg	Kilogram	
L	Length, ft, in, cm	
lb	Pound	
mm	Millimeter	
N_R	Reynolds number (dimensionless)	
P	Pressure, lb/in², g/cm²	
P'	Number of atmospheres (dimensionless)	
PCB	Printed circuit board	
psi	Pounds per square inch	
Q	Power dissipation, watts, Btu/hr, cal/sec	
q	Heat flow per unit length, Btu/hr ft, cal/sec cm	
R	Thermal resistance, hr °F/Btu, sec °C/cal, °C/watt, °C in/watt	
R	Gas constant, 53.3 ft/°R, 2924 cm/°K	
R	Radius, ft, in, cm	
°R	Absolute temperature, degrees Rankine	
RFI	Radio frequency interference	
rms	Root mean square	
RTV	Room temperature vulcanizing	
S	Slope, °F/hr, °C/hr	
sec	Second	
T	Absolute temperature, °R, °K	
t	Temperature, °F, °C	
tanh	Hyperbolic tangent	
T_i	Time, hr, min, sec	
torr	Atmospheric pressure, mm Hg	
V	Velocity, ft/min, cm/sec	
VP	Velocity pressure, in H_2O, cm H_2O	
W	Weight, lb, g	
W	Weight flow, lb/hr, lb/min, g/sec	
W	Width, ft, in, cm	
X	Coordinate axis	
Z	Air flow impedance, in $H_2O/(lb/min)^2$	

Greek Symbols

α	Solar absorptivity, (dimensionless)
β	Bulk modulus, 1/°R, 1/°K
Δ	Difference or change, temperature or pressure
δ	Fin thickness, ft, in, cm
η	Fin efficiency factor, percent
θ	Component resistance to junction, °F/watt, °C/watt
μ	Viscosity, lb/ft hr, g/cm sec
ρ	Density, lb/ft^3, lb/in^3, g/cm^3
σ	Stefan-Boltzmann constant, Btu/hr ft^2 °R^4, cal/sec cm^2 °K^4
σ	Ratio, air density to standard density (dimensionless)
τ	Time constant, hr, min, sec

Subscripts

a	Ambient
AG	Air gap
ALT	Altitude
av	Average
c	Convection, component, combined
cr	Critical
Cu	Copper
e	End, equivalent, epoxy, environment
eff	Effective
H	Heat, hydraulic
hs	Hot spot
i	Interface, internal
j	Junction
jc	Junction to case
L	Loss
Lm	Log mean
min	Minimum
n	Net
o	Initial condition
r	Radiation
s	Surface, static, solar
s-a	Surface to ambient
SL	Sea level
ss	Steady state
std	Standard
t	Total
trans	Transistor
v	Velocity

Cooling Techniques for
Electronic Equipment

1
Evaluating the Cooling Requirements

1.1 HEAT SOURCES

Electronic equipment relies on the flow and control of electrical current to perform a fantastic variety of functions, in virtually every major industry throughout the world. Whenever electrical current flows through a resistive element, heat is generated in that element. An increase in the current or resistance produces an increase in the amount of heat that is generated in the element. The heat continues to be generated as long as the current continues to flow. As the heat builds up, the temperature of the resistive element starts to rise, unless the heat can find a flow path that carries it away from the element. If the heat flow path is poor, the temperature may continue to rise until the resistive element is destroyed and the current stops flowing. If the heat flow path is good, the temperature may rise until it stabilizes at a point where the heat flowing away from the element is equal to the heat generated by the electrical current flowing in the element.

Heat is generated by the flow of electrical current in electronic component parts such as resistors, diodes, integrated circuits (ICs), hybrids, transistors, microprocessors, relays, dual inline packages (DIPs), large scale integrated circuits (LSIs), and very large scale integrated circuits (VSIs).

Figure 1.1 shows an electronic chassis that has heat exchangers (cold plates) on the top and bottom surfaces which rely on conditioned cooling air for controlling temperatures on plug-in circuit boards.

Electronic components and electronic systems are rapidly shrinking in size while their complexity and capability continue to grow at an amazing rate. In addition, the power has been increasing while the volume has been decreasing. This has produced a dramatic increase in the power density, resulting in rapidly rising temperatures and a large increase in the number of failures.

The temperatures must be controlled on every component to ensure a reliable electronic system. If the operating temperatures become too high, electronic malfunctions may occur. Malfunctions may produce a simple out

Figure 1.1 The author discussing design details with two associates. (Courtesy Kearfott Division, The Singer Co.)

of tolerance condition for a minor temperature increase, or a catastrophic failure for a major temperature increase.

Heat always flows from the hot area to the cool area. Since the electronic components are usually the source of the heat, the electronic components will usually be the hottest spots in an electronic system. (During transient conditions and temperature cycling tests, the electronic components may not necessarily be the hottest points in the system.) The basic heat transfer problem in electronic systems is, therefore, the removal of internally generated heat by providing a good heat flow path from the heat sources to an ultimate sink, which is often the surrounding ambient air.

1.2 HEAT TRANSMISSION

There are three basic methods by which heat can be transferred: conduction, convection, and radiation. The laws relating to these methods of heat transmission are of primary importance in the design and operation of electronic equipment.

1.2 Heat Transmission

Conduction is the transfer of kinetic energy from one molecule to another. In an opaque solid it is the only method of heat transfer, where heat flows from the hot areas to the cooler areas of the solid. Heat conduction also occurs in gases and liquids, but the amount of heat transferred is usually smaller for the same geometry.

Convection is the transfer of heat by the mixing action of fluids. When the mixing is due entirely to temperature differences within the fluid, resulting in different densities, the action is known as natural convection. When the mixing is produced by mechanical means, such as fans and pumps, the action is known as forced convection.

Thermal radiation is the transfer of energy by electromagnetic waves that are produced by bodies because of their temperature. A hot body radiates energy in all directions. When this energy strikes another body, the part that is absorbed is transformed into heat.

Most electronic systems make use of all three basic methods of heat transfer to some extent, even though one method may dominate the design. For example, an electronic box cooled by forced convection might utilize a fan to draw air over electronic components mounted on printed circuit boards (PCBs), as shown in Figure 1.2.

The greatest amount of heat is picked up by forced convection as the cooling air passes over the individual electronic components that are mounted on the PCB. However, some of the heat from the electronic components is conducted directly to the PCB under the component body, and some of the heat is conducted to the back side of the PCB, through the components electrical lead wires, as shown in Figure 1.3. Since the cooling

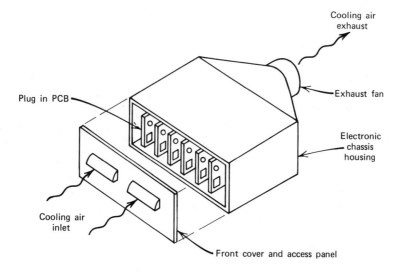

Figure 1.2 Electronic box cooled with an exhaust fan.

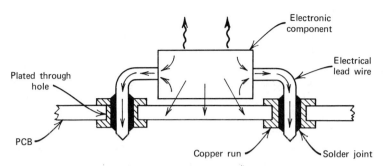

Figure 1.3 Heat conduction path from component, with heat flow through lead wires, to back side of PCB.

air passes over both surfaces of the PCB, the conduction of heat to the back side of the PCB provides additional surface area for improved cooling.

In addition, some of the heat is radiated from the hot components to the surrounding chassis walls and to the cooler spots on adjacent PCBs. This helps to reduce the component hot spot temperatures.

1.3 STEADY STATE HEAT TRANSFER

If an electronic system is turned on and left running for a very long period of time, and if the power requirements remain constant during that period, the temperatures of the electronic components and their mounting structures, such as PCBs, will usually become stable. Minor fluctuations in the line voltages, small changes in the physical properties of the individual components, and slight variations in the outside ambient conditions may have some small effects on the temperatures within the electronic system. For all practical purposes, however, the heat gained (or the power dissipated) by the electronic components is equal to the heat lost, so that the system has reached thermal equilibrium. The internal heat has found one or more thermal paths from the heat source to the ultimate heat sink. Usually, all three methods of heat transfer—conduction, convection, and radiation—are involved. When the thermal equilibrium condition has been reached, the rate of heat being transferred by each of the three methods remains constant. The temperature gradients are now fixed with the heat flowing from the hotter parts of the system to the cooler parts of the system, until the heat finally reaches the ultimate sink. These characteristics indicate that the system has reached the steady state heat transfer condition. Steady state conditions may develop in a matter of minutes for small components such as transistors and diodes. However, for large electronic consoles, it may take a full day of operation before steady state heat transfer conditions are reached.

1.4 TRANSIENT HEAT TRANSFER

When the rate of heat flow changes within an electronic system, it will normally produce a temperature change somewhere in that system. Also, when there is a temperature change within an electronic system, there will normally be a change in the heat flow rate somewhere in the system. Changes of these types are defined as transient heat transfer conditions, because the thermal equilibrium of the system is unbalanced. Transient heat transfer conditions develop, for example, when the power is first turned on in an electronic system. As the current flows through the electronic components, heat is generated and the temperatures within the components begin to rise, resulting in a transient, or changing condition.

Transient conditions will also occur in an electronic system when it is subjected to temperature cycling tests. Consider a system sitting in an environmental chamber where the ambient temperature is slowly being cycled between $-54°C$ ($-65°F$) and $+71°C$ ($+160°F$). In this case, the outside temperature often increases more rapidly than do the temperatures within the electronic box. Heat then flows from the outside of the box toward the interior, because heat always flows from the hot body to the cold body.

A satellite in orbit around the earth experiences transient heat transfer due to the constantly changing angle with respect to the sun and the earth. The intensity of the solar radiation may be constant, but the heat absorbed will vary along the surface because the angle the sun makes with respect to the surface is changing.

Sometimes it is necessary to use an auxiliary cooling device or technique for a short period, until the regular cooling system is available to take over the job. Consider the case where a missile is carried under the wing of an airplane. An auxiliary cooling cart is available to supply cooling to the electronic system within the missile while the airplane engine is being started and checked. The missile electronic system is normally cooled by the ram air during the captive flight phase and during the free flight phase after the missile is released from the airplane. No cooling air is provided for the missile electronics during the taxi and takeoff period, because of the extra weight and cost. Instead, the electronics must rely upon the thermal capacity or thermal inertia of the system to absorb the heat without developing excessive temperatures during this period. When a number of airplanes are lined up, waiting to take off, delays of 30 min may occur. This may cause the electronics to overheat. If the weight of the electronic system is increased, it will increase the thermal inertia and permit cooler operation for longer periods of time. For higher power systems, however, a very large mass may be required to keep the electronics cool for 30 min, so that a more sophisticated technique may be required.

Sometimes it is desirable to use the change of state from a solid to a liquid, or from a liquid to a gas, to absorb heat. A large amount of heat can be absorbed under these conditions. It is often possible to use hollow wall construction for the electronic chassis, which could be filled with wax that melts at a predetermined temperature. The change of state from a solid to a liquid may absorb enough heat to permit the electronics to survive a 30 min period with no cooling air. Once the airplane is flying, the ram air cools the wax, which returns to the solid state. If the missile is not fired, the melting wax permits the delaying cycle to be repeated over and over again.

1.5 ELECTRONIC EQUIPMENT FOR AIRPLANES, MISSILES, SATELLITES, AND SPACECRAFT

Electronic boxes used in airplanes, missiles, satellites, and spacecraft often have odd shapes that permit them to make maximum use of the volume available in odd-shaped structures. An odd-shaped box may require more time to design, because it is usually more difficult to provide the circuit cards with an efficient heat flow path, regardless of the cooling method used.

The trend in military and commercial airplanes and helicopters is toward a series of several standard sizes for plug-in types of electronic boxes that fit in racks. These are called ATR (air transport rack) boxes. They are of various widths, which are known as one quarter, one half, three quarters, and full width, each with a short and a long length. The electrical interface connectors are often at the rear of the box, with quick release fasteners at the front [1].*

Many of the electronic boxes are cooled by forced convection with bleed air from the jet engine compressor section. Since this air is at a high temperature and pressure, it is throttled (passed through the cooling turbine), cooled, and dried with a water separator before it is used. This air often enters the electronic box at the rear, adjacent to the electrical connectors. Rubber gaskets are used around the inlet ports at the air interface to provide an effective plug-in connection, which reduces the leakage at the cooling air interface.

Sometimes the conditioned cooling air is not completely dry because of excessive moisture in the air from humidity or a rainstorm. Small drops of water will often be carried into the electronics section together with the cooling air. If this water accumulates on PCBs or their plug-in connectors, electrical problems may develop. Therefore, many specifications do not permit external cooling air to come into direct contact with electronic components or circuits.

Air cooled heat exchangers, commonly called air cooled cold plates, which are being used more and more in airplanes, provide conditioned air for cooling the electronics. These heat exchangers are usually dip brazed

---

* *Numbers in brackets refer to references at the end of the book.*

1.5 Electronic Equipment for Airplanes, Missiles, Satellites, and Spacecraft

when many thin [0.006 to 0.008 in (0.15 to 0.20 mm)] aluminum plate fins are used. Pin fin aluminum castings are becoming very popular because of their low cost. There is usually a slight weight increase with pin fins because the walls and fins have to be thicker to permit the molten aluminum to flow [2, 3].

Electronic systems for missiles generally have two cooling conditions to consider, captive and free flight. If the missile flight duration is relatively short, the electronics can be precooled during the captive phase so that the system can function with no additional cooling during the flight phase. The electronic support structure would act as the heat sink, soaking up the heat as it is generated, to permit the electronic system to function during the free flight phase.

Some missiles, such as the Cruise missiles, have a very long free flight phase, so that the cooling system must be capable of cooling the electronics for several hours. If ram air is used at speeds near Mach 1, the ram temperature rise of the cooling air may exceed 100°F (55°C). Since Cruise missiles fly at low altitudes, where the surrounding ambient air temperatures can be as high as 100°F, the cooling air temperatures could reach values of 200°F (93°C) even before the cooling process begins. Since the maximum desirable component mounting surface temperature is about 212°F (100°C), the outside ambient air cannot be used directly for cooling.

Cruise missiles must carry a large supply of fuel for their long flights. The fuel is often pumped through liquid cooled cold plates to provide cooling for the electronics. Toward the end of the flight mission, when the fuel supply runs low, the temperatures may increase. At this point it may be necessary to use the thermal inertia in the electronics structure to keep the system cool enough to finish its flight.

Electronic systems for satellites and spacecraft generally rely upon radiation to deep space for all their cooling. Deep space has a temperature of absolute zero, −460°F or 0°Rankine (−273°C or 0°Kelvin). Temperatures this low can provide excellent cooling if the proper surface finishes are used [4].

Special surface finishes and treatments may be required for satellites and spacecraft to prevent them from absorbing large quantities of heat from the sun. This heat may be direct solar radiation plus solar radiation reflected from the various planets and their moons (reflected radiation is called albedo) [4, 5].

Liquid cooled cold plates are often used to support electronic systems. Pumps then circulate the cooling fluid from the cold plates, where the heat is picked up, to the space radiators, where the heat is dumped to space.

Conduction heat transfer is used extensively for cooling electronic equipment in space environments. In the hard vacuum conditions of outer space, flat and smooth surfaces must be utilized with high contact pressures, to minimize the temperature rise across each interface. Although air is not normally considered to be a good heat conductor, its presence will sharply reduce the contact resistance at most interfaces. Thermal greases are sometimes used to reduce the interface resistance in hard vacuum environments, by filling the small voids that would otherwise develop when the air is evacuated.

Air cooling can still be provided in a hard vacuum environment if a sealed and pressurized box is used. An internal fan can be used to circulate internal cooling air through a liquid cooled cold plate, which would carry away the heat. The heat from the fan must be added to the total heat load of the system. Also, a sealed box will have a large pressure differential across the surfaces of the box in the vacuum of outer space. The outer surfaces of the box must therefore be thick and stiff to resist excessive deformations and stresses caused by these pressures.

Natural convection cannot be used to transfer heat in satellites and spacecraft electronics. Natural convection requires a gravity field to permit the heated air to rise, because of its reduced density. In satellites and spacecraft, the effects of gravity are neutralized by the velocity and the continuous free fall characteristics of the flight path. Therefore, only radiation, conduction, and forced convection (in sealed boxes) should be considered for cooling electronic systems in space.

1.6 ELECTRONIC EQUIPMENT FOR SHIPS AND SUBMARINES

Large cabinets, consoles, or enclosures are normally used to support the electronic equipment used on ships and submarines. These cabinets are usually heavy and rugged, to provide protection for the electronics during storms and rough seas. Some cabinets may dissipate more than 2 kilowatts of heat, so liquid cooled cold plates and heat exchangers are often used to cool the equipment. The electronic components are often mounted on panels and sliding drawers. Panels are used to support displays, and drawers are used to support the heavy power supply units. Completely enclosed cabinets are often used by the Navy with special radio frequency interference (RFI) and electromagnetic interference (EMI) gaskets on heavy doors. This protects and shields the electronics from undesirable radiated and conducted electromagnetic waves, which can interfere with the operation of the equipment. Because these systems are completely enclosed, cooling problems may become quite severe [6].

Water is usually available on ships and submarines, so that it is natural to utilize it for cooling. Water cooled heat exchangers are often used, with external fins to permit cooling with forced air. Fans are used to force the air through the heat exchanger fins to cool the air, which is then circulated through the console. This type of forced convection cooling can be used with both closed loop and open loop systems. With a closed loop system, a cooling air supply plenum and a return plenum may be established within the side walls of the consoles. The side walls are often several inches deep, with ribs to provide rigidity from high shock loads, so that they can easily carry the cooling air to and from the electronics. If an open loop system is used for cooling, the air entering the console would be forced through the water cooled heat exchanger, which is usually at the base of the console. The conditioned air would be circulated through the electronics and then exhausted at the top of the console.

When the heat loads are not too high, natural convection techniques can often be used to cool the electronics. This works well on tall cabinets, which can use chimney effects to force the air through the system without the use of fans or pumps. Air enters at the bottom of this cabinet, where it first picks up heat from the electronics. The warmer air has a reduced density, so that it starts to rise through the chassis, picking up more heat as it rises. The cooling air finally exits at the top of the cabinet.

Large open spaces must be provided in a chassis that is cooled by natural convection. The flow resistance must be minimized to permit the cooling air to pass freely through the cabinet. This requires a low packaging density, with very few turns in the air flow path, to ensure the proper cooling air flow [7, 8].

Often designers package electronic equipment in a console and completely ignore the cooling air flow path. A large fan is then placed at the bottom of the unit to "blow air at the electronics," with the hope that somehow, somewhere, the air will find a path to the hot spots so that they will be adequately cooled. This brute force technique is usually justified on the grounds that it has been used many times by many people and it "seems to work." Needless to say, this is a poor policy, because there is no assurance that the blowing air will ever reach the spots that require cooling, and there is no way to determine what the cooling effects will be. It is best to plan ahead so that the air can be directed to the most critical electronic components for more efficient cooling. This will result in a smaller, lighter, and less expensive electronic system, with higher reliability.

1.7 ELECTRONIC EQUIPMENT FOR COMMUNICATION SYSTEMS AND GROUND SUPPORT SYSTEMS

Communication systems and ground support systems must both be capable of continuous operation for extended periods of time at high altitudes, in hot desert areas, in arctic areas, and in rain, sleet, and snow. Large systems have their electronics completely enclosed within shelters that resemble small barns and can hold several people for long periods of time. Small systems are often enclosed in transit cases, which may be carried on the back seat of a Jeep type of vehicle for rapid mobility. Small panel trucks are very popular for transporting moderate sized units.

Communication shelters are used all over the world by military personnel and by commercial television networks, to transmit and receive all forms of data. The shelters are usually lined from wall to wall with large electronic control consoles that can easily dissipate many kilowatts of heat.

Such shelters are insulated to protect them from both heat and cold. In hot climates the shelters are often equipped with exhaust fans to flush out hot air that has been trapped. This is important because the electronic consoles within the shelters use the local ambient air for cooling. This is done using forced convection or natural convection, depending upon the power dissipation and the location of the console. If the internal ambient temperatures within the shelter become too hot, refrigeration units may be provided, with auxiliary power units, to keep the shelters cool.

Ground support systems are generally used to perform functional checks of the electronic equipment for airplanes, missiles, ships, submarines, trains, trucks, and automobiles to ensure that they are operating properly. If malfunctions are discovered, replacements can be made to ensure successful completion of the mission or trip.

Ground support equipment is also used to supply auxiliary power or auxiliary cooling for electronic equipment that requires temporary boosters. These units could contain gasoline powered engines, with generators and regulated power supplies or large batteries as the power source.

Lightweight controls and displays are generally supported by light, dripproof, sheet metal enclosures. If heavy items such as large transformers and power supply components are required, they are mounted at the bottom of a rugged electronic enclosure, to reduce the dynamic loads that may be developed in a shock and vibration environment.

1.8 MINICOMPUTERS, MICROCOMPUTERS, AND MICROPROCESSORS

The computer industry has been changing very rapidly since the introduction of miniature semiconductor silicone chips. The first chips, which were introduced in about 1962, contained only about 15 to 20 diodes, transistors, and resistors on a substrate that measured about 0.150 by 0.200 in (0.381 by 0.508 cm). Component densities have since increased sharply, so that the same size chip can now incorporate over 30,000 components, and the costs have dropped just as fast. Large computers utilizing discrete diode, transistor, and resistor components were soon replaced with much smaller machines which had the same power at a much lower cost because they utilized the new semiconductor chip technology. These new, smaller machines became known as minicomputers because they were so small compared to the older and larger machines but had the same capability.

Minicomputers have found exciting new applications in many areas. Their size, flexibility, reduced costs, and improved memory storage have permitted more small businesses to make use of them. The power requirements of these machines have also dropped, so that exotic and expensive techniques are not required to cool the electronics.

Microcomputers are even smaller than minicomputers. A typical minicomputer will occupy one drawer of a filing cabinet. A typical microcomputer will fit on one plug-in PCB about 5 × 7 in (12.70 × 17.78 cm). Microcomputers are not as fast as minicomputers and are generally used where flexibility, size, and cost are more important than speed.

Microcomputers have been replacing systems that utilize custom-made large scale integrated circuits (LSIs) and hybrid circuits. Although these circuits have a high density and low power dissipation, they lack flexibility and cannot be modified or adapted for different applications.

Both the minicomputers and the microcomputers require some type of central processing unit (CPU) to control input and output, to perform mathematical operations, to decode, and to move information in and out of the memory. This CPU, which would normally require the mounting surface area of one plug-in PCB, can now be placed on a single small chip. This chip is called a microprocessor, and it is the most expensive part of the microcomputer.

Microprocessors are available in rectangular cases about 2.5 × 0.75 × 0.20 in (6.35 × 1.905 × 0.508 cm) with about 40 external wires, which perform all of the CPU functions. Microprocessors must be used with memory systems, which can also be expensive. Many memory systems use dual inline package (DIP) components, which are fabricated on silicon chips in much the same way that microprocessors are made. The semiconductor memories require less power than do core memories. However, core-type memories are able to operate at higher temperatures.

Floppy disk memories are rapidly becoming available with very large capacities. These thin disks are made of a soft or floppy material (hence the name) that is coded magnetically on one or both sides and enclosed in a square cardboard cover. These disks provide a large amount of memory at a very low price.

Bits are recorded on the surface of the rotating disk with a movable head. The surface of the disk is divided into sectors, shaped like a slice of pie, and tracks, shaped like concentric rings. The disk requires 7 to 15 watts of power for its operation. Many systems use two disks because the drive mechanism for two disks is only slightly more expensive than that for one disk.

Cassette tape memories are quite common in low cost systems. The cassette memory is the cheapest mass memory available. This type of memory can store large amounts of information, but the system is very slow and usually must be turned on and off manually.

1.9 COOLING SPECIFICATIONS FOR ELECTRONICS

A number of government and civil agencies issue documents that relate to the cooling of electronic equipment. A few of these are described below.

1. *MIL-E-16400 (Navy) Electronic, Interior Communication and Navigation Equipment, Naval Ship and Shore, General Specification*. This specification covers the general requirements applicable to the design and construction of electronic interior communication and navigation equipment intended for naval ship or shore applications. The specification defines the environmental conditions within which the equipment must operate satisfactorily and reliably. This includes the selection and application of materials and parts and means by which the assembled system will be tested to ensure that it is acceptable to the U.S. Navy.

2. *MIL-STD-202, Test Methods for Electronic and Electrical Component Parts*. This specification establishes uniform methods for testing electronic and electrical component parts. This includes basic environmental tests to determine resistance and capability of withstanding natural elements and conditions surrounding military operations. The specification includes items such as capacitors, resistors, switches, relays, transformers, and jacks. Only small parts are considered in this specification. The maximum weight of transformers and inductors is limited to 300 pounds or having a rms test voltage up to 50,000 volts.

3. *MIL-STD-810, Environmental Test Methods*. This document describes all test methods of a similar character that have appeared in the various joint or single service specifications. It also describes newly developed test methods that are feasible for use by more than one service.

4. *MIL-E-5400, Electronic Equipment, Airborne, General Specification*. This specification covers the general requirements for airborne electronic equipment for operation primarily in piloted aircraft. It includes various temperature and altitude conditions for different classes of equipment, with details on selecting materials, components, and fasteners.

5. *MIL-STD-781B, Reliability Tests: Exponential Distribution*. This standard outlines test levels and test plans for reliability qualification, production acceptance, and longevity tests. Included are temperature cycling tests for isolating weak components and poor designs.

6. MIL-HDBK-251 : THERMAL

7. MIL-HDBK-5 : STRUCTURAL

1.10 SPECIFYING THE POWER DISSIPATION

Power dissipations in electronic systems should be evaluated carefully, because the reliability and the mean time between failures (MTBF) can be sharply reduced by excessive component temperatures. It is better to be slightly conservative (slightly higher) in estimating power dissipations, to provide for future growth. Also, some compensation must be made for the normal tolerance variations in the component ratings and in the line voltages, so that a little safety factor should be added to the power dissipation in the system.

A safety factor will also provide a cushion to fall back upon when it is suddenly discovered that some critical components will not be available for

1.10 Specifying the Power Dissipation

production, or that substitute components will dissipate 10% more power. It is very expensive to change the cooling technique in a system after the design has been completed.

It is possible, of course, to be too conservative and to specify power dissipations that are far higher than would normally be expected. This condition will result in added expense, weight, and size. In some cases being too conservative may result in the use of a large fan when none is needed. A large and heavy liquid cooling system may be installed instead of a simple fan because someone was too conservative in specifying the power dissipation.

Some contracts may require a thermal analysis to be completed 3 months before the qualification test program. In order to prove the validity of the analysis, the contract may require the analysis to agree within $\pm 10\%$ of the actual temperatures achieved during the qualification test. This is a difficult situation to deal with, because the thermal design may be based upon a worst-case duty cycle or test condition, where high line voltages and component tolerance variations have been included in the total power dissipation. Higher power dissipations will result in higher component temperatures. During the actual qualification test, on the other hand, a completely different power dissipation may be used because a different duty cycle may be specified for the tests. This may result in poor agreement between the analysis and the tests [9].

It may be necessary to investigate many different sets of conditions to prove the thermal design of an electronic system. For example, a high altitude, high temperature environment may produce the highest temperatures in the logic section of the system rather than in the power supply section. Instead, a high altitude, low temperature environment may be the most severe for power supplies, because this condition draws the maximum heater power for the warm-up period. High current demands, even for short periods, may create hot spots in the electronic components that control the power [9].

Many electronic systems require special test sets to simulate test conditions similar to those that will be experienced in actual field operation. These test sets may be designed to exercise the electronic systems in many different ways, depending upon the manner in which the equipment designers interpret the test specifications. This can result in a test power dissipation that is completely different from the dissipation experienced in actual field operation. Therefore, it is important to make sure that the test equipment designers and the electronic system designers agree on the power requirements. This will reduce the chances of having a big difference in the design power dissipation and the test power dissipation.

The engineers responsible for determining the power dissipation in an electronic system should take the time required to determine these values accurately. The power dissipations may dictate the methods of heat removal, which, in turn, will determine the hot spot component temperatures.

It is worth the extra time required for a more accurate estimate of the power dissipation, because the reliability, size, weight, and cost of the system may depend upon it.

1.11 DIMENSIONAL UNITS AND CONVERSION FACTORS

The metric system for measuring is rapidly gaining acceptance in the United States, but the English system is still used as the standard by industry, by most lay persons, by many engineers. Both sets of dimensional units are therefore used in this book to permit engineers to use the system that is more convenient for them and to permit them to convert easily from one system to the other.

Power dissipation, or heat (Q), for electronic systems is expressed in three standard sets of units: watts, British thermal units per hour (Btu/hr), and calories per second (cal/sec). The watt is the standard unit of heat and power measure for electronic equipment in most countries throughout the world, including the United States.

The British thermal unit (Btu) is the quantity of heat required to raise the temperature of 1 pound mass of water 1 degree Fahrenheit (1°F) at standard pressure [10].

The calorie (cal) is the quantity of heat required to raise the temperature of 1 gram mass of water 1 degree Celsius (1°C) at standard pressure. This is sometimes written as gram-calorie (g-cal), which is called the small calorie [10].

A consistent set of units is always used in this book. When the unit of heat is the Btu, the length dimension will be feet (ft) and the time unit will be hours (hr). When the unit of heat is the calorie, the length dimension will be centimeters (cm) and the time unit will be seconds (sec).

Conversions can be made between the various units with the use of Table 1.1.

Table 1.1 Power/Heat Rate Conversions[a] Q

Multiply Number of → By ↘ to Obtain ↓	watts	Btu/hr	cal/sec
watts	1	0.293	4.187
Btu/hr	3.413	1	14.285
cal/sec	0.239	0.070	1

[a] 1 watt = 1 joule/sec.
  1 Btu = 1055 joules = 252 cal.

1.11 Dimensional Units and Conversion Factors

For example, to convert 10 Btu/hr to cal/sec using Table 1.1:

$$\left(10 \frac{\text{Btu}}{\text{hr}}\right)(0.070) = 0.70 \frac{\text{cal}}{\text{sec}} \tag{1.1}$$

Without the table, using dimension analysis,

$$\left(10 \frac{\text{Btu}}{\text{hr}}\right)\left(225 \frac{\text{cal}}{\text{Btu}}\right)\left(\frac{\text{hr}}{3600 \text{ sec}}\right) = 0.70 \frac{\text{cal}}{\text{sec}} \tag{1.2}$$

To convert 30 watts to Btu/hr using Table 1.1:

$$(30 \text{ watts})(3.413) = 102.39 \frac{\text{Btu}}{\text{hr}} \tag{1.3}$$

Thermal conductivity is expressed in English units and in metric units.

English units: $K = \dfrac{\text{Btu}}{\text{hr ft}^2 \, °\text{F/ft}} = \dfrac{\text{Btu ft}}{\text{hr ft}^2 \, °\text{F}} = \dfrac{\text{Btu}}{\text{hr ft} \, °\text{F}}$

Metric units: $K = \dfrac{\text{g cal}}{\text{sec cm}^2 \, °\text{C/cm}} = \dfrac{\text{cal cm}}{\text{sec cm}^2 \, °\text{C}} = \dfrac{\text{cal}}{\text{sec cm} \, °\text{C}}$

Conversions can be made between English units and metric units using Table 1.2.

Table 1.2 Thermal Conductivity Conversions K

Multiply Number of → By ↘ to Obtain ↓	$\dfrac{\text{Btu ft}}{\text{hr ft}^2 \, °\text{F}}$ or $\dfrac{\text{Btu}}{\text{hr ft} \, °\text{F}}$	$\dfrac{\text{cal cm}}{\text{sec cm}^2 \, °\text{C}}$ or $\dfrac{\text{cal}}{\text{sec cm} \, °\text{C}}$
$\dfrac{\text{Btu}}{\text{hr ft} \, °\text{F}}$	1	241.9
$\dfrac{\text{cal}}{\text{sec cm} \, °\text{C}}$	0.00413	1

For example, to convert 88 Btu/hr ft °F to cal/sec cm °C using Table 1.2:

$$\left(88 \frac{\text{Btu}}{\text{hr ft} \, °\text{F}}\right)(0.00413) = 0.363 \frac{\text{cal}}{\text{sec cm} \, °\text{C}} \tag{1.4}$$

The coefficient of heat transfer is expressed in English units and metric units in Table 1.3.

$$\text{English units:} \quad h = \frac{\text{Btu}}{\text{hr ft}^2 \, °\text{F}}$$

$$\text{Metric units:} \quad h = \frac{\text{cal}}{\text{sec cm}^2 \, °\text{C}}$$

Other conversion values are given in Tables 1.4 through 1.11.

Table 1.3   Coefficient of Heat Transfer Conversions h

Multiply Number of → By ↓ to Obtain ↓	$\dfrac{\text{Btu}}{\text{hr ft}^2 \, °\text{F}}$	$\dfrac{\text{cal}}{\text{sec cm}^2 \, °\text{C}}$
$\dfrac{\text{Btu}}{\text{hr ft}^2 \, °\text{F}}$	1	7373
$\dfrac{\text{cal}}{\text{sec cm}^2 \, °\text{C}}$	0.0001356	1

Table 1.4   Viscosity Conversions[a] μ

Multiply Number of → By ↓ to Obtain ↓	$\dfrac{\text{lb}}{\text{ft hr}}$	$\dfrac{\text{g}}{\text{cm sec}}$ (poise)	$\dfrac{0.01 \text{ g}}{\text{cm sec}}$ (centipoise)
$\dfrac{\text{lb}}{\text{ft hr}}$	1	242	2.42
$\dfrac{\text{g}}{\text{cm sec}}$ (poise)	0.00413	1	100
$\dfrac{0.01 \text{ g}}{\text{cm sec}}$ (centipoise)	0.413	0.01	1

[a] English units: $\mu = lb/ft \, hr$; metric units: $\mu = g/cm \, sec = poise$ (100 centipoise = 1 poise).

Table 1.5 Specific Heat Conversions C_p

Multiply Number of → By → to Obtain ↓	$\dfrac{\text{Btu}}{\text{lb °F}}$	$\dfrac{\text{cal}}{\text{g °C}}$
$\dfrac{\text{Btu}}{\text{lb °F}}$	1	1
$\dfrac{\text{cal}}{\text{g °C}}$	1	1

Table 1.6 Density Conversions ρ

Multiply Number of → By → to Obtain ↓	$\dfrac{\text{lb}}{\text{ft}^3}$	$\dfrac{\text{lb}}{\text{in}^3}$	$\dfrac{\text{g}}{\text{cm}^3}$
$\dfrac{\text{lb}}{\text{ft}^3}$	1	1728	62.43
$\dfrac{\text{lb}}{\text{in}^3}$	0.000579	1	0.036
$\dfrac{\text{g}}{\text{cm}^3}$	0.0160	27.68	1

Table 1.7 Pressure Conversions

Multiply Number of → By → to Obtain ↓	$\dfrac{lb}{in^2}$	in H$_2$O	$\dfrac{g}{cm^2}$	mm Hg
$\dfrac{lb}{in^2}$	1	0.0361	0.0142	0.0193
in H$_2$O	27.7	1	0.394	0.535
$\dfrac{g}{cm^2}$	70.37	2.538	1	1.361
mm Hg	51.71	1.869	0.735	1

Table 1.8 Weight Conversion

Multiply Number of → By → to Obtain ↓	lb	g	kg
lb	1	0.00220	2.205
g	453.6	1	1000
kg	0.4536	0.001	1

Table 1.9 Length Conversion

Multiply Number of → By → to Obtain ↓	in	ft	cm
in	1	12	0.3937
ft	0.0833	1	0.0328
cm	2.54	30.48	1

Table 1.10 Area Conversions

Multiply Number of → By → to Obtain ↓	in^2	ft^2	cm^2
in^2	1	144	0.155
ft^2	0.00694	1	0.00107
cm^2	6.452	929	1

Table 1.11 Volume Conversions

Multiply Number of → By ↓ to Obtain ↓	in³	ft³	cm³
in³	1	1728	0.0610
ft³	5.787×10^{-4}	1	3.531×10^{-5}
cm³	16.39	2.832×10^{4}	1

Figure 1.4 shows an investment cast aluminum chassis with the rear cover removed.

Figure 1.4 Investment cast electronic box, where the temperatures are controlled by conduction of heat to air cooled pin fin heat exchangers in the side walls. (Courtesy Litton Systems, Inc.)

2
Designing the Electronic Chassis

2.1 FORMED SHEET METAL ELECTRONIC ASSEMBLIES

Electronic systems normally consist of many different discrete electronic component parts, such as resistors, capacitors, diodes, transistors, hybrids, DIPs, LSIs, microprocessors, and transformers, which are enclosed within a support structure called the chassis. Power is dissipated by these different components, because the electronic system is not 100% efficient. Any power that is dissipated is rejected in the form of heat. This heat must be removed to prevent excessive temperatures from developing in these electronic components.

The purpose of the electronic chassis is to support the components while providing a low resistance thermal path to a heat sink, which will absorb this waste heat with a minimum rise in the temperature of the components. The heat sink may be the ambient air surrounding the chassis or a liquid cooled cold plate that is an integral part of the chassis wall.

Whatever method of heat transfer is selected, the technique should be as simple and as cost effective as possible. Many factors will have to be considered, such as the space available, the power requirements of the cooling system, the maximum allowable component temperatures, component sizes, power densities, and the heat sink temperature. Other factors, such as shock and vibration, may have to be considered together with the thermal environment to ensure an adequate chassis design [1, 11, 12].

Sheet metal structures are often used for many different types of electronic boxes because the manufacturing costs are so low. Thin gauge aluminum or steel sheets can be blanked and formed into a lightweight chassis with the use of rivets, spot welding, or arc welding. The final assembly is usually painted for protection and appearance. This type of construction can be used for a chassis 7 or 70 in high.

Light gauge steel sheet metal structures are usually not suitable for cooling high power electronic systems by means of conduction. The cross section areas are small because the metal is thin and the thermal conductance is low. This increases the thermal resistance, which also increases the component hot spot temperatures. Also, light gauge sheet metal structures are not generally capable of withstanding high vibration and shock levels, so that their use in these environments is very limited.

Lightweight sheet metal structures may tend to amplify any acoustic noise generated by cooling fans or pumps. If large, thin flat panels are used on an electronic box that is to be fan cooled, make sure that it is not too close to workers, who may object to the acoustic noise that is generated.

Structural epoxies are being used very successfully for assembling small electronic boxes. These epoxies have a high peel and shear strength, with a short cure time at 212°F (100°C), which makes them very cost effective. Lap joints with large surface areas must be used to obtain the necessary strength at the bonded interfaces.

Maintainability has recently become a big factor in the design of electronic equipment. Quick access is usually required to fault-isolate an electrical malfunction and to provide some means for a quick repair. This has resulted in the extensive use of plug-in types of circuit boards which can be removed and replaced very easily. These circuit boards are generally used to support small electronic components that do not dissipate a great deal of power. Large, heavy, or high power dissipating components are often mounted on the chassis walls with special support brackets.

A typical chassis would have two sheet metal dust covers on opposite sides of the chassis. One cover would provide access to the plug-in PCBs; the cover on the opposite side would provide access to the wiring harness or the master interconnecting board.

2.2 DIP BRAZED BOXES WITH INTEGRAL COLD PLATES

Dip brazed aluminum boxes are convenient to use for small, lightweight systems when the quantities are small, usually less than about eight or ten boxes. If the quantities are much larger, it is generally more cost effective to use aluminum investment castings or plaster mold castings, even though they require special tooling and pattern costs. Most electronic boxes will require shelves and brackets for mounting electronic components, ribs for stiffening the chassis to resist vibration, and cutouts for the cables and electrical harness. A dip brazed electronic chassis can often provide these features at a relatively low cost.

The size of the dip brazed assembly is usually limited by the size of the dip brazed tank, or salt bath, which is used to completely submerge the structure that is to be brazed. The individual parts of the chassis are joined together like a three dimensional jigsaw puzzle. Sometimes, a stainless steel fixture is used to hold the individual parts together during the dip brazing process. Sometimes the parts are tack welded, riveted, or screwed together with aluminum screws. An aluminum slurry or a brazing strip, with a melting

2.2 Dip Brazed Boxes with Integral Cold Plates

temperature slightly lower than the temperature of the salt bath, is used to join the individual structural members, which have a melting temperature slightly higher than that of the salt bath. As the slurry melts, capillary action draws the molten aluminum into the small voids between the individual parts and joins them rigidly.

Sometimes a structure may be too big to dip braze in one piece; in this case two separate dip brazed subassemblies are made. These can be joined later with screws and structural epoxy.

Thin plate fin types of heat exchangers are often used for the side walls of a chassis. The heat exchangers (or cold plates) are often dip brazed first as a subassembly and then cemented together to form a chassis. Sometimes the heat exchangers can be dip brazed as an integral part of the chassis without first forming a subassembly. The multiple fins provide a large surface area, which sharply increases the amount of heat that can be removed from the chassis with air or with liquids. A typical cross section through a chassis with finned side wall heat exchangers is shown in Fig. 2.1.

The plate fin material is usually about 0.006 in (0.0152 cm) or 0.008 in (0.0203 cm) thick so that many fins can be spaced close together. The typical spacing is about 14 to 18 fins per inch. Some companies can dip braze as many as 23 to 25 fins per inch.

When the fins get very close together, it becomes very difficult to clean out the salts left from the dip brazed tank, so that they may become trapped. This can block the fin passages and reduce the cooling effectiveness. Also, trapped salts can corrode the metal and weaken the structure.

It is desirable to have some method for checking the dip brazed heat exchangers to make sure that the fins are not blocked. A visual check is very valuable. If it is possible to look down the heat exchanger, blocked fins can be spotted and either rejected or cleaned. If a visual check is not possible, a pressure drop check should be made with a known flow passing

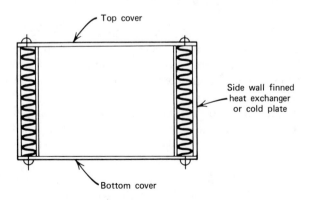

Figure 2.1 Chassis with side wall heat exchanger.

through the fins. If a blockage occurs, the pressure drop across the fins will be very high.

The fin material is generally very soft, so that it can be deformed easily. If fins in an air cooled cold plate are allowed to extend to the end of the chassis opening, they can become bent or deformed by foreign objects, such as bolts, nuts, screwdrivers, and even pencils. Therefore, for added protection, the ends of the fins should be recessed at least $\frac{1}{4}$ in.

2.3 PLASTER MOLD AND INVESTMENT CASTINGS WITH COOLING FINS

Plaster mold castings and investment castings are very popular for small electronic support structures or small electronic enclosures. Plaster mold castings and investment castings use plaster for the mold, into which a molten metal is poured. Typical metals are usually aluminum, magnesium, zinc, bronzes, and some steels. Wall thicknesses of 0.040 to 0.060 in (0.102 to 0.152 cm) can be readily obtained with both methods.

Investment piece part castings are generally slightly more expensive than plaster mold castings because an intermediate wax core is required. This core is made up in the exact detail required by the finished product. It is coated with several thin coats of plaster or refractory material and thoroughly dried. The coated core is heated to melt the wax, which is then drained, leaving the hollow mold. The mold is filled with molten metal while a vacuum is applied to remove tiny air pockets from the porous plaster or refractory outer shell. This permits the molten metal to fill in every small corner in the mold for excellent detail, surface finish, and accuracy. The outer shell must be broken away and destroyed to obtain the finished product.

Investment castings are somewhat limited in their size, depending upon the complexity. Intricate chassis castings 15 × 8 × 10 in (38.10 × 20.32 × 25.40 cm), with wall thicknesses of 0.070 in (0.178 cm), can be readily obtained.

Plaster mold casted piece parts are normally less expensive than investment castings because the process does not require the use of an intermediate disposable wax pattern. However, tooling costs for the plaster mold casting will be higher if permanent molds are used, because of the extra machining required.

Very large plaster mold castings can be made, up to 100 in (254 cm) in length if required, with considerable detail. However, plaster castings usually cannot produce small details as well as can investment castings.

Recent advances in casting technology have made it possible to produce thin wall aluminum castings that are pressure-tight for large molecule vapors such as nitrogen, or exotic liquids. The process, called the Antioch process, produces progressive freezing from the distant points in the casting toward the risers. The process uses a plaster mold that has the permeability of a sand mold. Impregnation with an epoxy resin is not required to obtain an airtight box. This is important for applications that require electroplated nickel, electroless nickel, or some other superior plating process for superior corrosion resistance.

Investment castings and plaster castings methods can be used to make very efficient, lightweight heat exchangers and cold plates with integrally cast pin fins. Typical pin fins are about 0.062 in (0.157 cm) in diameter, 0.50 in (1.27 cm) long, and spaced on 0.200 in (0.508 cm) centers. Plate fin types of heat exchangers and cold plates are much more difficult to cast. Continuous plate fins require cores that must be supported as the molten metal is poured. These cores can shift and crack if the ribs are very long. If plate fins are desired, it might be better to cast them in short lengths instead of in a continuous length, to provide a means for supporting the cores.

Cemented construction techniques can often be combined with castings to provide a plate fin heat exchanger or cold plate. Extruded plate fin sections can be cemented to cast plate fin sections to provide a plate fin heat exchanger, as shown in Figure 2.2.

2.4 DIE CAST HOUSINGS

The die casting process is capable of providing the lowest cost piece parts, with high quality and excellent appearance. However, tooling costs are very high, tools take a long time to fabricate, and modifications or design changes are very expensive. Die castings should therefore not be considered for production runs of less than about 1000 piece parts. This requires long range planning, scheduling, and coordinating to ensure a satisfactory product.

For large production runs it may even be possible to use investment castings or plaster mold castings as a buffer until the die casting tools have been fabricated, installed, and proven out.

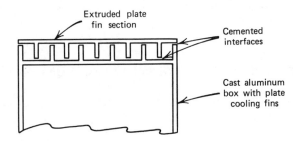

Figure 2.2 Plate fin extrusion cemented to a cast plate fin box to form a multiple fin heat exchanger.

2.5 LARGE SAND CASTINGS

Large electronic enclosures are often required for systems that must operate in severe vibration and shock environments. Rigid structures are required to withstand high acceleration levels, without fatigue failures, while providing an adequate support for electronic components. If weight is not too critical, fabrication costs can often be substantially reduced with the use of large aluminum sand castings. This type of cabinet is often suitable for ships and submarines, which must be capable of operating in rough seas during storms.

Mounting surfaces for the cabinet and for electronic subassemblies and components may have to be machined, because of poor casting tolerance control. However, with a little planning, the amount of machining can often be reduced to provide a cost effective structure.

2.6 EXTRUDED SECTIONS FOR LARGE CABINETS

Weight is often a problem with large cabinets that must be designed to withstand the Navy shock (MIL-S-901) and vibration (MIL-STD-167) requirements. Thick cast walls can provide the required rigidity, but with a high weight penalty. Under these circumstances extruded sections, with ribs or hollow cores, are capable of providing a rigid but relatively lightweight electronic enclosure. This type of structure can be welded, bolted, or riveted together to form a very rugged console. The hollow core type of extrusion is very convenient for ducting cooling air to various parts of the cabinet with fans or blowers. Large hollow core cross sections can carry large quantities of cooling air with a small pressure drop. Openings can be placed at various points in the extruded wall sections to direct the cooling air to hot spot areas. These openings can be blocked or reduced in size with plugs if the power distribution is changed at a later date.

Extruded side wall sections are convenient to use with closed forced air cooling systems, where a water cooled heat exchanger or a refrigeration unit cools the recirculated air. One vertical side wall can be used as the supply plenum, and the opposite side wall can be used as the return plenum, as shown in Figure 2.3.

2.7 HUMIDITY CONSIDERATIONS IN ELECTRONIC BOXES

Electronic equipment must often be capable of operating in very humid environments, where condensation will produce large amounts of water over a long period of time [13].

When high humidity environments are encountered, equipment designers must make a choice. They can seal the box against moisture or let the box

2.7 Humidity Considerations In Electronic Boxes

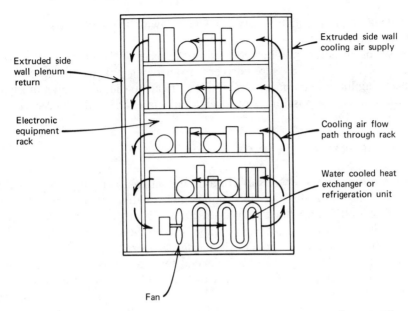

Figure 2.3 Closed cooling air recirculating system for electronic equipment rack.

breathe. Past experience with moisture problems shows that it is better to let the box breathe, because it is extremely difficult to seal an electronic box against moisture [14].

A sealed box is larger and heavier than a nonsealed box, especially if the system is required to operate at high altitudes. The walls of a sealed box must be capable of withstanding high forces due to high pressure differentials at high altitudes. The increased stiffness requirements will result in a weight increase and a size increase.

Humidity can cause serious problems in the electronic equipment when the internal circulating air is cooler than the outside ambient air. Moisture can condense on the electronic components, connectors, and circuit boards, producing short circuits or radical changes in the resistance between electronic components.

A moisture drainage path should be provided that permits the condensation to drain from the console walls, circuit boards, connectors, and wire harness to a drip pan at the bottom of the unit, where the moisture can evaporate or drain out of the system. Avoid moisture traps where the condensate can settle and cause corrosion. Drill holes, with a minimum diameter of 0.25 in, in the corners of horizontal bulkheads, away from the electronic components, to provide a moisture drainage path through the cabinet.

Use vertically oriented circuit boards and connectors, if possible, to provide a natural moisture drainage path away from the boards to the bottom of the chassis.

Offset drain holes may be required in the base of the chassis to prevent foreign objects, such as screwdrivers, from being poked into these holes and causing internal damage. Drain holes similar to the one shown in Figure 2.4 will reduce the possibility of foreign objects entering the chassis.

2.8 CONFORMAL COATINGS

There are four popular types of conformal coatings: acrylic, epoxy, polyurethane, and silicone. A thin conformal coating 0.003 to 0.005 in (0.0076 to 0.0127 cm) thick may be required to protect the circuit boards from moisture. However, these coatings should be applied only if they are absolutely necessary. In general, coatings are expensive to apply. They require special cleaning processes, special tools to apply the coatings, and the circuit boards are difficult to rework. In addition, many coatings tend to crack, chip, and peel, and they will contaminate the connector contacts unless they are masked during the application. Also, water vapor tends to creep under the coatings and condense, so that coatings can change the electrical resistance between the circuits they are supposed to help. Therefore, if it becomes necessary to apply a conformal coating for moisture protection, make sure that the cleaning and application procedures are carefully followed, or the coatings may create more problems than they solve.

The conformal coat should not be applied so that it bridges the strain relief on the component lead wires. The purpose of the strain relief is to reduce stresses in the wire and in the solder joints. If the strain relief is bridged (completely filled), it will act like a short circuit and will not provide strain relief during vibration as the PCB flexes. Also, a filled strain relief will restrict the relative motion that results during temperature cycling tests. This will increase the stresses in the solder joints, which will increase the chance of failure.

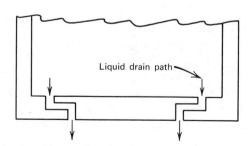

Figure 2.4 Offset drain holes in bottom of chassis.

2.9 SEALED ELECTRONIC BOXES

Electronic systems with high impedance circuits are usually very sensitive to humidity and moisture in the ambient air. Slight amounts of condensation on sensitive components, printed circuits, or electrical connectors can often produce large changes in the operating characteristics of the system. Therefore, to minimize potential problems resulting from humidity, moisture, and condensation, sensitive electronic systems are often packaged in sealed electronic boxes.

Sealed electronic boxes are also used for some air cooled electronic systems, which must be capable of operating in the hard vacuum environment of outer space. The sealed box is used to prevent the loss of air, which is required to cool the electronic components.

Many different types of seals can be used, depending upon the size of the unit, the cost, and the ease of repair. Solder seals are very effective for small covers on boxes, but repairs are inconvenient. O-ring seals are popular and easy to use for large or small boxes. A large, stiff, machined mounting flange with many high strength screws is required for the box and the cover, to provide an effective seal. A typical O-ring seal is shown in Figure 2.5.

Sealed electronic boxes, which are used for space applications, may require the use of a pressure relief valve to reduce the high internal pressure that will develop at high altitudes. If a relief valve is not used, the walls of the box will have to be made stiff enough to resist the high deflections and stresses that may result when the external atmosphere is no longer present.

The seal must do two things. First, it must prevent air (or any other gas, such as dry nitrogen) from leaking out of the box. Second, it must prevent water vapor from entering the box, because water vapor may condense and cause arcing or short circuits. Preventing air from leaking out of the box is

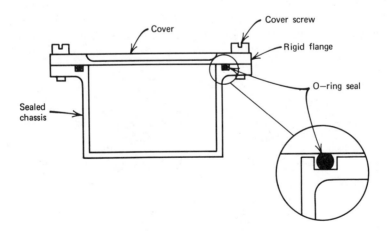

Figure 2.5 O-ring seal for electronic box.

not too difficult to accomplish, since many different types of gaskets will provide an effective seal for an airtight box. A vaportight box may be very difficult to achieve, however, if the wrong type of gasket is used.

Flat rubber gaskets are easy to use and cheap to fabricate. These gaskets are capable of providing an airtight seal for a large electronic box. A typical flat gasket seal is shown in Figure 2.6. A flat rubber gasket should not be used for boxes that require a water vapor seal. The interface pressure between the flat rubber gasket and its mating box and cover flanges is usually very low, because the rubber is free to deform as the cover screws are tightened. Although this gasket will seal in air, it will not seal out water vapor. The water vapor molecule is smaller than the air molecule, so that the water vapor module will pass through restricted openings more readily than the air molecule.

The water vapor molecule is smaller than the air molecule because the water vapor molecule has a smaller molecular weight. The size of the gas molecule is related to the molecular weight of the gas, so that a gas molecule with a high molecular weight will be larger than a gas with a low molecular weight. Air is primarily a mixture of oxygen and nitrogen gases. Oxygen (O_2) has a molecular weight of 32 and nitrogen (N_2) has a molecular weight of 28. Water vapor is still water (H_2O). Two atoms of hydrogen (H_2) have an atomic weight of 2, and one atom of oxygen (O) has an atomic weight of 16, for a total molecular weight of 18 [15].

The smaller size of the water vapor molecule therefore permits it to pass through openings that would normally restrict the flow of the air (oxygen and nitrogen) molecules. The O-ring type of gasket seal has a very high interface pressure, which makes it very effective for boxes that require a water vapor seal.

Water vapor is a gas and therefore obeys the normal gas laws. The force driving the water vapor migration is the difference in the partial pressure of the water vapor between two different areas. Water vapor will tend to flow from the high partial pressure areas to the low partial pressure areas. If an electronic box is assembled in a dry climate, the partial pressure of the

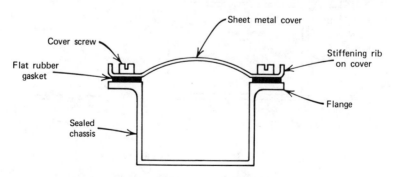

Figure 2.6 Flat rubber gasket for electronic box.

2.9 Sealed Electronic Boxes

water vapor within the box will be low. If that box is now transported to a humid area, such as Florida, the partial pressure of the water vapor in the atmosphere will be quite high. The water vapor will try to equalize the partial pressure, so that it will try to enter the box. It does not matter if the absolute air pressure within the box is 50 psia and the outside ambient pressure is only 14.7 psia. Since the partial pressure of the water vapor within the box is low, the tendency will be for the water vapor to try to enter the box, to equalize the partial pressures [14].

If the type of gasket used to seal an electronic box is not effective, water vapor will pass through the seal and condense within the electronic box. It does not take very long for half a pint of water to accumulate within a small chassis due to condensation. An electronic box 15 in × 10 in wide × 8 in high, with a flat rubber gasket seal under the top cover can accumulate that amount of condensation in a period of only about 3 months when it is continually exposed to humid air.

Cover gaskets are not the only areas that will permit water vapor to enter an electronic box. Other sources may be the connector interfaces, areas around fasteners, and the basic structure itself.

Connector interfaces on an electronic box must be adequately sealed if the box is to prevent the entry of water vapor. Solder seals are very effective for small connectors, but, again, repairs are inconvenient. O-ring gaskets are convenient, relatively easy to use, and can provide an effective water vapor seal.

A sealed electronic box should not have bolts or other fasteners passing through the outer structural wall into the interior of the chassis. Every hole in the chassis wall is the possible source of a leak. If bolts are required to pass through an external wall, gaskets or some other provision must be made to ensure an effective seal at that interface.

The basic structure of the electronic box itself may not be capable of providing an effective water vapor seal. Sheet metal structures are generally satisfactory. The sheet metal is made from rolled stock, so that it is not porous. However, the box corners may be welded and welds may leak.

Cast structures are often very porous. Some magnesium and aluminum castings 1 in thick will not hold air for more than 5 minutes. Structures of this type can often be impregnated with an epoxy resin. The most effective technique for applying the resin is by means of a vacuum. The porous structure is placed in a vacuum chamber, the air is evacuated, and the resin is applied to the surfaces to be impregnated. The vacuum is then released and the increased air pressure drives the resin into the porous surface, which seals the structure.

Dry nitrogen is very effective for removing water vapor from a sealed box. An inlet and outlet valve are placed at opposite ends of the chassis, which is purged with dry nitrogen to remove all traces of water vapor. The nitrogen valves are then closed so that the box is sealed with dry nitrogen. Since most of our normal atmosphere is nitrogen, the characteristics of the

32 Designing the Electronic Chassis

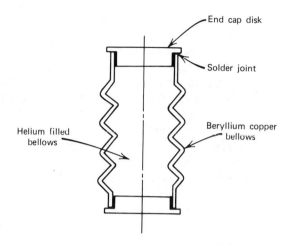

Figure 2.7 Cross section through helium filled sealed bellows with end caps.

gas within the box is very similar to the atmosphere, except that it contains no water vapor.

Desiccators may be placed within the electronic enclosure to absorb any water vapor that may seep into the sealed box over a long period of time. Materials such as silica jel or molecular sieve are quite effective for this purpose.

Care should be exercised in the selection of materials that are to be used for sealed packages. Extrusions should be avoided, if possible, because

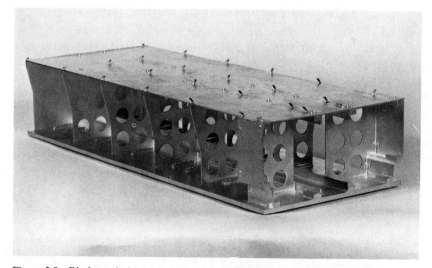

Figure 2.8 Dip brazed electronic box with self jigging tabs. (Courtesy Honeywell Co.)

2.10 Standard Electronic Box Sizes

Figure 2.9 Dip brazed electronic chassis with aluminum screws and tack welds for positioning piece parts. (Courtesy Kèarfott Division, The Singer Co.)

extruded sections can be very porous. For example, end caps are often used on sealed helium filled bellows, as shown in Figure 2.7. These bellows are used to take up the expansion of the liquid used in gyros, which must operate over a wide temperature range.

If the end caps are made from disk sections that have been sliced from an extruded rod, the end caps can be porous and the helium will leak out. To avoid this difficulty, the end caps should be stamped or machined from flat rolled sheet stock. This would provide end caps that have the grain running perpendicular to the axis of the bellows, so that the axial pores are sharply reduced.

2.10 STANDARD ELECTRONIC BOX SIZES

Electronic boxes can come in a wide variety of sizes and shapes, depending upon the application and environment. One organization, however, has

Table 2.1 ARINC Standard Rectangular Box Sizes

Description	Width (in)	Length (in)	Height (in)
Short one quarter ATR	2.250	12.5625	7.625
Long one quarter ATR	2.250	19.5625	7.625
Short three eights ATR	3.5625	12.5625	7.625
Long three eights ATR	3.5625	19.5625	7.625
Short half ATR	4.875	12.5625	7.625
Long half ATR	4.875	19.5625	7.625
Short three quarters ATR	7.50	12.5625	7.625
Long three quarters ATR	7.50	19.5625	7.625
One ATR	10.125	19.5625	7.625
One and one half ATR	15.375	19.5625	7.625

made an attempt to standardize the size, shape, and mounting for electronic boxes used in air transport equipment. This has become the standard known as ARINC, which is an acronym for Aeronautical Radio Inc. This specification, ARINC 404, Air Transport Equipment Cases and Racking (ATR), December 31, 1970, defines a group of rectangular plug-in types of electronic equipment cases, which have the outer dimensions listed in Table 2.1.

Figure 2.8 shows a dip brazed aluminum chassis with tabs that can be twisted or bent to lock piece parts together before brazing. In Figure 2.9, the piece parts are held by aluminum screws and tack welds prior to brazing.

3
Conduction Cooling for Chassis and Circuit Boards

3.1 CONCENTRATED HEAT SOURCES, STEADY STATE CONDUCTION

Electronic components are not 100% efficient, so that heat is generated in active components whenever the electronic system is in operation. As the heat is generated, the temperature of the component increases and heat attempts to flow through any path it can find. If the heat source is constant, the temperature within the component continues to rise until the rate of the heat being generated is equal to the rate of the heat flowing away from the component. Steady state heat transfer then exists—the heat flowing into the component is equal to the heat flowing away from the component.

Heat always flows from high temperature areas to low temperature areas. When the electronic system is operating, the electronic components are its hottest parts. To control the hot spot component temperatures, the heat flow path must be controlled. If this is not done properly, the component temperatures are forced to rise in an attempt to balance the heat flow. Eventually, the temperatures may become so high that the component is destroyed if a protective circuit or a thermal cutoff switch is not used.

Conduction heat transfer in an electronic system is generally a slow process. Heat flow will not occur until a temperature difference has been established. This means that each member along the heat flow path must experience a temperature rise before the heat will flow to the next point along the path. This process continues until the final heat sink is reached, which is often the outside ambient air. This process may take many hours, perhaps even many days for a very large system, depending upon its size.

The materials along the heat flow path sharply influence the temperature gradients developed along the path. Heat can be conducted through any material: solid, liquid, or gas. The ability of the heat to flow through any material is determined by the physical properties of the material and by the geometry of the structure. The basic heat flow relation for steady state conduction from a single concentrated heat load is shown in Eq. 3.1. This

Table 3.1 Units Used with Thermal Conductivity Equation[a]

Item	Symbol	English Units	Metric Units
Power dissipation	Q	$\dfrac{Btu}{hr}$	$\dfrac{cal}{sec}$
Thermal conductivity	K	$\dfrac{Btu}{hr\ ft\ °F}$	$\dfrac{cal}{sec\ cm\ °C}$
Cross section area	A	ft^2	cm^2
Temperature difference	Δt	°F	°C
Length	L	ft	cm

[a] *Power in watts must always be converted into Btu/hr or cal/sec before it can be used in the equation.*

relation can be used for both English and metric units (Table 3.1). Consistent sets of units must always be used, however, as defined below [15–17].

$$Q = KA\frac{\Delta t}{L} \qquad (3.1)$$

3.2 MOUNTING ELECTRONIC COMPONENTS ON BRACKETS

Electronic components are often mounted on brackets, because wall space on a typical chassis is very limited. The bracket must provide a good heat flow path from the component to the ultimate heat sink, which may be a cold plate on the chassis wall. The temperature rise that develops from the component to the chassis wall is determined by the size of the bracket, how the bracket is fastened to the wall, and how the component is fastened to the bracket. The actual temperature that is developed at the mounting surface of the component is simply the temperature of the wall plus the temperature rise from the wall to the surface of the component.

Structural epoxies are convenient for mounting brackets to chassis walls. Epoxies are strong, easy to apply, and cure rapidly. One note of caution: Avoid silicone greases in areas that will be cemented. Silicones tend to migrate. Even if the grease is not in contact with the cemented joint, the migration will carry it to the joint before the epoxy is applied. This prevents the epoxy from bonding properly, resulting in a weakened joint.

If the chassis will experience any vibration or shock, it is a good idea to use mechanical fasteners such as rivets or screws with the epoxy to improve the bond.

Cantilevered brackets should be avoided in shock and vibration environments because they have low resonant frequencies and high transmissibilities. This results in high dynamic stresses and a low fatigue life [1].

Table 3.2 Thermal Conductivity of Various Materials [18–23]

Material (room temperature)	English Units $\left(\dfrac{\text{Btu}}{\text{hr ft °F}}\right)$	Metric Units $\left(\dfrac{\text{cal}}{\text{sec cm °C}}\right)$
Metal		
Aluminum		
Pure	125	0.52
5052	83	0.34
6061 T6	90	0.37
2024 T4	70	0.29
7075 T6	70	0.29
356 T6	87	0.36
Beryllium	95	0.39
Beryllium copper	50	0.21
Copper		
Pure	230	0.95
Drawn wire	166	0.68
Bronze	130	0.54
Red brass	64	0.26
Yellow brass	54	0.22
5% phosphor bronze	30	0.12
Gold	170	0.70
Iron		
Wrought	34	0.14
Cast	32	0.13
Kovar	9	0.037
Lead	19	0.078
Magnesium		
Pure	92	0.38
Cast	41	0.17
Molybdenum	75	0.31
Nickel		
Pure	46	0.19
Inconel	10	0.041
Monel	15	0.062
Silver	242	1.00
Steel		
SAE 1010	34	0.14
1020	32	0.13
1045	26	0.11
4130	24	0.10
Tin	36	0.15
Titanium	9	0.037
Zinc	59	0.24

Table 3.2 (Continued)

Material (room temperature)	English Units $\left(\dfrac{\text{Btu}}{\text{hr ft °F}}\right)$	Metric Units $\left(\dfrac{\text{cal}}{\text{sec cm °C}}\right)$
Nonmetal		
Air	0.0153	0.000063
Alumina	17.0	0.070
Bakelite	0.11	0.00045
Beryllia (99.5%)	114.	0.47
Carbon	4.0	0.016
Epoxy		
No fill	0.12	0.00049
High fill	1.25	0.0051
Epoxy fiberglass	0.15	0.00062
Multilayer 0.005 in epoxy lamina with 0.0028 copper parallel to plane	20.	0.083
Perpendicular to plane (with plated throughholes)	2.	0.0082
Glass	0.50	0.0021
Glass wool	0.023	0.00009
Ice	1.23	0.0051
Mica	0.41	0.0017
Mylar	0.11	0.00045
Nylon	0.14	0.00058
Phenolic		
Plain	0.30	0.0012
Paper base	0.16	0.00066
Plexiglass	0.11	0.00045
Polystyrene	0.061	0.00025
Polyvinyl chloride	0.09	0.00037
Pyrex	0.73	0.0030
Rubber		
Butyl	0.15	0.00062
Hard	0.11	0.00045
Soft	0.08	0.00033
Silicone grease	0.12	0.00049
Silicone rubber	0.11	0.00045
Styrofoam	0.02	0.00008
Teflon	0.11	0.00045
Water	0.38	0.0016
Wood		
Maple	0.096	0.00039
Pine	0.067	0.00027

3.3 SAMPLE PROBLEM—TRANSISTOR MOUNTED ON A BRACKET

Determine the mounting surface temperature on the case of a transistor that is bolted to an aluminum bracket (5052 aluminum) as shown in Figure 3.1. The bracket is cemented to a heat sink wall, which is maintained at a temperature of 131°F (55°C). The transistor has a mica insulator at the mounting interface and the power dissipation is 7.5 watts. Radiation and convection heat transfer from the assembly are small and can be ignored.

SOLUTION

The heat transfer path from the transistor to the heat sink wall is broken into three parts, and each part is examined separately, using Eq. 3.1 written in a slightly different form, as shown in Eq. 3.2. The solution is shown in English units and metric units.

$$\Delta t = \frac{QL}{KA} \qquad (3.2)$$

Part 1. Temperature Rise across Cemented Interface

Given Q = concentrated heat load of 7.5 watts
  Q = 25.6 Btu/hr = 1.79 cal/sec (ref. Table 1.1)
  L = thickness of cemented interface
  L = 0.010 in = 0.000833 ft = 0.0254 cm
  K = thermal conductivity epoxy cement (ref. Table 3.2)
  K = 0.167 Btu/hr ft °F = 0.000689 cal/sec cm °C (ref. Table 1.2)
  A = cross section area along heat flow path
  A = (1.3)(0.75) = 0.975 in² = 0.00677 ft² = 6.29 cm²

Substitute into Eq. 3.2 using English units.

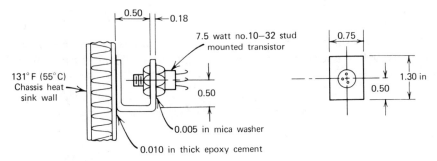

Figure 3.1 Transistor mounted on a bracket.

$$\Delta t_1 = \frac{(25.6 \text{ Btu/hr})(0.000833 \text{ ft})}{(0.167 \text{ Btu/hr ft °F})(0.00677 \text{ ft}^2)} = 18.9°\text{F} \qquad (3.3)$$

Substitute into Eq. 3.2 using metric units.

$$\Delta t_1 = \frac{(1.79 \text{ cal/sec})(0.0254 \text{ cm})}{(0.000689 \text{ cal/sec cm °C})(6.29 \text{ cm}^2)} = 10.5°\text{C} \qquad (3.3a)$$

Part 2. Temperature Rise along the Aluminum Bracket

Given Q = 7.5 watts = 25.6 Btu/hr = 1.79 cal/sec
 L = length of bracket from center of cemented face to center of transistor mount
 L = 0.50 + 0.50 + 0.50 = 1.5 in = 0.125 ft = 3.81 cm
 K = thermal conductivity 5052 aluminum bracket (ref. Table 3.2)
 K = 83 Btu/hr ft °F = 0.343 cal/sec cm °C
 A = cross section area of bracket
 A = (0.75)(0.18) = 0.135 in² = 0.000937 ft² = 0.871 cm²

Substitute into Eq. 3.2 using English units.

$$\Delta t_2 = \frac{(25.6)(0.125)}{(83)(0.000937)} = 41.1°\text{F} \qquad (3.4)$$

Substitute into Eq. 3.2 using metric units.

$$\Delta t_2 = \frac{(1.79)(3.81)}{(0.343)(0.871)} = 22.8°\text{C} \qquad (3.4a)$$

Part 3. Temperature Rise across Mica Insulator (Ref. Table 3.3)

Given Q = 7.5 watts $\qquad (3.5)$
 R = 2.4 °C/watt (10–32 stud; ref. Table 3.3)
 Δt_3 = (7.5 watts)(2.4 °C/watt)
 Δt_3 = 18°C = 32.4°F

The mounting surface temperature of the transistor is determined from the chassis wall temperature plus the temperature rises along the heat flow path.

$$t_{\text{trans}} = t_{\text{surf}} + \Delta t_1 + \Delta t_2 + \Delta t_3$$

In English units, the transistor temperature is

$$t_{\text{trans}} = 131 + 18.9 + 41.1 + 32.4 = 223.4°\text{F} \qquad (3.6)$$

3.4 Uniformly Distributed Heat Sources, Steady State Conduction

Table 3.3 Thermal Resistance R across Interface from Case to Sink for Stud Mounted Transistors and Diodes

Stud Size	Hex Size across Flats	Dry	With Thermal Grease	Across Mica Washer 0.005 in (0.0127 cm)
10–32	$\frac{7}{16}$ in	$0.60 \frac{°C}{watt}$	$0.40 \frac{°C}{watt}$	$2.4 \frac{°C}{watt}$
$\frac{1}{4}$–28	$\frac{9}{16}$ in	$0.45 \frac{°C}{watt}$	$0.30 \frac{°C}{watt}$	$2.0 \frac{°C}{watt}$

In metric units, the transistor temperature is

$$t_{trans} = 55 + 10.5 + 22.8 + 18.0 = 106.3°C \tag{3.6a}$$

3.4 UNIFORMLY DISTRIBUTED HEAT SOURCES, STEADY STATE CONDUCTION

Identical electronic components are often placed next to one another, on mounting brackets or on circuit boards, as shown in Figure 3.2. When each component dissipates approximately the same amount of power, the result will be a uniformly distributed heat load.

When any one strip of electronic components is considered, as shown in

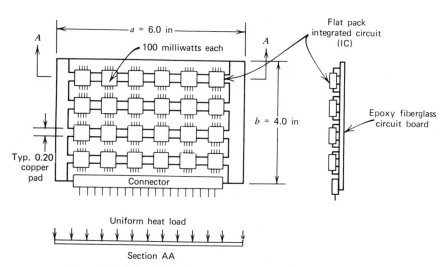

Figure 3.2 Printed circuit board with flat pack integrated circuits.

section AA, the heat input can be evaluated as a uniformly distributed heat load. On a typical printed circuit board (PCB), the heat will flow from the component to the heat sink strip under the component, and then to the outer edges of the PCB, where it is removed. The heat sink strip is usually made of copper or aluminum, which have high thermal conductivity. The maximum temperature will occur at the center of the PCB and the minimum temperature will occur at the edges. This produces a parabolic temperature distribution, as shown in Figure 3.3.

When only one side of one strip is considered, a heat balance equation can be obtained by considering a small element dx of the strip, along the span with a length of L. Then

$$dQ_1 + dQ_2 = dQ_3$$

where $dQ_1 = q\, dx$ = heat input

$$dQ_2 = -KA\frac{dt}{dx} = \text{heat flow}$$

$$dQ_3 = KA\frac{d}{dx}(t + dt) = \text{total heat}$$

Then

$$q\, dx - KA\frac{dt}{dx} = -KA\frac{dt}{dx} - KA\frac{d}{dx}(dt)$$

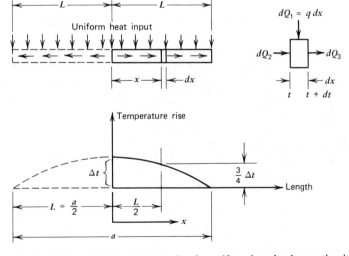

Figure 3.3 Parabolic temperature distribution for uniform heat load on a circuit board.

3.4 Uniformly Distributed Heat Sources, Steady State Conduction

or

$$\frac{d^2 t}{dx^2} = -\frac{q}{KA}$$

This is a second order differential equation, which can be solved by integrating twice. Integrating once yields

$$\frac{dt}{dx} = -\frac{qx}{KA} + C_1$$

Integrating twice yields

$$t = -\frac{qx^2}{2KA} + C_1 x + C_2$$

The constant C_1 is zero because at $x = 0$, the plate is adiabatic and no heat is lost. The slope of the temperature gradient is therefore zero.

The constant C_2 is determined by letting the temperature at the end of the plate be t_e; then

$$C_2 = t_e + \frac{qL^2}{2KA}$$

The temperature at any point along the plate (or strip) is

$$t = -\frac{qx^2}{2KA} + t_e + \frac{qL^2}{2KA}$$

Let the temperature at the center of the plate be t_0; then

$$t_0 - t_e = \frac{q}{2KA}(L^2 - x^2)$$

When $t_0 - t_e$ is represented as Δt,

$$\Delta t = \frac{q}{2KA}(L^2 - x^2) \tag{3.7}$$

This equation produces the parabolic shape for the temperature distribution, which is shown in Figure 3.3

When the full length of the strip is considered, the value of x in Eq. 3.7 is zero. The total heat flow along the length L then becomes

$$Q = qL \tag{3.8}$$

Substitute Eq. 3.8 into Eq. 3.7 when $x = 0$. This results in the equation for the maximum temperature rise in a strip with a uniformly distributed heat load.

$$\Delta t = \frac{QL}{2KA} \qquad \begin{array}{l} L = \tfrac{1}{2} \text{ TOTAL CARD LENGTH} \\ Q = \text{ " " POWER} \end{array} \qquad (3.9)$$

When the temperature rise at the midpoint along the heat sink strip of length L is desired, then $x = L/2$. Substituting this value into Eq. 3.7 will result in the temperature rise at the midpoint of the strip:

$$\text{midpoint } \Delta t = \frac{3QL}{8KA} \qquad (3.10)$$

Considering only the strip with a length of L, the ratio of the strip midpoint temperature rise to the maximum temperature rise is shown in the following relation:

$$\frac{\text{midpoint } \Delta t}{\text{maximum } \Delta t} = \frac{3QL/8KA}{QL/2KA} = \frac{3}{4} \qquad (3.11)$$

3.5 SAMPLE PROBLEM—COOLING INTEGRATED CIRCUITS ON A PCB

A series of flat pack integrated circuits are to be mounted on a multilayer printed circuit board (PCB) as shown in Figure 3.2. Each flat pack dissipates 100 milliwatts of power. Heat from the components is to be removed by conduction through the printed circuit copper pads, which have 2 ounces of copper [thickness is 0.0028 in (0.0071 cm)]. The heat must be conducted to the edges of the PCB, where it flows into a heat sink. Determine the temperature rise from the center of the PCB to the edge to see if the design will be satisfactory.

SOLUTION

The flat packs generate a uniformly distributed heat load, which results in the parabolic temperature distribution shown in Figure 3.4. Because of symmetry, only one half of the system is evaluated. Equation 3.9 is used to determine the temperature rise from the center of the PCB to the edge for one strip of components, using English units and metric units.

3.5 Sample Problem—Cooling Integrated Circuits on a PCB

Given $Q = 3(0.10) = 0.30$ watt heat input, one half strip
$Q = 1.02$ Btu/hr $= 0.0717$ cal/sec (ref. Table 1.1)
$L = 3.0$ in $= 0.25$ ft $= 7.62$ cm (length)
$K = 166$ Btu/hr ft °F $= 0.685$ cal/sec cm °C (Thermal conductivity of copper)
$A = (0.20)(0.0028) = 0.00056$ in² (cross section area)
$A = 3.89 \times 10^{-6}$ ft² $= 0.00361$ cm²

Substitute into Eq. 3.9 to obtain the temperature rise along the strip using English units.

$$\Delta t = \frac{(1.02)(0.25 \text{ ft})}{(2)(166)(3.89 \times 10^{-6} \text{ ft}^2)} = 197°F \qquad (3.12)$$

Substitute into Eq. 3.9 using metric units.

$$\Delta t = \frac{(0.0717)(7.62 \text{ cm})}{(2)(0.685)(0.00361 \text{ cm}^2)} = 110°C \qquad (3.12a)$$

The amount of heat that can be removed by radiation or convection for this type of system is very small. The temperature rise is therefore too high. By the time the sink temperature is added, assuming that it is 80°F, the case temperature on the component will be 277°F. Since the typical maximum allowable case temperature is about 212°F, the design is not acceptable.

If the copper thickness is doubled to 4 ounces, which has a thickness of 0.0056 in (0.014 cm), the temperature rise will be 98.5°F (55°C), which is still too high for any sink greater than about 115°F (46°C). For high temperature applications, the copper thickness will have to be increased to about 0.0112 in (0.0284 cm) for a good design.

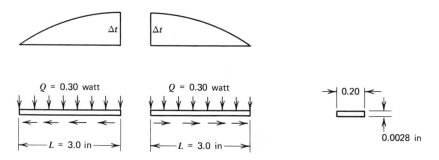

Figure 3.4 Uniformly distributed heat load on one copper strip.

3.6 CIRCUIT BOARD WITH AN ALUMINUM HEAT SINK CORE

Another method is often used for fabricating PCBs that will have a good conduction heat flow path. A thin circuit lamina is cemented to an aluminum plate, and flat pack integrated components are lap soldered to the lamina. Laminas can be cemented to both sides of the aluminum plate, so that both sides of the plate can be populated with integrated circuit components, as shown in Figure 3.5.

Twice as many integrated circuits can be mounted on this type of PCB. This means the total power dissipation will be twice that of a PCB which has components on one side only. If all the components have about the same power dissipation, a uniform heat load will be produced, so that the temperature rise distribution will still be parabolic, as shown in Figure 3.3.

3.7 SAMPLE PROBLEM—TEMPERATURE RISE ALONG A PCB HEAT SINK PLATE

Determine the temperature rise from the center of the aluminum heat sink plate to the side edges for the conductively cooled PCB shown in Figure 3.5. The heat flow path from the center of the heat sink to the side edges is shown in Figure 3.6. Components are mounted on both sides of the board, which results in a power dissipation of 2.4 watts per side, or a total power dissipation of 4.8 watts for the complete PCB.

SOLUTION

The temperature rise along the heat sink plate is determined from Eq. 3.9 considering English units and metric units.

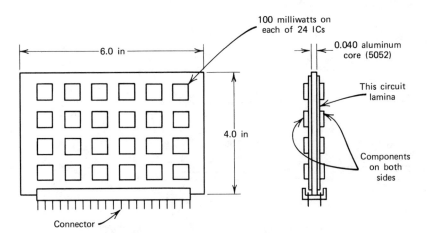

Figure 3.5 Circuit board with aluminum core for good heat conduction.

3.8 How to Avoid Warping on PCBs with Metal Heat Sinks

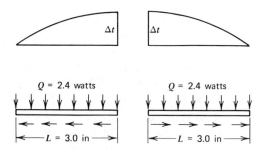

Figure 3.6 Uniformly distributed heat load on an aluminum board core.

Given $Q = 2.4$ watts $= 8.19$ Btu/hr $= 0.574$ cal/sec (ref. Table 1.1)
$L = 3.0$ in $= 0.25$ ft $= 7.62$ cm (heat sink length)
$K = 83$ Btu/hr ft °F $= 0.343$ cal/sec cm °C [aluminum (5052) conductivity]
$A = (0.04)(4.0) = 0.16$ in²
$A = 0.00111$ ft² $= 1.032$ cm² (cross section area)

Substitute the English units into Eq. 3.9.

$$\Delta t = \frac{(8.19)(0.25)}{(2)(83)(0.00111)} = 11.1°F \tag{3.13}$$

Substitute the metric units into Eq. 3.9.

$$\Delta t = \frac{(0.574)(7.62)}{(2)(0.343)(1.032)} = 6.2°C \tag{3.13a}$$

Note that these results do *not* include the temperature rise across the thin circuit lamina. The results are only for the aluminum heat sink plate.

With such a small temperature rise, this type of design is well suited for high temperature applications.

3.8 HOW TO AVOID WARPING ON PCBS WITH METAL HEAT SINKS

Aluminum and copper heat sink cores on circuit boards are capable of conducting away large amounts of heat if the proper construction techniques are used. Extreme care must be used if only one thin circuit lamina is to be bonded to a metal core. Improper bonding can result in warped boards, because of the big difference in the coefficient of thermal expansion between the metal and the epoxy fiberglass lamina.

Thin sheets of prepreg are often used to bond the epoxy fiberglass printed circuit laminas to the metal heat sink plate. This requires a high temperature and a high pressure to cure the prepreg adhesive. Since aluminum and copper both have a coefficient of thermal expansion that is about double that of the epoxy fiberglass, in the plane of the PCB the metal will expand more than the epoxy at the high curing temperature. When the cured assembly is brought down to room temperature, the metal will shrink more than the epoxy, causing the bonded assembly to warp.

The warping caused by nonuniform shrinking between the metal core and the epoxy fiberglass lamina can usually be reduced or eliminated by bonding a similar part on the opposite side of the plate.

When only one side of a metal plate is to be laminated with a printed circuit, a room temperature (or slightly higher) cementing process should be used instead of the high temperature bonding process. There are many epoxy cements that have a relatively low temperature curing cycle [22, 23].

3.9 CHASSIS WITH NONUNIFORM WALL SECTIONS

Electronic chassis always seem to require cutouts, notches, and clearance holes for assembly access, wire harnesses, or maintenance. These openings will generally cut through a bulkhead or other structural member, which is required to carry heat away from some critical high power electronic component. The cutouts result in nonuniform wall sections, which must be analyzed to determine their heat flow capability.

One convenient method for analyzing nonuniform wall sections is to subdivide them into smaller units that have relatively uniform sections. The heat flow path through each of the smaller, relatively uniform sections can then be defined in terms of a thermal resistance. A specific resistance value for each section can then be determined by its geometry and physical properties. This will result in a thermal analog resistor network, or mathematical model, which describes the thermal characteristics of that structural section of the electronic system.

The basic conduction heat flow relation shown by Eq. 3.1 can be modified slightly to utilize the thermal resistance concept. The thermal resistance for conduction heat flow is then defined by Eq. 3.14 [24].

$$R = \frac{L}{KA} \qquad (3.14)$$

Substituting this value into Eq. 3.1 results in the temperature rise relation when the thermal resistance concept is used, as shown in Eq. 3.15.

$$\Delta t = QR \qquad (3.15)$$

CONDUCTANCE = 1/R

3.9 Chassis with Nonuniform Wall Sections

Figure 3.7 Series flow resistor network.

The thermal resistance concept is very convenient for developing mathematical models of simple or complex electronic boxes. Analog resistor networks can be established for heat flow in one, two, and three dimensions with little effort. Complex shapes can often be modeled with many simple thermal resistors to provide an effective means for determining the temperature profile for almost any type of system.

Two basic resistance patterns, series and parallel, are used to generate analog resistor networks. A simple series pattern network is shown in Figure 3.7.

The total effective resistance (R_t) for the series flow network is determined by adding all the individual resistances, as shown in Eq. 3.16 [15, 17, 24].

$$R_t = R_1 + R_2 + R_3 + \cdots \qquad (3.16)$$

A simple parallel flow resistor network is shown in Figure 3.8.

The total effective resistance for the parallel flow network is determined by combining the individual resistances as shown in Eq. 3.17 [15, 17, 24].

$$\frac{1}{R_t} = \frac{1}{R_1} + \frac{1}{R_2} + \frac{1}{R_3} + \cdots \qquad (3.17)$$

A typical electronic system will normally consist of many different combinations of series and parallel flow resistor networks.

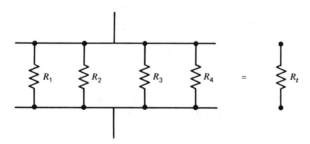

Figure 3.8 Parallel flow resistor network.

3.10 SAMPLE PROBLEM—HEAT FLOW ALONG NONUNIFORM BULKHEAD

An aluminum bulkhead is used to support a row of six power resistors. Each resistor dissipates 1.5 watts, for a total power dissipation of 9 watts. The bulkhead conducts the heat to the opposite wall of the chassis, which is cooled by a multiple fin heat exchanger. The bulkhead has two cutouts for connectors to pass through, as shown in Figure 3.9. Determine the temperature rise across the length of the bulkhead.

SOLUTION

A mathematical model with series and parallel thermal resistor networks can be established to represent the heat flow path, as shown in Figure 3.10a.

The thermal analog resistor network is simplified by first finding the equivalent effective single resistance for the parallel section represented by resistors R_2, R_3, and R_4 with the use of Eq. 3.17 and the geometry shown in Figure 3.9.

Determine resistor R_2:
Given L_2 = 1.5 in = 0.125 ft = 3.81 cm (length)
K_2 = 83 Btu/hr ft °F = 0.343 cal/sec cm °C [aluminum (5052) conductivity]
A_2 = (0.50)(0.060) = 0.030 in² — IGNORE .25 STEP; SMALL % ERROR
A_2 = 0.000208 ft² = 0.194 cm² (cross section area)

$$R_2 = \frac{0.125}{(83)(0.000208)} = 7.24 \frac{\text{hr °F}}{\text{Btu}} \text{ (English units)} \qquad (3.18)$$

$$R_2 = \frac{3.81}{(0.343)(0.194)} = 57.3 \frac{\text{sec °C}}{\text{cal}} \text{ (metric units)} \qquad (3.18a)$$

Determine resistor R_3:
Given L_3 = 1.5 in = 0.125 ft = 3.81 cm (length)
K_3 = 83 Btu/hr ft °F = 0.343 cal/sec cm °C (aluminum)
A_3 = (1.0)(0.060) = 0.060 in² (area)
A_3 = 0.000417 ft² = 0.387 cm²

$$R_3 = \frac{0.125}{(83)(0.000417)} = 3.61 \frac{\text{hr °F}}{\text{Btu}} \text{ (English units)} \qquad (3.19)$$

$$R_3 = \frac{3.81}{(0.343)(0.387)} = 28.7 \frac{\text{sec °C}}{\text{cal}} \text{ (metric units)} \qquad (3.19a)$$

3.10 Sample Problem—Heat Flow Along Nonuniform Bulkhead

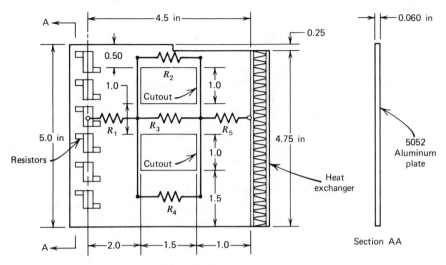

Figure 3.9 Bulkhead with two cutouts for connectors.

Determine resistor R_4:

Given $L_4 = 1.5$ in $= 0.125$ ft $= 3.81$ cm (length)
 $K_4 = 83$ Btu/hr ft °F $= 0.343$ cal/sec cm °C (aluminum)
 $A_4 = (1.5)(0.060) = 0.090$ in^2
 $A_4 = 0.000625$ ft$^2 = 0.581$ cm^2 (cross section area)

$$R_4 = \frac{0.125}{(83)(0.000625)} = 2.41 \frac{\text{hr °F}}{\text{Btu}} \text{ (English units)} \quad (3.20)$$

$$R_4 = \frac{3.81}{(0.341)(0.581)} = 19.1 \frac{\text{sec °C}}{\text{cal}} \text{ (metric units)} \quad (3.20a)$$

Resistors R_2, R_3, and R_4 are combined in parallel using Eq. 3.17, which results in resistor R_6.

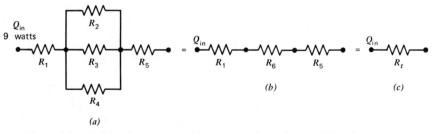

Figure 3.10 Bulkhead thermal model using a series and a parallel resistor network.

$$\frac{1}{R_6} = \frac{1}{7.24} + \frac{1}{3.61} + \frac{1}{2.41}$$

$$R_6 = 1.20 \frac{\text{hr °F}}{\text{Btu}} \text{ (English units)} \quad (3.21)$$

$$\frac{1}{R_6} = \frac{1}{57.3} + \frac{1}{28.7} + \frac{1}{19.1}$$

$$R_6 = 9.56 \frac{\text{sec °C}}{\text{cal}} \text{ (metric units)} \quad (3.21\text{a})$$

The mathematical model now appears as shown in Figure 3.10b. This can be further reduced by considering thermal resistance R_6 to be in series with resistors R_1 and R_5. Each resistance value is determined from Eq. 3.14. The geometry for these resistors is obtained from Figure 3.9.

Determine resistor R_1:
Given $L_1 = 2.0$ in $= 0.167$ ft $= 5.08$ cm (length)
$K_1 = 83$ Btu/hr ft °F $= 0.343$ cal/sec cm °C [aluminum (5052) conductivity]
$A = (5.0)(0.060) = 0.30$ in^2
$A = 0.00208$ ft$^2 = 1.935$ cm^2 (cross section area)

$$R_1 = \frac{0.167}{(83)(0.00208)} = 0.97 \frac{\text{hr °F}}{\text{Btu}} \text{ (English units)} \quad (3.22)$$

$$R_1 = \frac{5.08}{(0.343)(1.935)} = 7.65 \frac{\text{sec °C}}{\text{cal}} \text{ (metric units)} \quad (3.22\text{a})$$

Determine resistor R_5:

Given $L_5 = 1.0$ in $= 0.0833$ ft $= 2.54$ cm (length)
$K_5 = 83$ Btu/hr ft °F $= 0.343$ cal/sec cm °C (aluminum)
$A_5 = (4.75)(0.060) = 0.285$ in^2
$A_5 = 0.00198$ ft$^2 = 1.839$ cm^2 (cross section area)

$$R_5 = \frac{0.0833}{(83)(0.00198)} = 0.51 \frac{\text{hr °F}}{\text{Btu}} \text{ (English units)} \quad (3.23)$$

$$R_5 = \frac{2.54}{(0.343)(1.839)} = 4.03 \frac{\text{sec °C}}{\text{cal}} \text{ (metric units)} \quad (3.23\text{a})$$

The values of the three resistors in Figure 3.10b are now known, so that Eq. 3.16 can be used to determine the total resistance of the three resistors in series.

$$R_t = 0.97 + 1.20 + 0.51 = 2.68 \frac{\text{hr °F}}{\text{Btu}} \text{ (English units)} \quad (3.24)$$

$$R_t = 7.65 + 9.56 + 4.03 = 21.24 \frac{\text{sec °C}}{\text{cal}} \text{ (metric units)} \quad (3.24a)$$

The temperature rise across the length of the bulkhead is then determined from Eq. 3.15, in English units and metric units.

Given $Q = 9$ watts $= 30.72$ Btu/hr $= 2.15$ cal/sec (heat dissipated)
$R_t = 2.68$ hr °F/Btu $= 21.24$ sec °C/cal (thermal resistance)

$$\Delta t = (30.72)(2.68) = 82.3°F \text{ (English units)} \quad (3.25)$$

$$\Delta t = (2.15)(21.24) = 45.7°C \text{ (metric units)} \quad (3.25a)$$

3.11 TWO DIMENSIONAL ANALOG RESISTOR NETWORKS

The conduction heat flow path through a typical electronic chassis usually occurs in two or more directions at the same time. If there are several concentrated heat input sources along the path, the temperature distribution along the heat flow path may be difficult to evaluate. The problem may be simplified by breaking up the structure into a group of smaller elements, which are interconnected by thermal resistors. The mass of each small element would be concentrated at the geometric center of the element. This point is called the node point of the element and is shown in Figure 3.11. The network of resistors and nodes is called an analog resistor network, and it is based upon the electrical networks that are used so extensively by electrical and electronic engineers [15, 17, 24].

Very large two and three dimensional analog resistor networks are often used with high speed digital computers to determine the thermal profile of

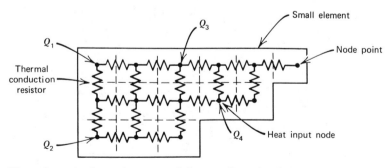

Figure 3.11 Analog resistor network for two dimensional heat flow in a plate.

complex structures. Temperatures in smaller models can often be determined with the use of the basic conduction heat flow relation shown in Eq. 3.1. A heat balance equation is written for each node point in the mathematical model. The simultaneous solution of these equations then results in the temperatures at each node point.

The heat balance equations are a little easier to use when the concept of thermal conductance is used instead of thermal resistance. Thermal conductance (k') is defined as the inverse of the thermal resistance (R), as shown in Eq. 3.26.

$$k' = \frac{1}{R} = \frac{KA}{L} \qquad (3.26)$$

Substitute Eq. 3.26 into Eq. 3.1. This results in the steady state heat flow relation using the concept of conductance, which is shown in Eq. 3.27.

$$Q = k' \, \Delta t = k'(t_2 - t_1) \qquad (3.27)$$

3.12 SAMPLE PROBLEM—TWO DIMENSIONAL CONDUCTION ON A POWER SUPPLY HEAT SINK

A cantilevered copper bracket is used to support part of a power supply, which consists of two power transistors and two wire wound resistors, with power dissipations as shown in Figure 3.12. The power supply heat sink is bolted to a cold plate, which is maintained at a temperature of 160°F (71°C). Determine the temperatures at the four component mounting points on the power supply heat sink.

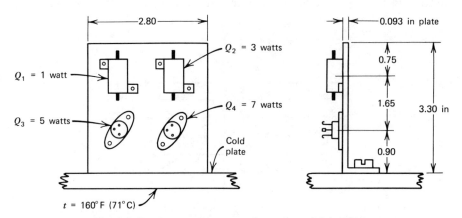

Figure 3.12 Components mounted on a heat sink bracket.

3.12 Sample Problem—Two Dimensional Conduction on a Power Supply Heat Sink

SOLUTION

The bracket is divided into small rectangular elements, with a node point at the geometric center of each element. The node points in each element are interconnected to form a rectangular grid pattern. This results in a mathematical model of the heat sink bracket, which is in the form of an analog conductance network representing a system with heat flow in two directions, as shown in Figure 3.13.

A heat balance equation must be written for each of the node points in the mathematical model to determine the temperature at each node point. To avoid the possibility of making an error in establishing a positive (+) or negative (−) heat flow direction within the heat sink bracket, a simple rule can be followed. Assume that all of the heat is flowing into each node as it is being examined. This will require changing the assumed heat flow direction in the various conductors, as the heat balance equations are written for different nodes. Starting at node 1, assume that the heat from node 2 flows into node 1, the heat from node 3 flows into node 1, and the heat from Q_1 flows into node 1. This is equivalent to assuming that the temperatures at node points 2 and 3 are higher than the temperature at node point 1. The heat flow relation shown by Eq. 3.27 is then used at each node.

The heat flow balance at node 1 is as follows:

$$k'_A(t_2 - t_1) + k'_B(t_3 - t_1) + Q_1 = 0$$

This leads to

$$(k'_A + k'_B)t_1 - k'_A t_2 - k'_B t_3 = Q_1 \tag{3.28}$$

The heat flow balance at node 2 is as follows:

$$k'_A(t_1 - t_2) + k'_C(t_4 - t_2) + Q_2 = 0$$

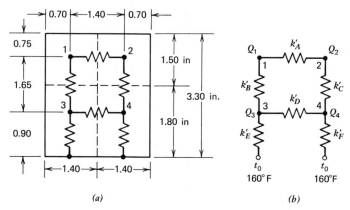

Figure 3.13 Mathematical model of the component bracket.

This leads to

$$-k'_A t_1 + (k'_A + k'_C) t_2 - k'_C t_4 = Q_2 \quad (3.29)$$

The heat flow balance at node 3 is as follows:

$$k'_B(t_1 - t_3) + k'_D(t_4 - t_3) + k'_E(t_0 - t_3) + Q_3 = 0$$

$$-k'_B t_1 + (k'_B + k'_D + k'_E) t_3 - k'_D t_4 = Q_3 + k'_E t_0 \quad (3.30)$$

The heat flow balance at node 4 is as follows:

$$k'_C(t_2 - t_4) + k'_D(t_3 - t_4) + k'_F(t_0 - t_4) + Q_4 = 0$$

$$-k'_C t_2 - k'_D t_3 + (k'_C + k'_D + k'_F) t_4 = Q_4 + k'_F t_0 \quad (3.31)$$

There are now four equations, 3.28 through 3.31, with four unknowns, t_1 through t_4, which can be solved simultaneously once all the conductances k'_A through k'_F have been determined. Each conductance can be determined using Figure 3.12 with Eq. 3.26 for English units and metric units. Start with conductance k'_A.

Given $K_A = 166$ Btu/hr ft °F = 0.686 cal/sec cm°C (copper conductivity)
$A_A = (1.5)(0.093) = 0.139$ in² (cross section area)
$A_A = 0.000965$ ft² = 0.897 cm²
$L_A = 1.40$ in = 0.117 ft = 3.56 cm (length)

In English units:

$$k'_A = \frac{(166 \text{ Btu/hr ft °F})(0.000965 \text{ ft}^2)}{0.117 \text{ ft}} = 1.37 \frac{\text{Btu}}{\text{hr °F}} \quad (3.32)$$

In metric units:

$$k'_A = \frac{(0.686 \text{ cal/sec cm °C})(0.897 \text{ cm}^2)}{3.56 \text{ cm}} = 0.173 \frac{\text{cal}}{\text{sec °C}} \quad (3.32a)$$

Calculate conductance k'_B.

Given $K_B = 166$ Btu/hr ft °F = 0.686 cal/sec cm°C (copper conductivity)
$A_B = (1.40)(0.093) = 0.130$ in² (cross section area)
$A_B = 0.000903$ ft² = 0.839 cm²
$L_B = 1.65$ in = 0.137 ft = 4.19 cm (length)

3.12 Sample Problem—Two Dimensional Conduction on a Power Supply Heat Sink

In English units:

$$k'_B = \frac{(166)(0.000903)}{0.137} = 1.09 \frac{\text{Btu}}{\text{hr °F}} \quad (3.33)$$

In metric units:

$$k'_B = \frac{(0.686)(0.839)}{4.19} = 0.137 \frac{\text{cal}}{\text{sec °C}} \quad (3.33a)$$

Calculate conductance k'_E.
Given $K_E = 166$ Btu/hr ft °F = 0.686 cal/sec cm °C (copper conductivity)
$A_E = (1.40)(0.093) = 0.130$ in² (cross section area)
$A_E = 0.000903$ ft² = 0.839 cm²
$L_E = 0.90$ in = 0.075 ft = 2.29 cm (length)

In English units:

$$k'_E = \frac{(166)(0.000903)}{0.075} = 2.00 \frac{\text{Btu}}{\text{hr °F}} \quad (3.34)$$

In metric units:

$$k'_E = \frac{(0.686)(0.839)}{2.29} = 0.251 \frac{\text{cal}}{\text{sec °C}} \quad (3.34a)$$

Calculate conductance k'_D on the copper plate.
Given $A_D = (1.80)(0.093) = 0.167$ in² (cross section area)
$A_D = 0.00116$ ft² = 1.077 cm²
$L_D = 1.40$ in = 0.117 ft = 3.56 cm (length)

In English units:

$$k'_D = \frac{(166)(0.00116)}{0.117} = 1.64 \frac{\text{Btu}}{\text{hr °F}} \quad (3.35)$$

In metric units:

$$k'_D = \frac{(0.686)(1.077)}{3.56} = 0.207 \frac{\text{cal}}{\text{sec °C}} \quad (3.35a)$$

Because of symmetry

$$k'_C = k'_B \quad \text{and} \quad k'_F = k'_E \quad (3.36)$$

The power dissipations at each of the four nodes are as follows (see Table 1.1 for power conversions):

$$Q_1 = 1 \text{ watt} = 3.41 \frac{\text{Btu}}{\text{hr}} = 0.239 \frac{\text{cal}}{\text{sec}}$$

$$Q_2 = 3 \text{ watts} = 10.24 \frac{\text{Btu}}{\text{hr}} = 0.717 \frac{\text{cal}}{\text{sec}} \quad (3.37)$$

$$Q_3 = 5 \text{ watts} = 17.06 \frac{\text{Btu}}{\text{hr}} = 1.19 \frac{\text{cal}}{\text{sec}}$$

$$Q_4 = 7 \text{ watts} = 23.89 \frac{\text{Btu}}{\text{hr}} = 1.67 \frac{\text{cal}}{\text{sec}}$$

Substitute Eqs. 3.32, 3.33, and 3.37 into Eq. 3.28 using English units.

$$2.46 t_1 - 1.37 t_2 - 1.09 t_3 = 3.41 \frac{\text{Btu}}{\text{hr}} \quad (3.38)$$

Substitute Eqs. 3.32, 3.36, and 3.37 into Eq. 3.29 using English units.

$$-1.37 t_1 + 2.46 t_2 - 1.09 t_4 = 10.24 \frac{\text{Btu}}{\text{hr}} \quad (3.39)$$

Substitute Eqs. 3.33, 3.35, 3.34, and 3.37 into Eq. 3.30 using English units.

$$-1.09 t_1 + 4.73 t_3 - 1.64 t_4 = 17.06 + (2.0)(160)$$

$$-1.09 t_1 + 4.73 t_3 - 1.64 t_4 = 337.06 \frac{\text{Btu}}{\text{hr}} \quad (3.40)$$

Substitute Eqs. 3.36, 3.35, and 3.37 into Eq. 3.31 using English units.

$$-1.09 t_2 - 1.64 t_3 + 4.73 t_4 = 23.89 + (2.0)(160)$$

$$-1.09 t_2 - 1.64 t_3 + 4.73 t_4 = 343.89 \frac{\text{Btu}}{\text{hr}} \quad (3.41)$$

Equations 3.38 through 3.41 must be solved simultaneously to obtain the temperatures at node points 1 through 4. Several different methods can be used if a calculator is not available. For example, using the substitution method, Eqs. 3.38 and 3.39 can be combined to eliminate t_1.

$$2.46 t_1 - 1.37 t_2 - 1.09 t_3 = 3.41 \quad \text{(ref. Eq. 3.38)}$$

$$\underline{-2.46 t_1 + 4.42 t_2 - 1.96 t_4 = 18.39} \quad \left(\text{ref. Eq. 3.39} \times \frac{2.46}{1.37}\right)$$

$$+ 3.05 t_2 - 1.09 t_3 - 1.96 t_4 = 21.80 \quad (3.42)$$

3.12 Sample Problem—Two Dimensional Conduction on a Power Supply Heat Sink

Next, combine Eqs. 3.39 and 3.40, eliminating t_4.

$$\begin{array}{r} 2.06t_1 - 3.70t_2 + 1.64t_4 = -15.41 \\ -1.09t_1 + 4.73t_3 - 1.64t_4 = 337.06 \\ \hline 0.97t_1 - 3.70t_2 + 4.73t_3 = 321.65 \end{array}$$

(3.43)

Combine Eqs. 3.40 and 3.41, eliminating t_3.

$$\begin{array}{r} -1.09t_1 + 4.73t_3 - 1.64t_4 = 337.06 \\ -3.14t_2 - 4.73t_3 + 13.64t_4 = 991.83 \\ \hline -1.09t_1 - 3.14t_2 + 12.00t_4 = 1328.89 \end{array}$$

(3.44)

Combine Eqs. 3.42 and 3.43, eliminating t_2.

$$\begin{array}{r} 3.70t_2 - 1.32t_3 - 2.38t_4 = 26.44 \\ 0.97t_1 - 3.70t_2 + 4.73t_3 = 321.65 \\ \hline 0.97t_1 + 3.41t_3 - 2.38t_4 = 348.09 \end{array}$$

(3.45)

Combine Eqs. 3.43 and 3.44, eliminating t_2.

$$\begin{array}{r} 0.97t_1 - 3.70t_2 + 4.73t_3 = 321.65 \\ 1.28t_1 + 3.70t_2 - 14.14t_4 = -1565.89 \\ \hline 2.25t_1 + 4.73t_3 - 14.14t_4 = -1244.24 \end{array}$$

(3.46)

Combine Eqs. 3.41 and 3.42, eliminating t_2.

$$\begin{array}{r} -3.05t_2 - 4.59t_3 + 13.23t_4 = 962.26 \\ 3.05t_2 - 1.09t_3 - 1.96t_4 = 21.80 \\ \hline -5.68t_3 + 11.27t_4 = 984.06 \end{array}$$

(3.47)

Combine Eqs. 3.45 and 3.46, eliminating t_1.

$$\begin{array}{r} -2.25t_1 - 7.91t_3 + 5.52t_4 = -807.42 \\ 2.25t_1 + 4.73t_3 - 14.14t_4 = -1244.24 \\ \hline -3.18t_3 - 8.62t_4 = -2051.66 \end{array}$$

(3.48)

Combine Eqs. 3.47 and 3.48, eliminating t_3. This will give the temperature at node point 4.

$$\begin{array}{r} -5.68t_3 + 11.27t_4 = 984.06 \\ +5.68t_3 + 15.40t_4 = 3664.60 \\ \hline 26.67t_4 = 4648.66 \\ t_4 = 174.3°F \end{array}$$

(3.49)

Substitute Eq. 3.49 into Eq. 3.48 for the temperature at mounting point 3, which is also node. 3.

$$-3.18t_3 - 8.62(174.3) = -2051.66$$

$$t_3 = 172.9°F \qquad (3.50)$$

Substitute Eqs. 3.49 and 3.50 into Eq. 3.46 for the temperature at mounting point 1.

$$2.25t_1 + 4.73(172.9) - 14.14(174.3) = -1244.24$$

$$t_1 = 178.8°F \qquad (3.51)$$

Substitute Eqs. 3.49 and 3.50 into Eq. 3.42 for the temperature at mounting point 2.

$$3.05t_2 - 1.09(172.9) - 1.96(174.3) = 21.80$$

$$t_2 = 181.0°F \qquad (3.52)$$

The same process can be repeated with the metric units by substituting Eqs. 3.32a through 3.36a and Eq. 3.37 into Eqs. 3.28 through 3.31. This will result in four equations with four unknowns, for temperatures in degrees Celsius.

3.13 HEAT CONDUCTION ACROSS INTERFACES IN AIR

The electronics industry makes extensive use of brackets, heat sinks, and circuit boards for mounting electronic components that are cooled by conduction. This often requires the transfer of heat across interfaces that may be bolted, riveted, or just clamped together. Extreme caution must be used if large amounts of heat will be conducted across these interfaces, because a poor thermal contact will produce a large thermal resistance, which means a high temperature rise. High temperatures will then result in more rapid electrical failures.

An enlarged view of two surfaces in contact will reveal that a very small portion of the total surface is actually in contact, as shown in Figure 3.14.

Most of the heat flow across the interface will take place between the various high points that contact each other. The number of high points per square inch of surface area that contact each other will generally depend upon the flatness, smoothness, hardness, stiffness, and interface pressure between the two mating surfaces [25–28].

The amount of heat that can be conducted across the interface will also depend upon the presence of another material at the interface, such as air, oil, or grease. Many manufacturers use special silicone greases, or pastes,

3.13 Heat Conduction Across Interfaces in Air

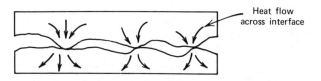

Figure 3.14 Enlarged view of two surfaces in contact.

which are filled with thermally conductive metal oxides to improve the overall heat transfer. These materials will not dry out or harden at temperatures well above the maximum operating temperatures of most electronic components. Water, which has a high thermal conductivity, is excellent as an interface conductor, but it evaporates too easily.

Sometimes large surfaces, which are not flat, must be mated using a dry interface. A soft, thin foil such as aluminum, indium, or copper about 0.002 in (0.0051 cm) thick will often improve the heat transfer across the interface.

Large surfaces should use stiff interface sheets to reduce the amount of warping. Large surfaces may also be improved with the use of an interface material made of a soft wire screen impregnated with a high carbon rubber. A thin beryllium copper sheet stamped with thousands of little spring fingers has worked well in many applications.

It is convenient to express the thermal interface characteristics in terms of a conductance coefficient h_i so that the heat transfer relation shown in Eq. 3.53 can be used [28].

$$Q = h_i A \, \Delta t \qquad (3.53)$$

Given

Item	Symbol	English Units	Metric units
Power dissipation	Q	$\dfrac{\text{Btu}}{\text{hr}}$	$\dfrac{\text{cal}}{\text{sec}}$
Interface conductance	h_i	$\dfrac{\text{Btu}}{\text{hr ft}^2 \, °\text{F}}$	$\dfrac{\text{cal}}{\text{sec cm}^2 \, °\text{C}}$
Area	A	ft^2	cm^2
Temperature difference	Δt	°F	°C

A considerable number of test data are available for the interface conductances of different types of joints, with different metals, different surface finishes, and different interface materials. These data are generally based upon the interface pressure between the two mating surfaces and the finish of these surfaces. The finish is measured by the rms (root mean square) index value of the small variations in the heights and depths of the many peaks and valleys found on all surfaces. Some typical rms values for various surfaces are shown in Table 3.4 [18, 19].

In general, a smooth surface is considered to have an rms finish less than about 15, and a rough surface is considered to have an rms finish greater than about 80. Some typical test data are shown in Figures 3.15, 3.16, and 3.17.

Table 3.5 shows typical interface conductances for various materials with an interface pressure of 10 psi.

An examination of the curves and tables shows that for a given surface finish and interface pressure, hard materials such as steel will have a lower interface conductance than will softer materials such as aluminum. Also, when the surface finish and surface flatness are improved, the interface conductance will rise substantially.

The interface conductance will increase when the clamping forces are increased. Low clamping forces will have lower conductance values than will high clamping forces. Smooth surfaces have high interface conductances with relatively low clamping forces. Increasing the clamping force between smooth surfaces does not greatly improve the interface conductance. Rough surfaces, on the other hand, have low interface conductances with low interface pressures. Increasing the clamping force between rough surfaces improves the interface conductance substantially [27, 28].

Figure 3.16 shows that for a given interface pressure and surface finish, a higher heat density will result in a higher interface conductance. Test data also show that the interface conductance rises substantially when the temperature is increased.

Most of the test data between two structural members, which are riveted,

Table 3.4 Typical Surface Finishes for Different Operations

Operation	Surface Finish (μin rms)
Lap	4
Grind	10
Rolled surface	15–20
Lathe, smooth cut	80
Lathe, rough cut	120
Milling machine, rough cut	125
Investment casting	125–240
File	240
Sand casting	240–300
Shaper	1000

3.13 Heat Conduction Across Interfaces in Air

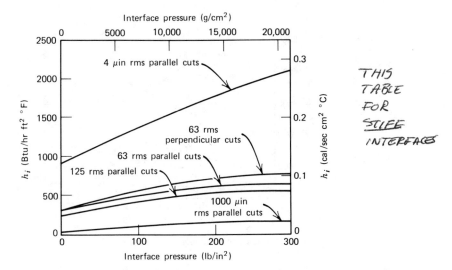

Figure 3.15 Interface conductance for parallel and perpendicular machine cuts for 18% carbon steel at 100°C (212°F) at sea level conditions.

(dry joint)

bolted, or spot welded together, will show a substantial amount of variation in the interface conductance values. Structural members that appear to be identical in all other respects will often have large variations in their interface conductances. These values are heavily dependent upon the surface contact area and interface pressure in the sections immediately under and around the riveted, bolted, and spot welded members. These parameters are difficult to control exactly, even with automatic machines. Therefore, the test data for structural members will show a great deal of variation [27, 28].

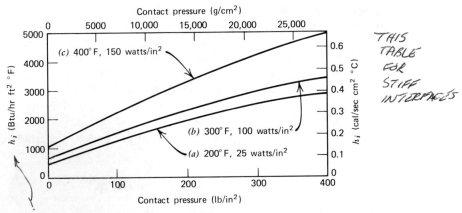

Figure 3.16 Interface conductance for 7075 T6 aluminum at sea level at various temperatures with 65 μin rms surface finish.

(dry joint)

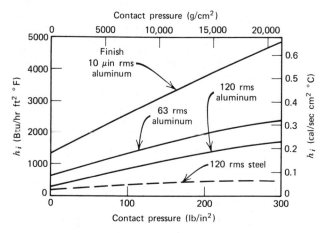

Figure 3.17 Interface conductance for various interface pressures at sea level.

3.14 SAMPLE PROBLEM—TEMPERATURE RISE ACROSS A BOLTED INTERFACE

Determine the temperature rise across the bolted interfaces of an aluminum bracket that is bolted to a chassis side wall. The bolts will produce an interface pressure of 200 lb/in² (14,074 g/cm²), as shown in Figure 3.18. The total power dissipation of the components on the bracket is 12 watts. The surface roughness at the bolted interfaces is about 65 μin rms. The bracket is in a confined area where very little heat will be lost by convection or radiation.

SOLUTION

Equation 3.53 can be used to determine the temperature rise across the bolted interfaces, together with Figure 3.16.

Given Q = 12 watts = 40.9 Btu/hr = 2.87 cal/sec (heat input)
A = 2[(0.50)(0.50) − (π/4)(0.150)²] = 0.464 in²
A = 0.00322 ft² = 2.99 cm² (interface area)
$\dfrac{Q}{A}$ = 12 watts/(0.464 in²) = 25.8 watts/in² = 4.01 watts/cm² (power density)
h_i = 2000 Btu/hr ft² °F = 0.270 cal/sec cm² °C (interface conductance; Figure 3.16, curve A)

Substitute into Eq. 3.53 for the temperature rise

$$\Delta t = \dfrac{40.9 \text{ Btu/hr}}{(2000 \text{ Btu/hr ft}^2 \text{ °F})(0.00322 \text{ ft}^2)} = 6.4°F \qquad (3.54)$$

3.15 Sample Problem—Temperature Rise Across a Small Air Gap

Table 3.5 Interface Conductance for Various Materials with an Interface Pressure of 10 psi

Material	Rms Surface 1	Rms Surface 2	Interface Conductance $\frac{\text{Btu}}{\text{hr ft}^2 \text{ °F}}$ Dry	$\frac{\text{Btu}}{\text{hr ft}^2 \text{ °F}}$ Oil[a]	$\frac{\text{cal}}{\text{sec cm}^2 \text{ °C}}$ Dry	$\frac{\text{cal}}{\text{sec cm}^2 \text{ °C}}$ Oil[a]
SAE 4141 steel	3	3	2200		0.297	
	70	85	400	1350	0.054	0.182
5051 aluminum	16	17	1800		0.243	
	60	60	1300	2000	0.175	0.270
	15	90	800	1600	0.108	0.216
Bronze AMS 4846	70	80	800	1200	0.108	0.162

[a] Oil thermal conductivity $K = 0.073$ Btu/hr ft °F = 0.00030 cal/sec cm °C.

In metric units:

$$\Delta t = \frac{2.87 \text{ cal/sec}}{(0.270 \text{ cal/sec cm}^2 \text{ °C})(2.99 \text{ cm}^2)} = 3.5°C \qquad (3.54a)$$

Air is a relatively good conductor of heat. Often, the air at the interface of two surfaces will carry a large part of the heat if the surfaces are very rough or the interface pressure is very low. This can be demonstrated by considering a slight modification of the preceding sample problem.

3.15 SAMPLE PROBLEM—TEMPERATURE RISE ACROSS A SMALL AIR GAP

In the preceding sample problem, there is dirt on the bracket, so that there is poor physical contact between the component bracket and the heat sink.

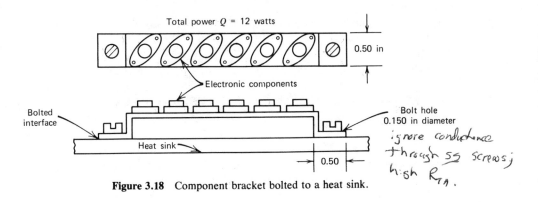

Figure 3.18 Component bracket bolted to a heat sink.

There is an air gap of 0.001 in (0.0025 cm) under the bracket. Determine the temperature rise across the air gap.

SOLUTION

The temperature rise across the air gap is determined with the use of the conduction relation shown in Eq. 3.1.

Given $Q = 12$ watts $= 40.9$ Btu/hr $= 2.87$ cal/sec (heat input)
$L = 0.001$ in $= 0.000083$ ft $= 0.0025$ cm (length of gap)
$K = 0.017$ Btu/hr ft °F $= 0.0000702$ cal/sec cm °C (air conductivity at 160°F)
$A = 0.464$ in² $= 0.00322$ ft² $= 2.99$ cm² (interface area)

In English units:

$$\Delta t = \frac{(40.9 \text{ Btu/hr})(0.000083 \text{ ft})}{(0.017 \text{ Btu/hr ft °F})(0.00322 \text{ ft}^2)} = 62.0°F \qquad (3.55)$$

In metric units:

$$\Delta t = \frac{(2.87 \text{ cal/sec})(0.0025 \text{ cm})}{(0.0000702 \text{ cal/sec cm°C})(2.99 \text{ cm}^2)} = 34.2°C \qquad (3.55a)$$

When a thermal grease is used at the interface of the mounting bracket, the temperature rise across the interface can be reduced. Thermal greases have thermal conductivities that generally range from about 0.2 to about 0.9 Btu/hr ft °F (0.00083 to 0.0037 cal/sec cm °C). For an average value of about 0.5 Btu/hr ft °F (0.0021 cal/sec cm °C), with an average thickness of 0.002 in (0.000166 ft), the temperature rise across the interface will be

$$\Delta t = \frac{(40.9)(0.000166)}{(0.5)(0.00322)} = 4.2°F \ (2.3°C) \qquad (3.56)$$

To get the same temperature rise across the bracket interface without the thermal grease, a better surface finish and a higher interface pressure will be required.

3.16 HEAT CONDUCTION ACROSS INTERFACES AT HIGH ALTITUDES

The amount of heat that can be transferred between two clamped surfaces is often sharply reduced at high altitudes. No heat will be conducted between two adjacent surfaces in a hard vacuum environment if they are not

3.16 Heat Conduction Across Interfaces at High Altitudes

in intimate contact with each other. Small amounts of heat may be radiated from one interface to another, depending upon the temperatures and the physical properties of the surfaces. However, conduction heat transfer requires some medium—solid, liquid, or gas—through which the heat can be transferred.

The change in the interface conductance between two adjacent surfaces with a low interface pressure, for various altitude conditions from sea level to a hard vacuum, is shown in Figure 3.19. This curve shows that special care must be used at the thermal interfaces of electronic assemblies that will be cooled by conduction in outer space applications. It also shows that there is very little change in the interface conduction for altitudes up to about 80,000 ft. Above that altitude, the interface conductance values begin to drop off very rapidly [25, 26].

Rigid thermal interfaces must be provided for surfaces that are expected to transfer heat in a vacuum environment. Test data show that typical sheet metal types of structures can have interface conductances that are only 10% of what they are at sea level. Some test data for different surface finishes are shown in Figure 3.20.

Most of the interface test data found in the literature are based upon very small and very rigid structural elements. This type of test sample is easy to control. The interface pressure can be established accurately, it is uniform across the entire surface, and different finishes are easy to obtain.

In the real world of electronic structures, the interface areas are often quite large, the interface pressures are not uniform, and the surfaces are not stiff and flat. Because of these factors, the actual interface conductances for large sheet metal structures will generally be much lower than most published test data show. Tests on large bolted sheet metal plates show that typical interface conductances will probably be in the range of a few hundred

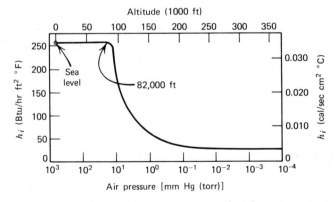

Figure 3.19 Interface conductance with contact pressure of 2 psi (140.7 g/cm^2). Copper, 0.011 TIR, 10 μin rms finish in contact with brass, 0.004 TIR, 22 μin rms finish, as a function of altitude.

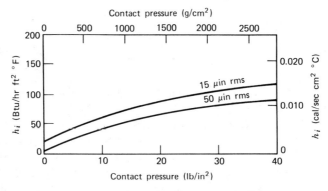

Figure 3.20 Interface conductance for 2024 T6 aluminum blocks at 10^{-4} torr.

rather than in the range of a few thousand (in English units). Some test data for large bolted sheet metal plates are shown in Table 3.6. Additional test data on various interfaces are shown in Tables 3.7 and 3.8.

When it is necessary to conduct heat across large nonrigid surfaces in a hard vacuum, some special heat transfer interfaces may have to be fabricated. One type of assembly that can provide an interface conductance of about 100 Btu/hr ft² °F (0.0135 cal/sec cm² °C) with an interface pressure of only 10 lb/in² (703.7 g/cm²) in a hard vacuum (10^{-6} torr) is shown in Figure

Table 3.6 Average Interface Conductance for Two As-Received Aluminum Plates 10 × 20 × 0.25 in Thick (25.4 × 50.8 × 0.635 cm) Bolted Only at the Four Corners and Using Various Interface Materials

	Interface Conductance (h_i)			
	In Air at Sea Level		Vacuum (10^{-6} torr)	
Interface Material	Btu hr ft² °F	cal sec cm² °C	Btu hr ft² °F	cal sec cm² °C
0.002 in (0.0051 cm) copper or aluminum foil	300	0.0405	100	0.0135
0.020 in (0.0508 cm) rubber-impregnated aluminum screen	200	0.027	40	0.0054
0.005 in (0.0127 cm) beryllium copper foil with 1000 small fingers	200	0.027	40	0.0054
No interface material	200	0.027	20	0.0027

Table 3.7 Average Interface Conductance for 5052 (1/4 Hard) Aluminum Strip 3.0 × 0.25 × 0.06 in Thick (7.62 × 0.635 × 0.152 cm) Clamped to a 0.25 in (0.635 cm) Thick Aluminum Plate[a]

	Interface Conductance (h_i)			
	Sea Level		10^{-5} Torr (mm hg)	
Interface Pressure	$\dfrac{\text{Btu}}{\text{hr ft}^2 \,°\text{F}}$	$\dfrac{\text{cal}}{\text{sec cm}^2 \,°\text{C}}$	$\dfrac{\text{Btu}}{\text{hr ft}^2 \,°\text{F}}$	$\dfrac{\text{cal}}{\text{sec cm}^2 \,°\text{C}}$
500 psi	350	0.0472	170	0.0230
1000 psi	520	0.0702	340	0.0459

[a] Surface finish on both members about 15 μin rms.

3.21. The interface assembly consists of a soft thin copper foil that is wrapped around a silicone rubber tube and soldered to a flat copper sheet.

3.17 OUTGASSING AT HIGH ALTITUDES

Many materials tend to release trapped gasses when the surrounding pressure is reduced. This applies to many common materials, such as rubbers, epoxies, greases, plastics, and even metals. More trapped gasses are released when the outside pressures are lower.

Outgassing may not be a problem unless there is optical equipment in the area or a hard vacuum condition is required for thermal isolation or electrical isolation.

Camera lenses and mirrors, which require a smooth clear surface, can often become coated if outgassing materials are present. The coating process may take a few days or a few weeks to develop, depending upon the

Table 3.8 Average Interface Conductance for a Small Chassis with a 5052 Aluminum Base Plate 4 × 6 × 0.125 in Thick (10.16 × 15.24 × 0.317 cm) Bolted to a 1.0 in (2.54 cm) Thick Cold Plate with Four No. 10–32 Screws at the Corners[a]

	Interface Conductance (h_i)			
	Sea Level		10^{-5} torr (mm hg)	
Interface Condition	English Units	Metric Units	English Units	Metric Units
Bolted at the four corners	220	0.0297	50	0.00675

[a] Surface finish on both members about 32 μin rms.

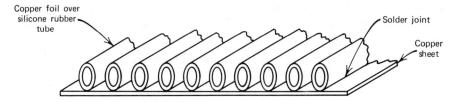

Figure 3.21 Resilient assembly for conducting heat across large nonrigid interfaces in a hard vacuum.

conditions. Once the optical surfaces are coated, however, the performance of the equipment will degrade substantially.

Outgassing can have the effect of increasing the pressure in a local area. Some electronic applications require thermal isolation to permit high temperatures to be developed with very low heating element power. If outgassing is present, it will substantially increase the heat transfer by gaseous conduction in small confined areas, so that the required high temperatures may never be reached.

In electrical applications, it is possible for arcing to develop if outgassing occurs in a vacuum environment. Arcing may not occur at sea level conditions or in a hard vacuum. However, with outgassing present, the effective altitude in the immediate area is changed so that ionization in the gas can occur and arcing will develop.

3.18 CIRCUIT BOARD EDGE GUIDES

Plug-in PCBs are often used with guides, which help to align the PCB connector with the chassis connector. These guides, which are usually fastened to the side walls of a chassis, grip the edges of the PCB as it is inserted and removed. If there is enough contact pressure and surface area at the interface between the edge guide and the PCB, the edge guide can be used to conduct heat away from the PCB. A typical installation is shown in Figure 3.22.

Plug-in PCBs must engage a blind connector, which is usually fastened to the chassis. If the chassis connector is rigid, with no floating provisions, the edge guide must provide that float, or many connector pins will be bent and broken. Rigid chassis connectors are generally used for electrical connections because wire wrap harnesses, flex tape harnesses, and multilayer master interconnecting mother boards are being used for production systems. These interconnections cannot withstand the excessive motion required for a floating connector system. Therefore, the floating mechanism is often built into the circuit board edge guides.

High pressures with dry, smooth interfaces must be provided in the edge connector if it is to be a good heat conductor. Thermal greases are unde-

3.18 Circuit Board Edge Guides

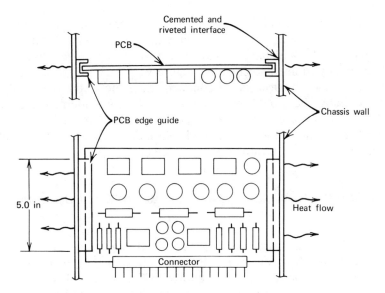

Figure 3.22 Plug-in PCB assembly with board edge guides.

sirable at these interfaces because they are very messy to handle and must be replaced every time a PCB is pulled for inspection or repair.

Many different types of circuit board edge guides are used by the electronics industry. The thermal resistance across a typical guide is rather difficult to calculate accurately, so that tests are used to establish these values. The units used to express the thermal resistance across an edge guide are °C in/watt. When the length of the guide is expressed in inches and the heat flow is in watts, the temperature rise is in degrees Celsius. Figure 3.23 shows four different types of edge guides, with the thermal resistance across each guide.

These board edge guides are generally satisfactory for applications at sea level or medium altitudes up to about 50,000 ft. At altitudes of 100,000 ft, test data show that the resistance values for the guides in Figure 3.23a–c will increase about 30%. The wedge clamp shown in Figure 3.23d will have about a 5% increase.

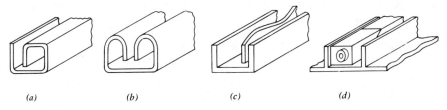

Figure 3.23 Board edge guides with typical thermal resistance. (a) G guide, 12°C in/watt; (b) B guide, 8°C in/watt; (c) U guide, 6°C in/watt; (d) wedge clamp, 2°C in/watt.

For hard vacuum conditions, where the pressure is less than 10^{-6} torr (mm hg), as normally experienced in outer space work, some of these edge guides may not be satisfactory, except for very low power dissipations. Outer space environments generally require very high pressure interfaces, with flat and smooth surfaces, to conduct heat effectively. Wedge clamps are very effective here because they can produce high interface pressures. Bolted interfaces are also used extensively because even small bolts can produce high forces. However, bolted interfaces lack the plug-in feature that is considered very desirable for maintenance.

3.19 SAMPLE PROBLEM—TEMPERATURE RISE ACROSS A PCB EDGE GUIDE

Determine the temperature rise across the PCB edge guide (from the edge of the PCB to the chassis wall) for the assembly shown in Figure 3.22. The edge guide is 5.0 in long, type c, as shown in Figure 3.23. The total power dissipation of the PCB is 10 watts, uniformly distributed, and the equipment must operate at 100,000 ft.

SOLUTION

Since there are two edge guides, half of the total power will be conducted through each guide. The temperature rise at sea level conditions can be determined from Eq. 3.57.

$$\Delta t = \frac{RQ}{L} \qquad (3.57)$$

Given $R = 6$ °C in/watt
$Q = 10/2 = 5$ watts (on half of the PCB)
$L = 5$ in (length of one guide)

$$\Delta t = \frac{(6 \text{ °C in/watt})(5 \text{ watts})}{5 \text{ in}} = 6°C \text{ (rise at sea level)}$$

At an altitude of 100,000 ft, test data and Figure 3.24 show that the resistance across the edge guide will increase about 30%. The temperature rise at this altitude will then be

$$\Delta t = 1.30(6°C) = 7.8°C \text{ (rise)} \qquad (3.58)$$

3.20 HEAT CONDUCTION THROUGH SHEET METAL COVERS

Lightweight electronic boxes often use sheet metal covers to enclose the chassis. Sometimes these covers are expected to conduct the heat away

3.21 Radial Heat Flow

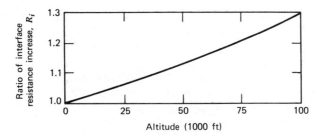

Figure 3.24 Change of board edge guide thermal interface resistance as a function of altitude (types *a–c*, Figure 3.23).

(R_i FOR IERC ZIF, WEDGE CLAMPS IS CONSTANT W/R TO ALTITUDE

from the chassis for additional cooling. To effectively conduct heat from the chassis to the cover, a flat, smooth, high pressure interface must be provided. The normal surface contact obtained at the interfaces of sheet metal members will not provide an effective heat conduction path. However, if a lip can be formed in the cover or on the side walls of the chassis, the heat conduction path can be substantially improved. Figure 3.25 shows how lips may be formed in the chassis and cover to improve the heat transfer across the interface.

3.21 RADIAL HEAT FLOW

Electronic support structures are not always rectangular in shape. If an electronic system is enclosed within a cylindrical structure, such as a small missile, it would be a waste of space to use a rectangular box for the electronics. The packaging form factor would probably conform to the natural cylindrical shape. Electronic components mounted on the inside surface of a cylindrical chassis would require a radial heat flow path to remove the heat when the heat sink is on the outer surface of the structure. The temperature rise from the inside surface to the outside surface of the

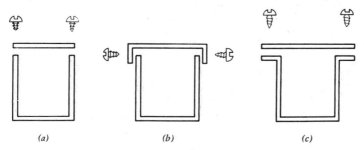

Figure 3.25 Different types of sheet metal covers on electronic boxes. (*a*) Poor; (*b*) good; (*c*) good.

cylindrical structure can be determined from the standard heat flow relation shown by Eq. 3.2 for a concentrated heat load. However, in a radial heat flow system, the cross section area is constantly changing as the radius changes, as shown in Figure 3.26.

An average cross section area can be used to obtain a rough approximation of the temperature rise through the cylindrical wall. If a more accurate evaluation of the temperature rise is desired, the logarithmic mean area must be used, as shown by Eq. 3.59 [15, 17, 24].

$$A_{lm} = \frac{A_{out} - A_{in}}{\log_e(A_{out}/A_{in})} \qquad (3.59)$$

When the ratio of A_{out}/A_{in} is less than 2, the average cross section area is within 4% of the logarithmic area.

3.22 SAMPLE PROBLEM—TEMPERATURE RISE THROUGH A CYLINDRICAL SHELL

A hollow steel cylindrical sheel has a group of resistors mounted on the inside surface, as shown in Figure 3.27. Heat is removed from the outside surface of the cylinder, so that the heat flow path is radial through the wall of the cylinder. Determine the temperature rise through the steel cylinder when the power dissipation is 10 watts.

SOLUTION

The logarithmic mean area is determined from Eq. 3.59.

Given $\quad A_{out} = \pi D_{out} a = (3.14)(4.2)(1.5) = 19.79 \text{ in}^2 = 0.1374 \text{ ft}^2$
$\quad\quad\quad\quad A_{in} = \pi D_{in} a = (3.14)(2.0)(1.5) = 9.42 \text{ in}^2 = 0.0654 \text{ ft}^2$
$\quad\quad\quad\quad Q = 10 \text{ watts} = 34.13 \text{ Btu/hr (heat input)}$
$\quad\quad\quad\quad K = 24 \text{ Btu/hr ft °F (steel conductivity)}$

$$A_{lm} = \frac{0.1374 - 0.0654}{\log_e(0.1374/0.0654)} = 0.0970 \text{ ft}^2 \text{ (logarithmic mean area)}$$

Substitute into Eq. 3.2 for the temperature rise through the steel cylin-

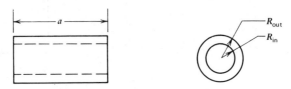

Figure 3.26 Hollow cylinder.

3.22 Sample Problem—Temperature Rise Through a Cylindrical Shell

Figure 3.27 Resistors mounted on the inside wall of a hollow cylinder.

drical wall. The length of the heat flow path is the thickness of the steel wall, 1.1 in = 0.0916 ft.

$$\Delta t = \frac{(34.13)(0.0916)}{(24)(0.0970)} = 1.34°F \qquad (3.60)$$

If the average cross section area for radial heat flow through the cylinder is used:

$$A_{av} = \frac{A_{out} + A_{in}}{2} = \frac{0.1374 + 0.0654}{2} = 0.1014 \text{ ft}^2$$

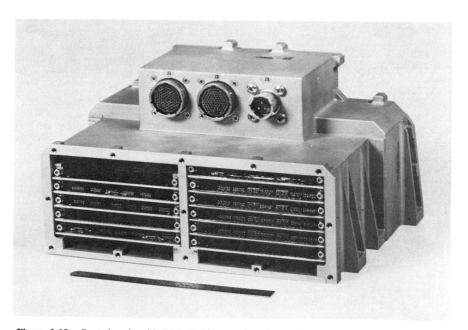

Figure 3.28 Cast chassis with PCBs held by wedge clamps for conduction cooling. (Courtesy Litton Systems, Inc.)

Substitute into Eq. 3.2 for the temperature rise.

$$\Delta t = \frac{(34.13)(0.0916)}{(24)(0.1014)} = 1.28°F \qquad (3.61)$$

In this case there is very little difference in the temperature rise using the logarithmic mean cross section area or the average cross section area for the radial heat flow through the cylinder.

Figure 3.28 shows a rugged cast aluminum chassis with conduction-cooled circuit boards held by wedge clamps.

4

Mounting and Cooling Techniques for Electronic Components

4.1 VARIOUS TYPES OF COMPONENTS

Plug-in types of printed circuit boards (PCBs) have become the standard in the electronics industry. This type of equipment is easy to troubleshoot and maintain. A considerable amount of time and effort has therefore been expended in finding ways to improve the performance and to reduce the cost of PCBs. Many different construction techniques and many different cooling schemes have been devised by the electronic companies involved in the manufacturing of this hardware.

One area that has received a considerable amount of attention is that of mounting electronic components on PCBs. Many different types of components are involved with various shapes and electrical lead wire arrangements. Dual inline packages (DIPs), large scale integrated circuits (LSIs), hybrids, and most recently the microprocessors have been replacing the discrete resistors, capacitors, transistors, and diodes. However, many discrete components are still being used because of their low costs [29].

No matter which groups of components are used for an electronic system, the mounting techniques must provide sufficient cooling to permit the device to operate effectively in its environment. The best method for determining the effectiveness of any cooling system appears to be in the measurement of the case and the junction temperatures of the individual electronic components. Experience has shown that excessive component case temperatures will result in rapid electronic failures. Reliability studies have shown that it is not a good practice to exceed case temperatures of 212°F (100°C) for long periods on electrically operating equipment with any of the components mentioned above, or the failure rates will show a large increase.

When the electronic components are to be cooled by conduction, a good heat flow path must be provided from each component to the ultimate heat sink. The ultimate heat sink may simply be the outside ambient air, or it may be a sophisticated liquid cooled heat exchanger. Each segment along the heat flow path must be examined in detail to be sure that the thermal resistance is low enough to ensure the proper cooling [30].

4.2 MOUNTING COMPONENTS ON PCBS

The electronic components are usually the major heat source in electronic systems. For effective cooling of these components, it is necessary to plan in advance the heat flow mechanism and heat flow path from the component to the sink. The method used for mounting components to PCBs cooled by conduction should provide a low thermal resistance from the component to the PCB. Metal heat sinks are often laminated to the PCB because metals have a high thermal conductivity. The two most popular metals are aluminum and copper, which can be laminated in several different ways, as shown in Figure 4.1.

Some types of components, such as DIPs, LSIs, hybrids, and microprocessors, have a large number of electrical lead wires extending from their cases. About half of the heat dissipated by these components may be conducted through the wires, and about half may be conducted through the bodies of each component. Therefore, when provisions are made for mount-

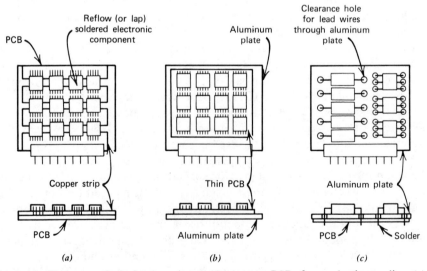

Figure 4.1 Various methods for mounting components on PCBs for conduction cooling. (*a*) Copper strips over PCB and under components; (*b*) aluminum plate bonded to a thin PCB; (*c*) clearance holes in aluminum plate with PCB on back side.

4.2 Mounting Components on PCBs

ing these components on PCBs, adequate attention should be given to the heat flow through the wires. For example, if the mounting method shown in Figure 4.1b is used, plated throughholes should be located in every pad that has a soldered lead connection. Plated throughholes will reduce the thermal resistance through the PCB, which will reduce the temperature rise from the component to the aluminum heat sink plate. Plated throughholes will also prevent the pads from lifting off the PCB, when the components are soldered to the pads on a lap-soldered joint.

The body of the component with a high heat dissipation (greater than about 4 watts/in^2) should be fastened to the PCB so that there is a low thermal resistance path from the body of the component to the heat sink plate. Many different adhesives can be used, such as RTV cements, double-back tape (Mylar or Kapton), epoxies, or just the adhesive quality of a conformal coating if it is used to protect the PCB. The heat flow from the components to the ultimate heat sink should then follow a predetermined path similar to the one shown in Figure 4.2 [9, 29-31].

Some type of adhesive should be used under high power components mounted on the PCB. If no adhesive is used and only an air gap exists, it is very difficult to control the size of the gap. A small air gap may grow into a large air gap after the PCB has been handled by several people. As the uncontrolled air gap grows, the thermal resistance also grows. This increases the temperature rise across the gap, which increases the component case temperatures.

Tests were run on several different types of carbon composition resistors, which were mounted on heat sinks using different mounting methods. The thermal resistance from the top surface of the resistors to the aluminum heat sink immediately under the resistors is shown in Table 4.1.

Tests were also run on $\frac{1}{4} \times \frac{1}{4}$ in flat pack integrated circuits mounted on a 0.005 in thick PCB which was laminated to an aluminum heat sink using a 0.003 in thick prepreg cement, making a total PCB thickness of 0.008 in. The results are shown in Table 4.2.

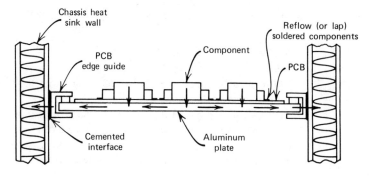

Figure 4.2 Conduction heat flow path from component to heat sink.

Table 4.1 Thermal Resistance for Resistors Mounted on an Aluminum Heat Sink

Component	Interface Cemented with Humiseal Conformal Coating	Resistance with No Cement
$\frac{1}{4}$ watt Resistor RC 07	46.2°C/watt	75°C/watt
$\frac{1}{2}$ watt Resistor RC 20	34.2°C/watt	58°C/watt
1 watt Resistor RC 32	19.1°C/watt	26°C/watt

Epoxy adhesives under components should be avoided if extensive repairs are anticipated on the PCB. Components cemented with epoxy adhesives are extremely difficult to remove without damaging the PCB. Other adhesives, such as RTVs, should be used if repairs are required.

Double-sided Mylar tapes work well for holding components to the PCB. However, cleaning solvents normally used to clean the solder flux off of the lamina will dissolve some of the Mylar adhesive and wash it away. At least half of the adhesive can be lost due to cleaning solvents. Pull tests should be run on components that use double-back Mylar tape, to make sure most of the solvent has not been washed away. A minimum pull force of at least 15 psi should be required in the vertical direction (perpendicular to the PCB) for acceptance. This means that a component with a body size of 1 by 1 in should be capable of withstanding a straight vertical pull force of 15 pounds perpendicular to the plane of the PCB.

Table 4.2 Thermal Resistance from Integrated Circuit Case, through 0.008 in Thick Epoxy Fiberglass Lamina, to Aluminum Heat Sink Plate

Component	Resistance with Cement under Component	Resistance with Air Gap under Component
$\frac{1}{4} \times \frac{1}{4}$ in flat pack	20°C/watt with 0.001 in Humiseal conformal coat	—
$\frac{1}{4} \times \frac{1}{4}$ in flat pack	29°C/watt with 0.003 in thick double-sided Mylar tape	60°C/watt with 0.005 in air gap
$\frac{1}{4} \times \frac{1}{4}$ in flat pack	30°C/watt with 0.002 in thick RTV adhesive over 0.004 in thick Stycast epoxy, which insulates copper runs on PCB	35°C/watt with 0.001 in air gap over 0.004 in thick Stycast epoxy, which insulates copper runs on PCB

4.3 Sample Problem—Hot Spot Temperature of an Integrated Circuit on a Plug-In PCB

Extra plated throughholes should be used under the bodies of high power dissipating components mounted on PCBs that are cooled by conduction. The plated throughholes have a small cylindrical ring of copper, which is a good heat conductor. This can decrease the thermal resistance through the PCB lamina as much as 25% in many cases. A lower thermal resistance means a lower component case temperature.

Low power dissipating components may not require any type of adhesive for fastening the components to the PCB. It may be possible to use an air gap under the components if the height of the gap can be controlled to a value less than about 0.005 in. Test data in Table 4.2 show the thermal resistance across an air gap of 0.005 in (including an 0.008 in thick epoxy fiberglass lamina) is 60°C/watt.

When the power dissipation of a $\frac{1}{4} \times \frac{1}{4}$ inch integrated circuit is only 0.10 watt, the temperature rise across the 0.005 in air gap (and the 0.008 in lamina) will only be about 6°C. If this is acceptable, a substantial sum of money may be saved in the fabrication of these PCBs [32].

PCBs with air gaps under the components must not be used for very high altitude or for space applications. There is no heat conduction across an air gap in a vacuum.

4.3 SAMPLE PROBLEM—HOT SPOT TEMPERATURE OF AN INTEGRATED CIRCUIT ON A PLUG-IN PCB

An electronic chassis, with many plug-in types of PCBs, is to be bolted to a liquid cooled cold plate because the system must be capable of operating at an altitude of 100,000 ft (30,480 m). Conduction cooling is the only reliable method of cooling, because natural convection is sharply reduced at high altitudes. In addition, there are other warm electronic boxes in the same equipment bay, which will prevent effective cooling by radiation. The PCBs will be populated with flat pack integrated circuits. These will be uniformly distributed, with a total power dissipation of 8 watts for each board, as shown in Figure 4.3. Components are mounted on both sides of a PCB which has an aluminum heat sink core at the center. The maximum allowable component case temperature is 212°F (100°C). Determine if the proposed design is satisfactory.

SOLUTION

The hot spot case temperature on the flat pack IC can be determined by calculating the temperature rise along individual segments of the heat flow path from the components to the liquid cooled cold plate heat sink. The heat flow path is broken up into five individual segments as follows:

Δt_1 = temperature rise from integrated circuit (IC) case through an 0.008

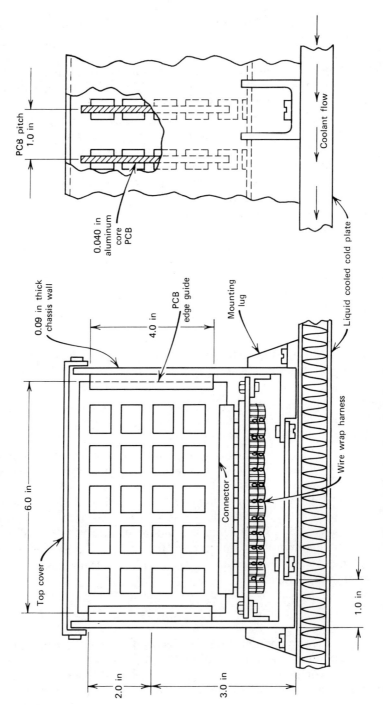

Figure 4.3 Cross section through a chassis with flat pack integrated circuits on PCBs.

4.3 Sample Problem—Hot Spot Temperature of an Integrated Circuit on a Plug-In PCB

in (0.020 cm) thick PCB lamina, plus an air gap of 0.005 in (0.013 cm) under IC case

Δt_2 = temperature rise from center of aluminum heat sink core on circuit board to sides of heat sink by edge guide

Δt_3 = temperature rise across board edge guide from heat sink core to chassis side wall, including cemented interface from guide to wall at 100,000 ft

Δt_4 = temperature rise down along the chassis side wall from the board edge guide to the base of the chassis

Δt_5 = temperature rise across the bolted interface from the base of the chassis to the liquid cooled cold plate at 100,000 ft

The various temperature rises are determined for each of the five segments, using English units and metric units.

All the PCBs in the chassis are very similar and all are on a 1 in pitch. For convenience, therefore, at 1.0 in (2.54 cm) chassis width is analyzed together with one typical PCB.

Δt_1: This represents the temperature rise across the 0.008 in thick PCB lamina to the aluminum core for a single integrated circuit (IC) component that dissipates 0.20 watt. To reduce manufacturing costs, the ICs will *not* be cemented to the PCB. Instead, a 0.005 in air gap will be allowed under the component. This will be controlled by forming the electrical lead wires in a forming die. The ICs will be lap soldered to the printed circuits on the laminated board. The thermal resistance from the component to the aluminum center core, which includes the 0.008 in lamina and the 0.005 in air gap, is obtained from Table 4.2.

$$\Delta t_1 = 60 \frac{°C}{watt} \times 0.20 \text{ watt} = 12°C = 21.6°F \qquad (4.1)$$

Δt_2: This represents the temperature rise along the aluminum core from the center of the board to the sides, for a uniformly distributed heat load of 8.0 watt, on both sides of the PCB. Since the heat load is symmetrical, each half of the heat sink can be considered separately, as shown in Figure 4.4. The following information is required.

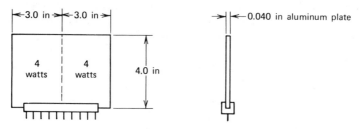

Figure 4.4 Aluminum PCB heat sink core.

Given $Q = 4.0$ watts $= 13.65$ Btu/hr $= 0.956$ cal/sec (heat input on half a board)
$L = 3.0$ in $= 0.25$ ft $= 7.62$ cm (length)
$K = 83$ Btu/hr ft °F $= 0.343$ cal/sec cm °C (aluminum conductivity)
$A = (4)(0.04) = 0.16$ in² $= 0.00111$ ft² $= 1.03$ cm² (area)

Substitute into Eq. 3.9 for the Δt from the center of the board heat sink to the side edge, in English units.

$$\Delta t_2 = \frac{(13.65 \text{ Btu/hr})(0.25 \text{ ft})}{(2)(83 \text{ Btu/hr ft °F})(0.00111 \text{ ft}^2)} = 18.5°F \qquad (4.2)$$

In metric units:

$$\Delta t_2 = \frac{(0.956 \text{ cal/sec})(7.62 \text{ cm})}{(2)(0.343 \text{ cal/sec cm °C})(1.03 \text{ cm}^2)} = 10.3°C \qquad (4.2a)$$

Δt_3: This represents the temperature rise across the board edge guide from the aluminum PCB core to the side wall of the chassis, using the Birtcher-type edge guide shown in Figure 3.23a.

$$\Delta t_3 = \left(12 \frac{°C \text{ in}}{\text{watt}}\right)\left(\frac{4 \text{ watts}}{4 \text{ in}}\right) = 12°C \text{ (at sea level)}$$

At an altitude of 100,000 ft this resistance across the edge guide increases about 30%, as shown in Figure 3.24.

$$\Delta t_3 = 1.3(12°C) = 15.6°C = 28.1°F \qquad (4.3)$$

Δt_4: This represents the temperature rise along the side of the chassis wall down to the bottom surface of the chassis. The temperature rise can be calculated as (a) a concentrated heat load down the side of the chassis or as (b) a combination of a uniformly distributed heat load and a concentrated heat load down the side of the chassis. Both methods will give the same results. The PCBs are on a 1.0 in (2.54 cm) pitch, so that a 1.0 in section of the chassis is analyzed.

(a) *Concentrated Heat Load.* For the concentrated heat load down the side of the chassis, the length of the heat flow path is taken from the center of the PCB to the center of the bottom flange, as shown in Figure 4.5.

$Q = 4$ watts $= 13.65$ Btu/hr $= 0.956$ cal/sec (heat; ref. Table 1.1)
$L = 3.5$ in $= 0.292$ ft $= 8.89$ cm (length)
$K = 83$ Btu/hr ft °F $= 0.343$ cal/sec cm °C (thermal conductivity; ref. Table 3.2)
$A = (1.0)(0.09) = 0.090$ in² $= 0.000625$ ft² $= 0.581$ cm² (area)

4.3 Sample Problem—Hot Spot Temperature of an Integrated Circuit on a Plug-In PCB

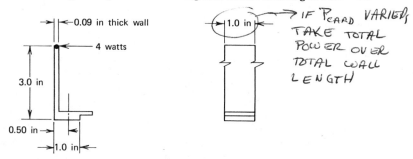

Figure 4.5 Concentrated heat load on side wall of the chassis.
(hot spot at middle of card)

IF P_{card} VARIES, TAKE TOTAL POWER OVER TOTAL WALL LENGTH

Substitute into Eq. 3.2 for the temperature rise in English units.

$$\Delta t_{4a} = \frac{(13.65)(0.292)}{(83)(0.000625)} = 76.8°F \qquad (4.4)$$

In metric units:

$$\Delta t_{4a} = \frac{(0.956)(8.89)}{(0.343)(0.581)} = 42.6°C \qquad (4.4a)$$

(b) *Uniform Heat Load with Concentrated Heat Load.* For the combination of a uniformly distributed heat load and a concentrated heat load down the side of the chassis, the uniform load acts only on the upper section of the chassis, as shown in Figure 4.6.

Using Eqs. 3.9 for the uniform heat load, and 3.2 for the concentrated

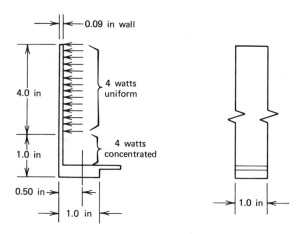

Figure 4.6 Uniform heat load combined with a concentrated heat load.
(hot spot at top of card)

heat load:

$$\Delta t_{4b} = \frac{(13.65)(4.0/12)}{(2)(83)(0.000625)} + \frac{(13.65)(1.5/12)}{(83)(0.000625)}$$

$$\Delta t_{4b} = 43.9 + 32.9 = 76.8°F \tag{4.5}$$

The results agree with Eq. 4.4

Δt_5: This represents the temperature rise across the interface from the chassis to the liquid cooled cold plate, for a section of the chassis that is 1 in wide. If the mounting bolts are spaced far apart on sheet metal structures, the average interface pressure will be quite low, so that the average interface conductance will also be quite low. For an air interface at sea level conditions, Table 3.6 shows a value of 200 Btu/hr ft² °F. At an altitude of 100,000 ft, the temperature rise will increase about 30%. (h_i does not increase)

$Q = 4$ watts $= 13.65$ Btu/hr $= 0.956$ cal/sec (heat)
$h_i = 200$ Btu/hr ft² °F $= 0.027$ cal/sec cm² °C (interface conductance)
$A = (1.0)(1.0) = 1.0$ in² $= 0.00694$ ft² $= 6.45$ cm² (area)

Substitute into Eq. 3.53 for an altitude of 100,000 ft, in English units.

$$\Delta t_5 = \frac{(1.30)(13.65)}{(200)(0.00694)} = 12.8°F \tag{4.6}$$

In metric units at the high altitude:

$$\Delta t_5 = \frac{(1.30)(0.956)}{(0.027)(6.45)} = 7.1°C \tag{4.6a}$$

The hot spot temperature at the mounting interface of the integrated circuits (ICs) at the center of the PCB can be determined from the 80°F (26.6°C) cold plate temperature and the sum of all the individual temperature rises in the heat flow path from the ICs to the cold plate. In English units:

$$t_{IC} = 80 + \Delta t_1 + \Delta t_2 + \Delta t_3 + \Delta t_4 + \Delta t_5 \tag{4.7}$$

$$t_{IC} = 80 + 21.6 + 18.5 + 28.1 + 76.8 + 12.8 = 237.8°F \tag{4.8}$$

In metric units:

$$t_{IC} = 26.6 + 12 + 10.3 + 15.6 + 42.6 + 7.1 = 114.2°C \tag{4.8a}$$

For a reliable electronic system, the maximum component surface temperature should not exceed 212°F (100°C). The design is therefore unsatisfactory, so that some design changes must be made. If the power dissipation cannot be reduced, the mechanical design of the structure must be changed.

An examination of the individual temperature rises shows that Δt_4 has the highest value of 76.8°F (42.6°C), which is shown in Eq. 4.4. The easiest way to reduce the temperature rise is to simply increase the wall thickness. If weight is not too critical, the wall thickness could be increased from 0.090 in to 0.120 in (0.230 to 0.305 cm). This increase would result in a new temperature rise, which is obtained with a direct ratio of the thickness.

$$\Delta t'_4 = 76.8 \left(\frac{0.090}{0.120}\right) = 57.6°F\ (32.0°C) \qquad (4.9)$$

This will result in a component surface temperature of 218.6°F (103.6°C), which is still too high, so that another change must be made. If the ICs are cemented to the PCB, Table 4.2 shows that the temperature rise will be reduced but the cost of the assembly will be increased. If a different PCB edge guide is used, the temperature rise can be reduced. For example, if the edge guide shown in Figure 3.23b is used, the thermal resistance across the guide will be 8°C in/watt. The new temperature rise across the guide can be obtained with a direct ratio of the resistance, using Eq. 4.3 for an altitude of 100,000 ft.

$$\Delta t'_3 = 28.1 \left(\frac{8}{12}\right) = 18.7°F\ (10.4°C) \qquad (4.10)$$

This will now result in a component surface temperature of 209.2°F (98.4°C), which is slightly less than the maximum allowable temperature for the 100,000 ft condition, so that the design is now satisfactory.

4.4 HOW TO MOUNT HIGH POWER COMPONENTS

High power components require special care for mounting, because temperatures can rise very rapidly and failures can occur quickly if the mounting interface conditions change [9]. Stud mounted diodes and transistors, which use insulating washers, must be mounted with care. Some types of insulating washers will deform and cold-flow after temperature cycling tests. This can result in a reduced mounting interface pressure between the component and the heat sink. This loose condition may not be noticed during assembly or inspection if a conformal coating is used, because the coating will tend to hold the parts in place. However, in any vibration or shock environments, the conformal coating may crack and the component will become loose. Without a good heat sink mount, these high power components will fail rapidly.

The insulating washers must be made of materials that will not crack or cold flow under high interface pressures. They must be capable of withstanding punctures from small sharp foreign particles and must have a high dielectric strength to prevent shorting. The insulating washers should also be easy to manufacture and to install.

Flat and smooth interfaces must be provided for the component and the heat sink mounting surfaces. All parts must be free from burrs and foreign or dirt particles, which would interfere with good heat transfer. The nut on stud mounted devices should be torqued to some predetermined value to ensure a consistent high interface pressure. Some manufacturers have been using Belleville-type spring washers under the nuts to compensate for minor cold flowing in the insulators. These spring washers tend to maintain a high component interface pressure under all conditions. A typical installation is shown in Figure 4.7.

Thermal greases, such as Dow Corning No. 340 and G.E. No. 2850, can increase the interface conductance values as much as 30%. When repairs are made, this thermal grease must be cleaned off and replaced carefully to ensure proper thermal operation. Caution must be exercised with thermal greases when cements or conformal coatings are used. Thermal greases that have a silicone base tend to migrate. In addition, these greases are very difficult to clean off. As a result, they often tend to contaminate clean surfaces, which interferes with the adhesion of cements and conformal coatings.

Junction temperatures must be controlled in high power components to prevent rapid failures. Most component manufacturers will specify the in-

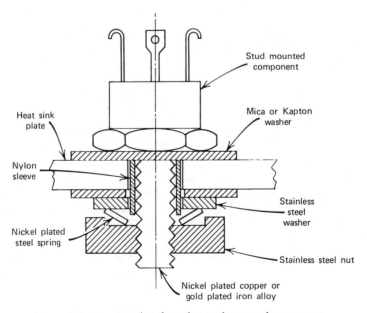

Figure 4.7 Cross section through a stud mounted component.

4.5 Sample Problem—Mounting High Power Transistors on a Heat Sink Plate

ternal thermal resistance from the junction of the device to the case using terms such as R_{jc} or Θ_{jc}. For high power components, this value is generally less than 1.0 °C/watt. These manufacturers often rate their components as 175°C or 200°C devices, which means that the junction temperatures are permitted to operate continuously at these values. For high reliable applications, however, the maximum allowable junction temperatures for power devices are usually specified as 125°C. When values this low are hard to achieve for a specific application, it may be possible for the manufacturer to reduce the internal resistance of the device so that the junction temperature will also be reduced for the same power dissipation. If this cannot be done, look for a similar device that has a lower internal resistance before making changes in the component mounting.

It is not always easy to control the junction or chip temperature within a power component. Many interfaces must be controlled to accomplish this. A typical cross section through such a device, showing these internal interfaces, is shown in Figure 4.8 [32].

The mounting position of the high power device within the chassis can often control the hot spot temperature. Sometimes it is relatively easy to reduce the hot spot component temperature just by changing the position of the component. For example, when a high power transistor is mounted at the center of the plug-in PCB, which is cooled by conduction, it will be much hotter than if the same component is mounted at the edge of the same PCB. This is demonstrated in the following sample problem.

4.5 SAMPLE PROBLEM—MOUNTING HIGH POWER TRANSISTORS ON A HEAT SINK PLATE

Several power transistors, which dissipate 5 watts each, are mounted on a power supply circuit board that has a 0.093 in (0.236 cm) thick 5052 alu-

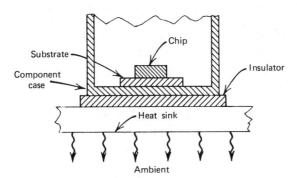

Figure 4.8 Cross section through a power device showing various heat transfer interfaces. The interfaces shown are: chip to substrate; substrate to case; case to insulator; insulator to heat sink; heat sink to ambient.

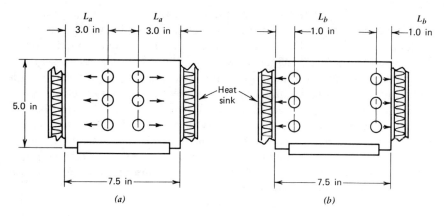

Figure 4.9 Power transistors mounted on an aluminum heat sink plate. (*a*) Old design; (*b*) new design.

minum heat sink plate, as shown in Figure 4.9. Determine how much lower the case temperatures will be when these components are mounted close to the edge of the PCB, as shown in Figure 4.9*b*, instead of the center, as shown in Figure 4.9*a*.

SOLUTION

Since both plug-in PCBs are symmetrical about the center, consider each half of the board for the analysis.

Given $Q = 3 \times 5 = 15$ watts $= 51.2$ Btu/hr $= 3.58$ cal/sec (heat)
$L_a = 3.0$ in $= 0.25$ ft $= 7.62$ cm (length of old design)
$L_b = 1.0$ in $= 0.0833$ ft $= 2.54$ cm (length of new design)
$K = 83$ Btu/hr ft °F $= 0.343$ cal/sec cm °C (conductivity 5052 aluminum)
$A = (5.0)(0.093) = 0.465$ in² $= 0.00323$ ft² $= 3.0$ cm² (area)

Substitute into Eq. 3.2 for the temperature rise at the two different mounting locations, in English units.

$$\Delta t_a = \frac{(51.2 \text{ Btu/hr})(0.25 \text{ ft})}{(83 \text{ Btu/hr ft °F})(0.00323 \text{ ft}^2)} = 47.7°F \quad (4.11)$$

In metric units:

$$\Delta t_a = \frac{(3.58 \text{ cal/sec})(7.62 \text{ cm})}{(0.343 \text{ cal/sec cm °C})(3.0 \text{ cm}^2)} = 26.5°C \quad (4.11a)$$

For the mounting position near the edge of the PCB, in English units:

$$\Delta t_b = \frac{(51.2)(0.0833)}{(83)(0.00323)} = 15.9°F \qquad (4.12)$$

In metric units:

$$\Delta t_b = \frac{(3.58)(2.54)}{(0.343)(3.00)} = 8.8°C \qquad (4.12a)$$

This shows that moving the transistors closer to the edge of the PCB can reduce the component surface mounting temperature by 47.7 − 15.9 = 31.8°F (17.7°C).

4.6 ELECTRICALLY ISOLATING HIGH POWER COMPONENTS

High power components, such as transistors and diodes, are often mounted on brackets, which are then mounted on heat sinks. When several different components are mounted on the same sink and the component cases are electrically hot, it is obvious that each component must be electrically isolated from the other. This is usually accomplished with the use of electrical isolators under each component, which are relatively good heat conductors. Various type of insulators are available ranging from a soft silicone rubber filled with thermally conductive metal oxides, to rigid but brittle beryllium oxide washers.

The soft type of silicone rubber gasket will deform when the mounting bolts on stud mounted components are tightened. The soft gasket tends to cold-flow into the small voids on the mating surfaces, which eliminates the air pockets on the component and the bracket. The thermal conductivity of the soft silicone gasket is not very high, but its ability to flow into small voids and its ability to deform to fit warped surfaces help to reduce the thermal resistance across the mounting interface.

Beryllium oxide washers provide a high thermal conductivity, which is equal to that of aluminum, while maintaining good electrical insulation. These washers are very brittle and may crack when mounting bolts are tightened if the mounting interfaces are not flat. If the mounting interfaces have small voids, the washer will not deform to fill them. Therefore, for high power applications and for easy replacement of parts, the washer is usually cemented to the heat sink and the component is usually cemented to the washer with an RTV cement, which can be removed easily.

There are conditions where the power requirements are so critical that no temperature gradient can be permitted between the component and the heat sink. Under these circumstances, the component is usually soldered directly to the heat sink so that the interface temperature rise is eliminated.

Hard anodized aluminum is a good electrical insulator and a good heat conductor. Often, components that have electrically hot cases or noncommon grounds may be mounted on the same heat sink plate, with no other insulation except the anodized finish for protection. This is not a good practice because experience has shown that the anodized finish is very hard and brittle and chips easily. This has often resulted in electrical shorts, which have caused electronic systems to malfunction.

Sometimes the thermal design of an electronic system will provide a component mounting surface temperature of 100°C, but the design is still not satisfactory, because the junction temperature is too high. Many component manufacturers make similar power devices which have different internal thermal resistances from the junction to the case (R_{jc}). Because of this, it is possible to have several different transistors or diodes with case temperatures of 100°C and junction temperatures that may range from 125°C to 150°C or higher. When the maximum allowable junction temperature is specified instead of the case temperature and the analysis shows that the resulting case temperature must be well below 100°C to provide the proper junction temperature, try looking for another transistor or diode which has a lower thermal resistance. If this is not available, a mechanical design change must be made to reduce the component case temperature.

It is extremely important to provide a low thermal resistance heat flow path from the high power dissipating components to the heat sink. When several different interfaces are involved, it may be possible to shift the location of the critical interface (the one with the highest temperature rise) so that the overall temperature rise to the component is reduced.

For example, in the sample problem shown in Figure 3.1, the transistor can be mounted directly to the aluminum bracket, without a mica washer. This will sharply reduce the temperature rise across the component interface, but it will make the bracket electrically hot. The bracket must now be electrically isolated from the heat sink. A thin strip of epoxy fiberglass can be used for the insulator. This can be made much larger than the mica washer because the interface from the bracket to the heat sink is much larger, so that the overall resistance is much lower. This results in a much lower temperature rise. This idea is illustrated in the following sample problem.

4.7 SAMPLE PROBLEM—MOUNTING A TRANSISTOR ON A HEAT SINK BRACKET

Figure 3.1 shows a sample problem where a transistor is mounted on an aluminum bracket with a mica washer insulator. Determine the component case temperature of the transistor when the mica washer is removed and the transistor is mounted directly to the aluminum bracket. Since the bracket will now be electrically hot, add a 0.002 in thick epoxy fiberglass insulator

4.7 Sample Problem—Mounting a Transistor on a Heat Sink Bracket

between the bracket and the heat sink. Compare the case temperatures of the transistor with and without the mica washer insulator.

SOLUTION

The heat transfer path from the transistor to the heat sink wall is broken up into three parts, and each part is examined separately, using Eq. 3.2.

Part 1. Temperature Rise across Cemented Epoxy Fiberglass

Given $L = 0.002$ in epoxy fiberglass + 0.002 in (epoxy cement)
$L = 0.004$ in = 0.000333 ft = 0.0101 cm (thickness)
$Q = 7.5$ watts = 25.6 Btu/hr = 1.79 cal/sec (heat dissipated)
$K = 0.167$ Btu/hr ft °F (thermal conductivity epoxy PCB + cement)
$A = (1.3)(0.75) = 0.975$ in² = 0.00677 ft² (insulator area)

Substitute into Eq. 3.2 using English units.

$$\Delta t_1 = \frac{(25.6)(0.000333)}{(0.167)(0.00677)} = 7.5°F \qquad (4.13)$$

Part 2. Temperature Rise along Aluminum Bracket

This will be the same as the value shown by Eq. 3.4 because the conditions are exactly the same.

$$\Delta t_2 = 41.1°F \qquad \text{(ref. Eq. 3.4)}$$

Part 3. Temperature Rise across Transistor Interface without Mica Washer

Table 3.3 is used to determine the temperature rise across the transistor interface, considering a No. 10-32 bolt stud.

$R = 0.60$ °C/watt (dry interface; ref. Table 3.3)
$Q = 7.5$ watts

$$\Delta t_3 = (7.5 \text{ watts})\left(0.60 \frac{°C}{\text{watt}}\right) = 4.5°C = 8.1°F \qquad (4.14)$$

The transistor case temperature is determined by adding the heat sink temperature of 131°F to the three Δt's just computed.

$$t_{\text{trans}} = 131 + 7.5 + 41.1 + 8.1 = 187.7°F \qquad (4.15)$$

Comparing Eqs. 4.15 and 3.6 shows there can be a net saving of 35.7°F by removing the mica washer and reducing the thickness of the epoxy cement while adding an insulator at the bracket interface to the heat sink.

4.8 POTTED MODULES

Electronic components with high heat densities, or components that must operate in severe vibration and shock environments, are often encapsulated (or potted) in various types of foams, epoxies, and silicone compounds. Thermally conductive additives, such as aluminum powder and metal oxides, may also be used to improve the overall thermal conductivity of the potting material.

Aluminum powder additives can increase the thermal conductivity of an epoxy by a factor of 10. However, experience with aluminum powder has shown that shorts can develop under certain circumstances. Tests on potted modules, which are filled with aluminum powder, show that when they are machined with sharp cutters, there will be no shorts between two points on the machined surfaces. However, when a dull tool, such as the tip of a screwdriver, is drawn across the machined surface, there will be electrical continuity between any two points within the groove. Apparently, a sharp machine cutting tool generates a clean cut on the small aluminum particles. A dull tool, on the other hand, smears the soft particles so that they overlap and join one another to form an electrically conductive path, which can cause short circuits.

High power transformers and chokes are often potted in thermally conductive epoxies to improve their heat transfer. These devices have a large amount of iron and copper, which are good heat conductors themselves. However, it is still necessary to provide a low resistance thermal path from the copper and iron to the ultimate heat sink, to prevent overheating in severe operating environments.

Discrete electronic components such as resistors, capacitors, diodes, and transistors are often potted in small modules to improve their vibration and shock resistance or to provide a single compact package that can be handled and mounted like a single component. Heat removal can be a problem, even for packages with relatively low power dissipations, unless proper attention is given to the packaging of these components.

The orientation of the component electrical lead wires within the potted module can sharply influence the heat transfer characteristics. This is because as much as 50% of the heat generated within some components may be conducted through the metal lead wires. In addition, the thermal conductivities of the various cylindrical component bodies on the resistors, capacitors, and diodes are much higher than the thermal conductivity of the plain potting material. For the lowest temperature rise along a given length, the heat flow path should therefore be parallel to the cylindrical axes of the component bodies.

4.9 Sample Problem—Temperature Rise in a Potted Module

Internal heat sinks can be used effectively in potted modules. When several components have high power dissipations, they can be mounted on insulated aluminum or copper strips before the potting compound is added. The aluminum or copper strips should then be brought close to the heat transfer surface for efficient heat removal. These strips will reduce the internal thermal resistance and reduce local hot spot temperatures.

When transistor heat sinks are required within a potted module, a good heat flow path can often be provided by mounting the transistors upside down, with their hat sections on metal heat sink strips, as shown in Figure 4.10.

When the module is to be mounted to a cold plate, it may be necessary to machine the mounting surface flat to ensure a good thermal interface for efficient heat removal.

Many potting compounds have a high coefficient of thermal expansion, which can cause problems in electronic systems that must be capable of operating over a wide temperature range. Most potting compounds will expand and contract through large displacements over a temperature range from $-65°F$ ($-54°C$) to $+185°F$ ($+85°C$). To avoid cracked components and broken solder joints, a thin resilient conformal coating should be applied to the entire module before it is potted.

The effective thermal conductivity of the potted module can be increased by spreading a number of thin copper wires throughout the module. This will increase the heat flow in the areas adjacent to the wires, which can help to reduce hot spot temperatures. This technique is demonstrated in the following sample problem.

4.9 SAMPLE PROBLEM—TEMPERATURE RISE IN A POTTED MODULE

Determine the temperature rise from the mounting surface to the top edge of the components in the potted module, which is shown in Figure 4.11, for

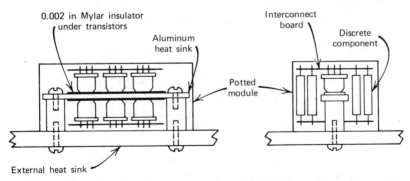

Figure 4.10 Cross section through a potted module showing an internal heat sink.

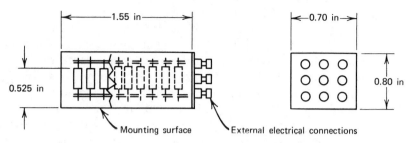

Figure 4.11 Potted module with a uniform power dissipation.

two different conditions:

1 Using an epoxy compound filled with aluminum oxide.
2 Using the compound in (1), but with 12 0.031 in (0.0787 cm) diameter copper wires distributed through the module adjacent to the components.

The power dissipation of the module is 2.0 watts and it is uniformly distributed through the module.

SOLUTION

Condition 1. Epoxy Potting Filled with Aluminum Oxide

The following information is required:

Given $Q = 2.0$ watts $= 6.83$ Btu/hr $= 0.478$ cal/sec (heat) (see Table 1.1 for power conversions)
$L_e = 0.525$ in $= 0.0437$ ft $= 1.333$ cm (epoxy length)
$K_e = 1.25$ Btu/hr ft °F $= 0.00516$ cal/sec cm °C (epoxy conductivity; see Table 3.2 for conductivity and Table 1.2 for conversions to English and metric units)
$A_e = (1.55)(0.70) = 1.085$ in² $= 0.00753$ ft² $= 7.0$ cm²
$A_e =$ (cross section area along heat flow path)

Substitute into Eq. 3.2 for the temperature rise in epoxy.

$$\Delta t_1 = \frac{(6.83)(0.0437)}{(1.25)(0.00753)} = 31.7°F \qquad (4.16)$$

Condition 2. Epoxy Potting with Copper Wires

A parallel heat flow path exists through the module. One path is through the epoxy, and one path is through the copper wires, as shown in Figure 4.12.

4.9 Sample Problem—Temperature Rise in a Potted Module

The following information is required:

Given $L_{cu} = 0.525$ in $= 0.0437$ ft $= 1.333$ cm (copper length)
$K_{cu} = 166$ Btu/hr ft °F $= 0.685$ cal/sec cm °C (copper conductivity)
$A_{cu} = (\pi/4)(0.031)^2(12) = 0.00906$ in² $= 0.0000629$ ft²
$A_{cu} = 0.0584$ cm² (copper cross section area)

Net area of epoxy without copper wires:

$$A_n = A_e - A_{cu} = 0.00753 - 0.0000629 = 0.00747 \text{ ft}^2$$

Substitute into Eq. 3.14 for the epoxy thermal resistance.

$$R_e = \frac{0.0437 \text{ ft}}{(1.25 \text{ Btu/hr ft °F})(0.00747 \text{ ft}^2)} = 4.68 \frac{\text{hr °F}}{\text{Btu}} \qquad (4.17)$$

Substitute into Eq. 3.14 for the copper thermal resistance.

$$R_{cu} = \frac{0.0437}{(166)(0.0000629)} = 4.18 \frac{\text{hr °F}}{\text{Btu}} \qquad (4.18)$$

Substitute Eqs. 4.17 and 4.18 into Eq. 3.17 for the combined thermal resistance of the epoxy and copper in condition 2 for the parallel flow path.

$$\frac{1}{R_2} = \frac{1}{4.68} + \frac{1}{4.18}$$

$$R_2 = 2.21 \frac{\text{hr °F}}{\text{Btu}} \qquad (4.19)$$

Substitute into Eq. 3.15 to obtain the temperature rise through the module with the potting and the copper wires.

$$\Delta t_2 = QR_2 = \left(6.83 \frac{\text{Btu}}{\text{hr}}\right)\left(2.21 \frac{\text{hr °F}}{\text{Btu}}\right) = 15.1 \text{°F} \qquad (4.20)$$

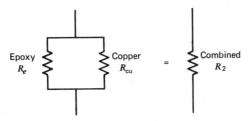

Figure 4.12 Parallel heat flow path through a potted module.

Comparing Eq. 4.16 with Eq. 4.20 shows that adding 12 thin copper wires to the potted module, parallel to the heat flow path, can reduce the average temperature rise about 50%.

4.10 COMPONENT LEAD WIRE STRAIN RELIEF

Electronic components have a wide variety of lead wire sizes, shapes, materials, and arrangements. One thing they all have in common is the potential for failure if the components are mounted without the proper strain relief in the lead wire. The lack of strain relief, or improper strain relief, can cause broken solder joints, lifted printed circuit pads, cracked components, cracked glass headers, short circuits, and many other strain related problems in systems that must operate over a wide temperature range. Special attention must be given to the component mounting techniques for electronic systems that will be exposed to temperatures that range from $-65°F(-54°C)$ to $+185°F(+85°C)$. This type of environment will generate high strains in elements that have different coefficients of expansion. High strains lead to high creep rates in solder joints, which can produce failures in a very short period of time. Solder joint stresses should be limited to a maximum value of 400 lb/in^2 (28,148 g/cm^2) for the foregoing temperature range to prevent temperature induced strain failures [20, 21, 33-35].

The trend in electronic packages is toward increased component density, so that the size of the lead wire strain relief has been constantly shrinking. A large strain relief means more space per component, which reduces the number of components that can be placed on a circuit board. However, when the strain relief gets too small, the problems begin to multiply rapidly.

Strains and stresses that are developed in the electronics are often the result of expansions and contractions in the structural elements of the system due to temperature changes. Expansion coefficients, which vary substantially for different materials, should be evaluated for their effects on the strain relief, to prevent excessive stresses from developing.

Cylindrical components such as resistors, capacitors, and diodes with axial leads should be provided with a minimum strain relief of about 0.100 in (0.254 cm), as shown in Figure 4.13a for lap and dip solder joints. Experience has shown that this type of strain relief will provide sufficient safety over the temperature range from $-65°F$ ($-54°C$) to $+185°F$ ($+85°C$) even with aluminum heat sinks.

Large rectangular components, such as transformers and chokes, generally have large wires, often with diameters of about 0.032 in (0.0813 cm). A large strain relief, such as shown in Figure 4.13b and c, may be required to prevent printed circuit pads from lifting on lap soldered joints.

Components are often placed closer together in high density packages, so that there is less room for a good strain relief in the lead wires. A longer

4.10 Component Lead Wire Strain Relief

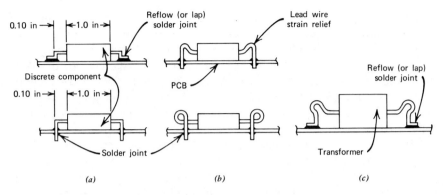

Figure 4.13 Typical lead wire strain relief for electronic components.

strain relief can be obtained by looping the lead wire, as shown in Figure 4.13b, to reduce the overall length of the component.

Transistors in TO-5 and TO-18 cans are often mounted on PCBs with the base of the transistor flush with the board, as shown in Figure 4.14a. This is a poor practice, because the lack of a strain relief in the lead wires can result in fractured solder joints as the circuit board thickness expands and contracts with temperature changes.

Flush mounted components may also produce trapped gas pockets in the electrical lead wire areas when the component is flow soldered to the PCB.

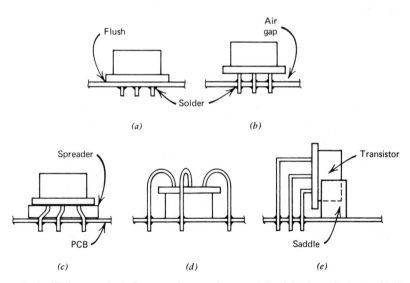

Figure 4.14 Various methods for mounting transistors. (a) Straight through (poor); (b) air gap (good low power); (c) spreader (good); (d) waterfall (good); (e) side saddle (good).

If the heated air is not permitted to escape from the small area around the lead wire, the increased air pressure will prevent the solder from wicking up the lead wire in the plated throughhole. This will result in a small solder fillet only at the bottom of the PCB. The outer appearance of the solder joint will seem normal, but the joint may have only 25% of the required amount of solder, which will result in a very weak solder joint that can fail very rapidly. Several techniques for mounting transistors are shown in Figure 4.14.

DIPs must also be carefully mounted, because they have very stiff lead wires with almost no strain relief. DIPs are made to be plugged into circuit boards, using the lead wires as electrical connectors, so that the lead wires must be rigid. When the dip leads are soldered to the PCB, this rigidity may produce solder joint failures when the body of the dip is mounted flush against the PCB. During temperature cycling, the expansion of the DIP body and the expansion of the epoxy fiberglass circuit board may be sufficient to produce an excessive strain in the solder joint, producing failures after several dozen cycles over a wide temperature range.

For conduction cooled systems, the body of the DIP should be cemented to the heat sink with a thermally conductive material that is resilient and will yield under the loads developed by the temperature cycle. RTV cements work well for applications of this type if the cement can be controlled to a thickness of about 0.005 to 0.010 in (0.0127 to 0.0254 cm). When the thickness of the cement exceeds 0.010 in, the thermal resistance from the component to the PCB may be too high, resulting in a high junction temperature. When the cement thickness is less than about 0.001 in (0.00254 cm), there will not be enough room for expansion at high temperatures, and the solder joints may again experience high stresses, resulting in joint failures after several dozen cycles.

DIPs have shoulders in their lead wires, which normally keeps the body of the DIP raised above the PCB. When metal strips are used under the component body for conduction cooled systems, the metal strip should be thick enough to bridge the resulting gap under the component body on high power dissipating DIPs. This should include any cements or tapes for holding the DIP to the heat sink.

When the power dissipation on a DIP is relatively low, below about 0.200 watt, it may not be necessary to cement the DIP to the heat sink in conductively cooled systems. Instead, a controlled air gap may be used under the DIP body. Air has a relatively low resistance for a small air gap. If the gap can be controlled to about 0.010 in (0.0254 cm), it may not result in extremely high junction temperatures. Some typical mounting installations for DIPs are shown in Figure 4.15.

The approximate temperature rise across the 0.010 in air gap for a power dissipation of 0.200 watt can be determined by considering only the heat conducted from the body of the DIP and ignoring the heat that is conducted

4.10 Component Lead Wire Strain Relief

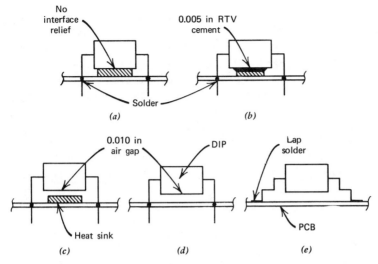

Figure 4.15 Various types of strain reliefs for DIPs mounted on PCBs. (*a*) No strain relief (poor); (*b*) RTV strain relief (good); (*c*) air gap strain relief (good low power); (*d*) air gap strain relief no sink (good low power); (*e*) air gap strain relief no sink (good low power).

through the lead wires. The following information is required:

Q = 0.200 watt = 0.683 Btu/hr (power dissipation)
L = 0.010 in = 0.000833 ft (length of air gap)
K = 0.018 Btu/hr ft °F (air conductivity at 200 °F; ref. Figure 6.28)
A = (0.30)(0.70) = 0.21 in² = 0.00146 ft² (area of DIP)

Substitute into Eq. 3.2 for the temperature rise across the DIP mounting interface.

$$\Delta t = \frac{(0.683)(0.000833)}{(0.018)(0.00146)} = 21.6°F \qquad (4.21)$$

If the air gap is doubled to 0.020 in (0.0508 cm), the temperature rise across the mounting interface will also double, to 43.2°F (24°C).

One of the problems with an air gap is that it is difficult to control. The air gap under a DIP can change when the PCB warps as the system heats. Also, rough handling may sharply change the size of the air gap, which would affect the heat transfer characteristics.

The thermal resistance of a small air gap will not change much, even at altitudes as high as 75,000 ft. Therefore, an air gap can still be used to cool low power dissipating DIPs at these altitudes. Air gaps will not provide

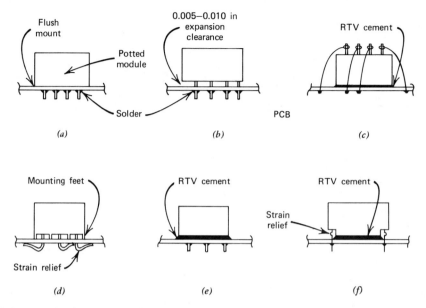

Figure 4.16 Various methods for mounting potted modules to PCBs. (*a*) No strain relief, no outgassing vents (poor); (*b*) air gap strain relief (good); (*c*) waterfall strain relief (good); (*d*) strain relief with outgassing provisions (good); (*e*) strain relief but no outgassing provisions (poor); (*f*) strain relief with outgassing provisions (good).

cooling in the hard vacuum environment of outer space, where there are no gases present to conduct away the heat.

Potted modules are often made with the electrical lead wires extending from one face of the unit. This face may then be mounted flush to a PCB and the lead wires will extend through plated through holes where they are soldered. If no strain relief is provided in the wires, the relatively high coefficients of expansion in the PCB and the potting may produce solder joint failures after a few dozen temperature cycles.

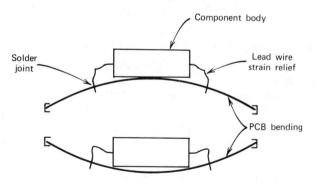

Figure 4.17 Circuit board flexing in a vibration environment.

4.10 Component Lead Wire Strain Relief

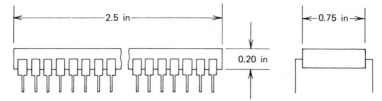

Figure 4.18 Typical microprocessor package.

Some type of strain relief should always be provided in the mounting geometry, to allow a small amount of relative motion to occur, to reduce the loads and strains developed in the solder joints. The maximum allowable thermal stresses in these solder joints should be limited to a value of about 400 psi under the most extreme temperature condition, to ensure a safe design. Some typical mounting configurations for potted modules are shown in Figure 4.16.

Conformal coatings are often used to protect the circuits and components from dirt and moisture. These coatings must not be permitted to fill in or to bridge the strain relief in the electrical lead wires or to fill in the strain

Figure 4.19 Composite aluminum PCB with side wedge clamps. (Courtesy Litton Systems, Inc.)

relief gap under large flat components that have leads that extend from the bottom surface. The bridging will increase the solder joint stresses because of the high coefficient of expansion of most conformal coatings, especially at the high tempertatures.

Circuit boards that must operate in severe vibration and shock environments should have electronic components mounted with an adequate lead wire strain relief. The type of lead wire and solder joint stresses developed in a vibration and shock environment are substantially different from the stresses developed in a thermal cycling environment. The number of stress reversals developed in a vibration environment can run into several hundred million. Therefore, stress concentrations such as nicks, scratches, and small bend radii become important and must be avoided, if possible [1].

Dynamic stresses in the lead wires are developed as the circuit board is forced to bend back and forth in a vibration environment. The condition is

Figure 4.20 Plug-in memory module designed for conduction cooling, uses side wedge clamps that provide high pressures at the thermal mounting interfaces. (Courtesy Litton Systems, Inc.)

4.10 Component Lead Wire Strain Relief

most severe during the PCB resonance, when the displacement amplitudes can become quite large. As the PCB bends back and forth, it forces the electrical lead wires on the components to bend back and forth, as shown in Figure 4.17. This can result in lead wire and solder joint failures when the stress levels and the number of fatigue cycles are high enough [1].

Microprocessors are particularly susceptible to solder joint and lead wire vibration faiiures. Microprocessors are manufactured in the form of a large DIP package. Typical dimensions are generally about 2.5 × 0.75 × 0.20 in (6.35 × 1.905 × 0.508 cm), as shown in Figure 4.18. The long component body length will result in an increase in the solder joint and lead wire stresses during vibration when the component is mounted at the center of the PCB. A high resonant frequency is required to keep the dynamic displacements low during resonant conditions, to provide a long fatigue life for the lead wires and solder joints. See reference [1] for more information on mounting electronic components that must operate in severe vibration environments.

Figure 4.19 shows an aluminum core composite circuit board with lap soldered electronic components.

Figure 4.20 shows a memory module made up of several circuit boards bolted together.

5

Practical Guides for Natural Convection and Radiation Cooling

5.1 HOW NATURAL CONVECTION IS DEVELOPED

Heat transfer by natural convection will occur when there is a change in the density of the fluid, which can be a gas or a liquid. Fluids tend to expand as they are heated, which results in a reduced density. In a gravity field, the fluid, which has a lower density, is lighter and therefore rises, creating a movement in the fluid which is called convection. This movement or convection permits the fluid to pick up heat and carry it away. When the fluid movement is produced only by a difference in the fluid density, the process is called natural or free convection. When the fluid movement is mechanically induced by means of pumps or fans, the process is called forced convection or forced circulation.

The driving force for natural convection is not very great because it depends upon the density change in the fluid. Therefore, any small obstacle or resistance in the flow path will sharply reduce the fluid flow rate, and therefore the cooling rate. Heat transfer by natural convection is more efficient if a well-defined, clear, unobstructed flow path is made available. Otherwise, the amount of heat carried away may be much less than anticipated and the resulting temperatures may be much higher than expected.

Natural convection heat transfer is extremely easy to use because no auxiliary equipment is required. It is also very reliable, because there are no auxiliary pumps or fans to malfunction or wear out.

The basic relation for free convection in a fluid (gas or liquid) is shown by three dimensionless ratios involving the Nusselt number, the Grashof

5.1 How Natural Convection Is Developed

number, and the Prandtl number, as shown in Eq. 5.1 [15, 17, 24]:

$$\underbrace{\frac{hL}{K}}_{\text{Nusselt number}} = C \underbrace{\left(\frac{L^3 \rho^2 g \beta \Delta t}{\mu^2}\right)^m}_{\text{Grashof number}} \underbrace{\left(\frac{C_p \mu}{K}\right)^n}_{\text{Prandtl number}} \quad (5.1)$$

where C = constant based upon geometry of surface
L = length along heat flow path (ft)
ρ = density of fluid (gas or liquid) (lb/ft^3)
g = acceleration of gravity = 4.17×10^8 (ft/hr^2)
β = volumetric expansion = 1/°R (or bulk modulus)
Δt = temperature difference (°F)
μ = viscosity (lb/ft hr)
C_p = specific heat of fluid (Btu/lb °F)
K = thermal conductivity (Btu/hr ft °F)
h = convection coefficient (Btu/hr ft^2 °F)

Experiments have shown that the exponents m and n are very nearly equal. Free convection calculations for electronic equipment can be made using an exponent value of 0.25.

Several items in Eq. 5.1 can be combined to simplify the convection relation as follows. Let

$$a = g\beta\rho^2 \frac{C_p \mu}{K}$$

Then

$$h = C\frac{K}{L}(aL^3 \Delta t)^{0.25} \quad (5.2)$$

The relation $Ka^{0.25}$ is fairly constant over a wide temperature range from 0°F to 200°F, with a value of about 0.52. Substituting this value into Eq. 5.2 results in the simplified natural convection equation for the laminar flow range [7, 8].

$$h = 0.52C \left(\frac{\Delta t}{L}\right)^{0.25} \quad (5.3)$$

Some typical values for C, which have been determined by experiments in free air, are shown in Table 5.1 [7, 8].

The characteristic length (L) is determined by the flow path the cooling air takes as it passes over the heated surface. Some typical values are shown in Table 5.2 [3, 7, 8].

The characteristic dimension for a flat horizontal rectangular plate is

Table 5.1 Values for Constants Based upon the Surface Geometry

Shape and Position	Value of C
Vertical plates	0.56
Long vertical cylinders and pipes	0.55
Horizontal cylinders and pipes	0.52
Horizontal plates facing upward	0.52
Horizontal plates facing downward	0.26
Spheres (where L = radius)	0.63
Small parts, also wires, (L = diameter)	1.45
Components on circuit boards	0.96
Small components in free air	1.39
Components in cordwood modules	0.48

obtained by using the geometric mean of the length and width. Considering a rectangular plate 12 × 6 in, Table 5.2 shows that the characteristic length would be as follows:

$$L = \frac{2(12)(6)}{12 + 6} = \frac{144}{18} = 8.0 \text{ in} = 0.606 \text{ ft}$$

For square plates, the characteristic dimension is the length of one side.

Table 5.2 Characteristic Length for Various Surfaces

Surface	Position	Characteristic Length (ft)
Flat plane	Vertical	Height, limited to 2 ft[a]
Flat plane (nonrectangular)	Vertical	$\frac{\text{Area}}{\text{Horizontal width}}$
Flat plane (circular)	Vertical	$\frac{\pi}{4}$ × diameter
Cylinder	Vertical	Height, limited to 2 ft
Flat plane	Horizontal	$\frac{2(\text{Length} \times \text{width})}{\text{Length} + \text{width}}$
Cylinder	Horizontal	Diameter
Sphere	Any	Radius

[a] When the actual plate height L exceeds 2 ft, use a maximum value of 2 ft in Eq. 5.3.

5.3 Natural Convection for Flat Horizontal Plates

For a 9.0 in square plate, Table 5.2 gives the correct value as follows:

$$L = \frac{2(9)(9)}{9+9} = \frac{162}{18} = 9.0 \text{ in} = 0.75 \text{ ft}$$

5.2 NATURAL CONVECTION FOR FLAT VERTICAL PLATES

The simplified laminar flow natural convection equation for a heated flat vertical plate at ordinary temperatures and sea level conditions is obtained with Eq. 5.3 and Table 5.1, as shown in Eq. 5.4. The temperature difference Δt must be in °F and the vertical height L must be in feet.

$$h_c = 0.29 \left(\frac{\Delta t}{L}\right)^{0.25} \tag{5.4}$$

The average natural convection coefficient h_c defines the thermal characteristics of the air film, which clings to the plate and restricts the flow of heat from the plate to the surrounding ambient air. A cross section through the plate and air film is shown in Figure 5.1.

Laminar flow occurs along the lower edges of a tall plate. The transition from laminar to turbulent flow generally starts at a distance of about 2 ft from the bottom edge of the plate. Turbulent flow occurs at distances greater than approximately 2 ft, depending upon the temperature difference from the plate to the ambient.

5.3 NATURAL CONVECTION FOR FLAT HORIZONTAL PLATES

The simplified laminar flow natural convection equation for a heated horizontal flat plate facing upward or a cooled plate facing downward is obtained

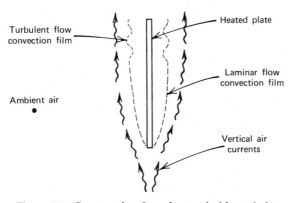

Figure 5.1 Cross section through a vertical heated plate.

from Eq. 5.3 and Table 5.1, as shown in Eq. 5.5. The temperature difference Δt must be in °F and the L dimension must be in feet.

$$h_c = 0.27 \left(\frac{\Delta t}{L}\right)^{0.25} \tag{5.5}$$

Cooling air flows in along the sides around the perimeter of the horizontal plate to replace the heated air, which rises as shown in Figure 5.2.

When the heated horizontal plate faces downward, the cooling air must flow along the bottom surface to the outer edges of the plate before it can rise, as shown in Figure 5.3. This increases the flow resistance, which reduces the cooling effectiveness.

The convection coefficient for a heated plate facing downward or a cooled plate facing upward is shown in Eq. 5.6. Its value is about half that of a heated plate facing upward.

$$h_c = 0.13 \left(\frac{\Delta t}{L}\right)^{0.25} \tag{5.6}$$

5.4 HEAT TRANSFERRED BY NATURAL CONVECTION

Once the convection coefficient h_c has been established, the amount of heat that can be carried away is determined from Eq. 5.7 [3, 7, 8].

$$Q = h_c A \, \Delta t \tag{5.7}$$

The units are the same as the relations shown in Eq. 3.53 except that the A represents the area of the heated surface, Δt represents the temperature difference between the heated surface and the surrounding ambient air, and h_c represents the convection coefficient.

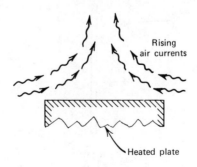

Figure 5.2 Heated horizontal surface facing upward.

5.5 Sample Problem—Vertical Plate Natural Convection

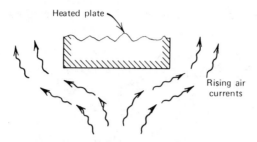

Figure 5.3 Heated horizontal surface facing downward.

5.5 SAMPLE PROBLEM—VERTICAL PLATE NATURAL CONVECTION

Determine the amount of heat that can be carried away from the vertical plate shown in Figure 5.4, considering only natural convection.

SOLUTION

Assuming laminar flow conditions (it is not known at this point if the flow is laminar or turbulent), the natural convection coefficient is determined from Eq. 5.4.

$$h_c = 0.29 \left[\frac{200 - 80 \ °F}{4 \ in/(12 \ in/ft)} \right]^{0.25} = 1.26 \ \frac{Btu}{hr \ ft^2 \ °F}$$

$$A = \frac{(8.0)(4.0) \ in^2}{144 \ in^2/ft^2} \ (two \ surfaces) = 0.444 \ ft^2$$

$$\Delta t = 200 - 80 = 120°F$$

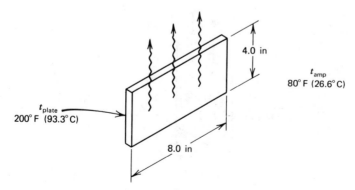

Figure 5.4 Vertical heated aluminum plate.

Substitute into Eq. 5.7 for the amount of heat convected away, using English units.

$$Q = \left(1.26 \frac{\text{Btu}}{\text{hr ft}^2 \, °\text{F}}\right) (0.444 \text{ ft}^2)(120°\text{F}) = 67.1 \frac{\text{Btu}}{\text{hr}} \quad (5.8)$$

Since 1 watt = 3.413 Btu/hr,

$$Q = \frac{67.1}{3.413} = 19.6 \text{ watts}$$

Tables 1.3 and 1.10 can be used to convert from English units to metric units.

$$h_c = \left(1.26 \frac{\text{Btu}}{\text{hr ft}^2 \, °\text{F}}\right)(0.000135) = 0.000170 \frac{\text{cal}}{\text{sec cm}^2 \, °\text{C}}$$

$$A = (0.444 \text{ ft}^2)(929) = 412.5 \text{ cm}^2$$

$$Q = \left(0.000170 \frac{\text{cal}}{\text{sec cm}^2 \, °\text{C}}\right)(412.5 \text{ cm}^2)(66.7°\text{C}) = 4.68 \frac{\text{cal}}{\text{sec}} \quad (5.9)$$

Since 1 watt = 0.239 cal/sec,

$$Q = \frac{4.68}{0.239} = 19.6 \text{ watts}$$

The amount of heat lost by natural convection will be different if the plate is rotated 90°, as shown in Figure 5.5. The vertical height L is now 8 in

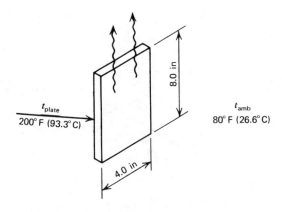

Figure 5.5 Vertical heated plate rotated 90°.

5.6 Turbulent Flow with Natural Convection

instead of 4 in. Using Eq. 5.4 once again, we obtain

$$h_c = 0.29 \left(\frac{120}{8/12}\right)^{0.25} = 1.06 \frac{\text{Btu}}{\text{hr ft}^2 \, °F}$$

Using Eq. 5.7 with the same plate area and temperature difference yields

$$Q = (1.06)(0.444)(120) = 56.5 \frac{\text{Btu}}{\text{hr}} \qquad (5.10)$$

Since 1 watt = 3.413 Btu/hr,

$$Q = 16.5 \text{ watts}$$

When the plate is rotated 90°, the amount of heat lost by natural convection is reduced about 15.8% because of the increased plate height, which reduces the natural convection coefficient. *Boundary layer is thicker, hence lower h_c, less dissipation*

5.6 TURBULENT FLOW WITH NATURAL CONVECTION

Turbulent flow conditions can develop during natural convection conditions when the temperature difference between the heated surface and the surrounding ambient air is high enough. The convection coefficient for turbulent conditions can be approximated by considering only the temperature difference. For heated vertical plates, the natural convection coefficient for turbulent flow is shown by Eq. 5.11. For heated horizontal plates facing upward, the natural convection coefficient for turbulent flow is shown by Eq. 5.12. The temperature difference must be in °F for these conditions [15, 17, 24, 36].

$$\text{Vertical plates:} \qquad h_c = 0.19 \Delta t^{0.333} \qquad (5.11)$$

$$\text{Horizontal plates:} \qquad h_c = 0.22 \Delta t^{0.333} \qquad (5.12)$$

The temperature difference required to change from laminar flow to turbulent flow for a vertical plate 1 ft high can be determined by setting Eq. 5.4 equal to Eq. 5.11.

$$0.29 \left(\frac{\Delta t}{1.0}\right)^{0.25} = 0.19 (\Delta t)^{0.333}$$

$$\frac{(\Delta t)^{0.333}}{(\Delta t)^{0.25}} = 1.526$$

$$(\Delta t)^{0.083} = 1.526$$

$$\Delta t = 163°F \qquad (5.13)$$

Δt actually less than this; laminar flow

This means that convection heat transfer for a vertical plate 1 ft high will change from laminar flow to turbulent flow when the temperature difference between the plate and the ambient is about 163°F. If the plate is less than 1 ft high, the temperature difference must be greater than 163°F to produce turbulent flow conditions.

The external surfaces of electronic boxes are often available to cool the equipment using natural convection. The heat dissipated by electronic components within the box may be conducted to the outside surfaces, where it is removed by convection. The amount of heat that can be removed is sharply influenced by the temperature difference between the surface of the box and the ambient air and also the surface area of the box.

5.7 SAMPLE PROBLEM—HEAT LOST FROM AN ELECTRONIC BOX

Determine the amount of heat that is lost by natural convection from the external surfaces of the box shown in Figure 5.6. The surface temperature is relatively uniform at 130°F and the ambient temperature is 80°F at sea level conditions. The box is aluminum with an iridited (or chromate) finish that has a low emissivity, so that heat lost by radiation is small and is ignored here.

SOLUTION

The natural convection coefficients for the vertical surfaces is obtained from Eq. 5.4 and Table 5.2.

$$h_c = 0.29 \left(\frac{\Delta t}{L}\right)^{0.25}$$

Given $\Delta t = 130 - 80 = 50°F$
$L = 6.0 \text{ in} = 0.50 \text{ ft}$

$$h_c = 0.29 \left(\frac{50}{0.50}\right)^{0.25} = 0.92 \frac{\text{Btu}}{\text{hr ft}^2 \text{ °F}} \tag{5.14}$$

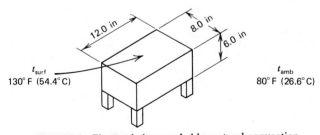

Figure 5.6 Electronic box cooled by natural convection.

5.7 Sample Problem—Heat Lost from an Electronic Box

The heat lost from the vertical sides of the box is determined from Eq. 5.7. Conversions from English units to metric units are made using Tables 1.3 and 1.10.

$$Q = h_c A \, \Delta t$$

Given $h_c = 0.92$ Btu/hr ft² °F $= 0.000124$ cal/sec cm² °C

$$A = \frac{[(12)(2) + (8)(2)](6) \text{ in}^2}{144 \text{ in}^2/\text{ft}^2}$$

$A = 1.667$ ft² $= 1548.6$ cm²
$\Delta t = 130 - 80 = 50°F = 27.8°C$

In English units:

$$Q_{\text{sides}} = (0.92)(1.667)(50) = 76.7 \frac{\text{Btu}}{\text{hr}} \qquad (5.15)$$

In metric units:

$$Q_{\text{sides}} = (0.000124)(1548.6)(27.8) = 5.34 \frac{\text{cal}}{\text{sec}} \qquad (5.15a)$$

The natural convection coefficient for the top horizontal surface is obtained from Eq. 5.5 and Table 5.2.

$$h_c = 0.27 \left(\frac{\Delta t}{L}\right)^{0.25}$$

Given $\Delta t = 130 - 80 = 50°F$

$$L = \frac{2(12)(8)}{12 + 8} = 9.6 \text{ in} = 0.80 \text{ ft}$$

$$h_c = 0.27 \left(\frac{50}{0.80}\right)^{0.25}$$

$$h_c = 0.76 \frac{\text{Btu}}{\text{hr ft}^2 \text{ °F}} = 0.000103 \frac{\text{cal}}{\text{sec cm}^2 \text{ °C}} \qquad (5.16)$$

The heat lost from the top surface of the box is determined from Eq. 5.7.

Given $A = \dfrac{(12)(8) \text{ in}^2}{144} = 0.667$ ft² $= 619.6$ cm²

In English units:

$$Q_{top} = (0.76)(0.667)(50) = 25.3 \frac{\text{Btu}}{\text{hr}} \qquad (5.17)$$

In metric units:

$$Q_{top} = (0.000103)(619.6)(27.7) = 1.77 \frac{\text{cal}}{\text{sec}} \qquad (5.17a)$$

The natural convection coefficient for the bottom horizontal surface is obtained from Eq. 5.6 and Table 5.2. Conversions to metric units are made with Tables 1.3 and 1.10.

$$h_c = 0.13 \left(\frac{\Delta t}{L}\right)^{0.25}$$

Given $\Delta t = 130 - 80 = 50°F$

$$L = \frac{2(12)(8)}{12 + 8} = 9.6 \text{ in} = 0.80 \text{ ft}$$

$$h_c = 0.13 \left(\frac{50}{0.80}\right)^{0.25}$$

$$h_c = 0.36 \frac{\text{Btu}}{\text{hr ft}^2 \, °F} = 0.0000486 \frac{\text{cal}}{\text{sec cm}^2 \, °C} \qquad (5.18)$$

The heat lost from the bottom surface is determined from Eq. 5.7.

Given $A = \frac{(12)(8) \text{ in}^2}{144} = 0.667 \text{ ft}^2 = 619.6 \text{ cm}^2$

In English units:

$$Q_{bot} = (0.36)(0.667)(50) = 12.0 \frac{\text{Btu}}{\text{hr}} \qquad (5.19)$$

In metric units:

$$Q_{bot} = (0.0000486)(619.6)(27.8) = 0.84 \frac{\text{cal}}{\text{sec}} \qquad (5.19a)$$

The total heat lost is the sum of the sides, top, and bottom.

5.8 Finned Surfaces for Natural Convection Cooling

In English units:

$$Q_{\text{total}} = 76.7 + 25.3 + 12.0 = 114.0 \frac{\text{Btu}}{\text{hr}} \tag{5.20}$$

Since 1 watt = 3.413 Btu/hr,

$$Q = 33.4 \text{ watts}$$

In metric units:

$$Q_{\text{total}} = 5.34 + 1.77 + 0.84 = 7.95 \frac{\text{cal}}{\text{sec}} \tag{5.20a}$$

Since 1 watt = 0.239 cal/sec,

$$Q = 33.3 \text{ watts}$$

5.8 FINNED SURFACES FOR NATURAL CONVECTION COOLING

The amount of heat that can be removed from an electronic box that is cooled by natural convection will be substantially increased if the surface area of the box can be substantially increased. One convenient method for increasing the surface area is to add fins with a low thermal resistance. The temperature of the fins will then be nearly equal to the surface temperature of the electronic box. The additional heat transferred to the atmosphere will then be approximately proportional to the increase in the surface area.

Fins will increase the size and weight of the electronic box. This may be a small penalty to pay if the cost is reduced and the reliability is increased by eliminating the need for a cooling fan.

The effectiveness of the finned surface will depend upon the temperature gradient along the fin as it extends from the surface of the electronic box. When the fin has a small temperature gradient, the temperature at the tip of the fin will be nearly equal to the temperature at the base of the fin or the chassis surface, and the fin will have a high efficiency. The natural convection coefficient h_c must be corrected when the fin efficiency is less than 100%. The fin efficiency is shown by Eq. 5.21 [15, 17, 24].

$$\eta = \frac{\tanh md}{md} = \text{fin efficiency} \tag{5.21}$$

where $m = \sqrt{\dfrac{2h_c}{K\delta}}$

h_c = convection coefficient
   = Btu/hr ft² °F or cal/sec cm² °C
K = thermal conductivity of fin material
   = Btu/hr ft °F or cal/sec cm °C
δ = fin thickness
   = ft, in, cm
d = fin height (distance fin extends from surface of box)
   = ft, in, cm

Fins should not be spaced too close together on the surface of a box, because this may result in pinching or choking the natural flow of cooling air. Closely spaced fins may increase the flow resistance and reduce the effective cooling available from the fins. When several fins are stacked together, there should be sufficient space to permit two convection films to develop, as shown in Figure 5.8.

The fin efficiency can also be determined from Figure 5.7 if the value of md is known.

If the fins are spaced closer together than about 0.60 in (1.52 cm) in perfectly still air at sea level, some pinching or choking may occur. However, under some natural convection conditions there is often additional air movement generated by the local ventilation system, so that the fins can be spaced closer together. Tests have shown that fins can be spaced about 0.50 in (1.27 cm) apart under these circumstances without choking the natural convection film coefficient. If the environmental conditions are unknown, a minimum clearance of 0.60 in is recommended to avoid trouble.

Fins can be cast integral with an electronic chassis, or they can be riveted, screwed, or cemented to the chassis structure. A low thermal interface

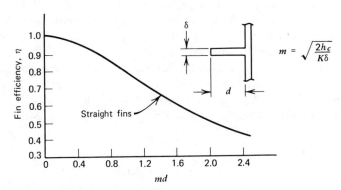

Figure 5.7 Fin efficiency factor for a single fin.

5.9 Sample Problem—Cooling Fins on an Electronic Box

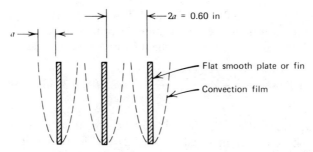

Figure 5.8 Convection film for several fins stacked together.

resistance must be provided between the fins and the chassis to ensure good heat transfer characteristics.

5.9 SAMPLE PROBLEM—COOLING FINS ON AN ELECTRONIC BOX

Determine the amount of heat that can be removed by natural convection from the electronic box in the preceding problem, when cooling fins are added as shown in Figure 5.9.

SOLUTION

The geometry of the electronic box is the same as Figure 5.6; only the vertical fins on the long sides have been added. The spacing between the

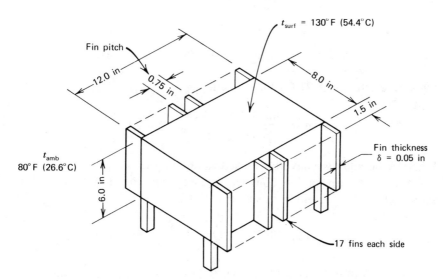

Figure 5.9 Electronic box with fins to improve natural convection cooling.

fin surfaces is 0.75 − 0.05 = 0.70 in (1.78 cm); therefore, no choking or pinching of the natural convection film will occur, even in perfectly still air.

The additional area supplied by the 17 fins on both sides is as follows:

$$A_{\text{fin}} = \frac{2(17)(6)(1.5)(2 \text{ sides})}{144 \text{ in}^2/\text{ft}^2} = 4.25 \text{ ft}^2 = 3948.3 \text{ cm}^2 \quad (5.22)$$

The temperature along the fin will not be constant, so that the fin efficiency η must be determined using Eq. 5.21 or Figure 5.7. The convection coefficient for the vertical sides of the box shown by Eq. 5.14 is used because the fins have the same height as the box.

Given h_c = 0.92 Btu/hr ft² °F = 0.000124 cal/sec cm² °C (ref. Eq. 5.14)
K = 83 Btu/hr ft °F = 0.343 cal/sec cm °C (aluminum fin conductivity)
δ = 0.05 in/12 = 0.00416 ft = 0.127 cm (fin thickness)
d = 1.5 in = 0.125 ft = 3.81 cm (fin height)

In English units:

$$m = \sqrt{\frac{2(0.92)}{(83)(0.00416)}} = 2.308 \frac{1}{\text{ft}} \quad (5.23)$$

$$md = (2.308)(0.125) = 0.288 \text{ (dimensionless)}$$

In metric units:

$$m = \sqrt{\frac{(2)(0.000124)}{(0.343)(0.127)}} = 0.0755 \frac{1}{\text{cm}} \quad (5.24)$$

$$md = (0.0755)(3.81) = 0.288 \text{ (dimensionless)}$$

Substitute into Eq. 5.21 for fin efficiency.

$$\eta = \frac{\tanh 0.288}{0.288} = \frac{0.280}{0.288} = 97.2\% \quad (5.25)$$

Figure 5.7 could also have been used to obtain a fin efficiency of 97.2%.

The effective fin natural convection coefficient can now be determined with the fin efficiency factor using Eq. (5.14)

$$h_{\text{eff}} = \eta h_c = (0.972)(0.92) = 0.89 \frac{\text{Btu}}{\text{hr ft}^2 \text{°F}} \text{ (English units)}$$

$$h_{\text{eff}} = (0.972)(0.000124) = 0.000121 \frac{\text{cal}}{\text{sec cm}^2 \text{°C}} \text{ (metric units)}$$

The additional heat lost by the fins only using natural convection is determined from Eq. 5.7.

$$Q = h_{eff} A \, \Delta t$$

In English units:

$$Q = (0.89)(4.25 \text{ ft}^2)(50°\text{F}) = 189.1 \frac{\text{Btu}}{\text{hr}} \qquad (5.26)$$

In metric units:

$$Q = (0.000121)(3948.3 \text{ cm}^2)(27.8°\text{C}) = 13.28 \frac{\text{cal}}{\text{sec}} \qquad (5.26\text{a})$$

The total heat loss from the box, in English units, is obtained by adding Eq. 5.20 to Eq. 5.26.

$$Q_{total} = 114.0 + 189.1 = 303.1 \frac{\text{Btu}}{\text{hr}} = 88.8 \text{ watts} \qquad (5.27)$$

The total heat lost from the box, in metric units, is obtained by adding Eq. 5.20a to Eq. 5.26a.

$$Q_{total} = 7.95 + 13.28 = 21.23 \frac{\text{cal}}{\text{sec}} = 88.8 \text{ watts} \qquad (5.27\text{a})$$

The fins thus increase the heat transfer capability of the box from 33.3 watts to 88.8 watts for the same temperature conditions, but with a slight increase in the volume and weight requirements.

5.10 NATURAL CONVECTION ANALOG RESISTOR NETWORKS

Computers are often used to solve heat transfer problems involving natural convection. This usually requires some type of mathematical model to simulate the physical characteristics of the system. Some analog resistor network models are shown in Figures 3.9 through 3.13.

The natural convection resistance is defined by Eq. 5.28.

$$R = \frac{1}{h_c A} \qquad (5.28)$$

Natural convection coefficients h_c for heated vertical plates can be determined from Eq. 5.4. Equation 5.5 can be used for heated horizontal plates facing upward. The area A represents the physical size of the plane surface in consistent units of ft^2 for English units or cm^2 for metric units.

Considering only single surfaces for the plates shown in Figure 5.10, for a temperature rise of 75°F, the thermal resistances would be as follows:

Vertical plate:

$$h_c = 0.29 \left(\frac{\Delta t}{L}\right)^{0.25} = 0.29 \left(\frac{75}{3/12}\right)^{0.25} = 1.21 \frac{\text{Btu}}{\text{hr ft}^2 \text{ °F}}$$

$$A = \frac{(3)(6) \text{ in}^2}{144 \text{ in}^2/\text{ft}^2} = 0.125 \text{ ft}^2$$

$$R_{\text{vert}} = \frac{1}{h_c A} = \frac{1}{(1.21)(0.125)} = 6.61 \frac{\text{hr °F}}{\text{Btu}} \tag{5.29}$$

Horizontal plate:

$$h_c = 0.27 \left(\frac{\Delta t}{L}\right)^{0.25}$$

$$L = \frac{L \times w}{(L + w)/2} = \frac{(3)(6)}{(3 + 6)/2} = 4.0 \text{ in} = 0.333 \text{ ft}$$

$$h_c = 0.27 \left(\frac{75}{0.333}\right)^{0.25} = 1.04 \frac{\text{Btu}}{\text{hr ft}^2 \text{ °F}}$$

$$R_{\text{hor}} = \frac{1}{h_c A} = \frac{1}{(1.04)(0.125)} = 7.69 \frac{\text{hr °F}}{\text{Btu}} \tag{5.30}$$

When a mathematical model for natural convection of a large vertical plate is being established, the conduction resistance between each node will be required, as well as the convection resistance to the ambient, as shown in Figure 5.11.

The small plate sections a through g are used to compute the conduction

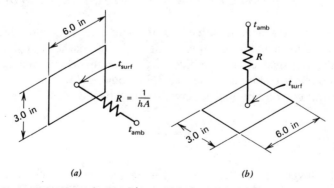

Figure 5.10 Natural convection models for vertical and horizontal plates.

5.11 Natural Convection Cooling for PCBs

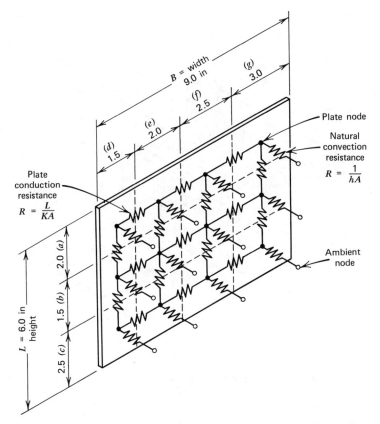

Figure 5.11 Mathematical model of a plate that combines conduction and natural convection thermal resistors.

resistances between each node point on the plate. However, the natural convection coefficient h_c depends upon the full height of the plate. Therefore, the full plate height of 6.0 in must be used for computing the natural convection coefficient, not the individual plate sections of a, b, and c.

5.11 NATURAL CONVECTION COOLING FOR PCBs

Printed circuit boards (PCBs) are often mounted within chassis that are completely open at the top and bottom to allow the free flow of cooling air. These PCBs should be mounted vertically with a minimum space of about 0.75 in (1.90 cm) free flow distance between the components and the adjacent PCB, to prevent pinching or choking of the natural convection flow, as shown in Figure 5.12.

Radiation cooling effects must be ignored here because the PCBs "see" each other. This simply results in a radiation heat interchange with no real

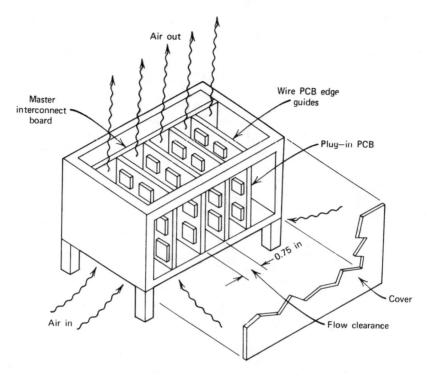

Figure 5.12 Plug-in PCBs cooled by natural convection.

cooling, except perhaps for the PCBs on each end of the chassis. These cooler end slots should be reserved for the highest-power-dissipating PCBs.

In a 160°F (71°C) ambient, a power dissipation of 0.10 watt/in² (0.0155 watt/cm²) will result in a component surface temperature of about 212°F (100°C) for a small PCB.

This considers only one side of the PCB, when electronic components are uniformly distributed with a uniform heat load along the surface of the PCB. When electronic components are mounted on both sides of the PCB with a uniform heat distribution, the total amount of power that can be dissipated from one PCB is double the power that can be dissipated from only one side of the PCB for a 30°C temperature rise. (54°F rise)

The foregoing approximation can be verified by considering a 6 by 9 in PCB, similar to Figure 5.11, with electronic components uniformly distributed on one side. The temperature rise from the component surface to the still ambient is 54°F (30°C). The natural convection coefficient is determined from Eq. 5.4.

$$h_c = 0.29 \left(\frac{54}{6/12} \right)^{0.25} = 0.93 \frac{\text{Btu}}{\text{hr ft}^2 \, °F}$$

5.12 Natural Convection Coefficient for Enclosed Air Space

The amount of heat lost by convection from one side of the PCB is determined from Eq. 5.7.

$$Q = 0.93 \left(\frac{6 \times 9 \text{ in}^2}{144 \text{ in}^2/\text{ft}^2} \right) (54°F) = 18.8 \frac{\text{Btu}}{\text{hr}} = 5.5 \text{ watts}$$

$$\text{power density} = \frac{Q}{A} = \frac{5.5 \text{ watts}}{6 \times 9 \text{ in}^2} = 0.10 \frac{\text{watt}}{\text{in}^2} \quad (5.31)$$

Equation 5.31 shows that a value of 0.10 watt/in² is a good approximation for the allowable power dissipation of a PCB cooled by natural convection.

Electronic components are usually mounted on one side of a PCB with the lead wires passing through the PCB to the opposite side for wave or flow soldering. Some of the heat dissipated within the electronic components will flow through the electrical lead wires to the back side of the PCB, as shown in Figure 1.3. When the back side of the PCB has a large number of copper printed circuit runs, they will act as heat spreaders. The back side of the PCB will then be capable of providing additional surface area for heat removal by natural convection. Thermal tests on PCBs with plated throughholes and flow-soldered electronic components, show that the back side of the PCB can increase the total effective heat transfer area about 30%.

If the electronic components are mounted on one side of the PCB, and lap solder joints are used (the electrical lead wires do not extend through the PCB with lap solder joints), the heat will not flow through to the back side of the PCB as easily. The back side of the PCB will not provide additional surface area for cooling under these circumstances.

5.12 NATURAL CONVECTION COEFFICIENT FOR ENCLOSED AIR SPACE

Printed circuit boards that are to be cooled by natural convection are sometimes placed so close to a wall that the natural flow of cooling air is blocked. The resulting clearance is too small to allow air movement, so that the clearance becomes an enclosed air space. Although natural convection is suppressed, heat transfer may still take place by conduction across the air gap. An equivalent convection coefficient can then be defined, based upon the heat conducted across the air gap as shown by Eq. 3.1.

$$Q = KA \frac{\Delta t}{L} \quad \text{(ref. Eq. 3.1)}$$

For convection, the general heat transfer relation is shown by Eq. 5.7.

$$Q = h_c A \, \Delta t \quad \text{(ref. Eq. 5.7)}$$

When the heat transferred is the same in both cases, the equations must be equal so that an air gap convection coefficient h_{AG} can be obtained as shown in Eq. 5.32.

$$h_{AG} A \, \Delta t = KA \frac{\Delta t}{L}$$

$$h_{AG} = \frac{K}{L} \tag{5.32}$$

where K = thermal conductivity of air gap
 L = thickness of air gap

The use of Eq. 5.32 can be demonstrated with a sample problem.

5.13 SAMPLE PROBLEM—PCB ADJACENT TO A CHASSIS WALL

An electronic chassis was designed for natural convection cooling, so that a clearance of 0.75 in (1.905 cm) was provided between the PCBs and components. However, a design change required the addition of another PCB, which might reduce the clearance too much unless the new PCB is placed very close to the side wall of the chassis, with a clearance of only 0.20 in (0.51 cm). The PCB measures 6 by 9 in and dissipates 5.5 watts. The electronic chassis must operate at sea level conditions in a maximum ambient temperature of 110°F (43.3°C). The maximum allowable component surface temperature is 212°F (100°C) with the chassis shown in Figure 5.13. The aluminum chassis has a polished finish that has a low emissivity, so that the heat lost by radiation is small. Determine if the design is adequate.

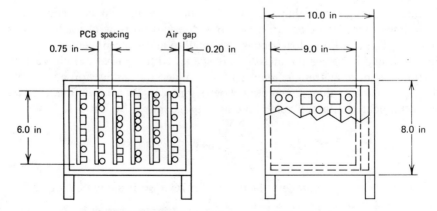

Figure 5.13 PCB spaced close to an end bulkhead.

5.13 Sample Problem—PCB Adjacent to a Chassis Wall

SOLUTION

The previous sample problem, shown in Figure 5.12, demonstrated that a clearance of 0.75 in will result in a temperature rise of 54°F (30°C) for the conditions given. This will result in a component surface temperature of 164°F, which is satisfactory. Therefore, only the PCB mounted close to the end wall has to be examined.

Heat from the components must flow to the outside ambient. This will require the heat to flow across two major resistance areas, the internal air gap of 0.20 in (R_1) and the external convection film (R_2), as shown in Figure 5.14.

The thermal conductivity of the air in the air gap is unknown, so that an average air temperature of 175°F is assumed and verified later. Determine the resistance with the convection coefficient for the 0.20 in air gap.

$$R_1 = \frac{1}{h_{AG}A} \quad \text{(ref. Eq. 5.28)}$$

Given $K = 0.017$ Btu/hr ft °F (ref. Figure 6.28)(page 202)

$$L = \frac{0.20}{12} = 0.0167 \text{ ft (air gap)}$$

$$h_{AG} = \frac{K}{L} = \frac{0.017}{0.0167} = 1.018 \frac{\text{Btu}}{\text{hr ft}^2 \text{°F}} \text{ (air gap convection)}$$

$$A = \frac{6 \times 9 \text{ in}^2}{144} = 0.375 \text{ ft}^2 \text{ (PCB area)}$$

$$R_1 = \frac{1}{(1.018)(0.375)} = 2.62 \frac{\text{hr °F}}{\text{Btu}} \tag{5.33}$$

The external convection coefficient must be estimated because the tem-

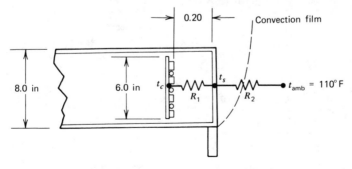

Figure 5.14 Thermal resistances in the heat flow path from PCB components to the outside ambient.

perature rise from the surface of the chassis to the ambient is unknown. In general, natural convection coefficients for this type of structure will range from about 0.6 to about 1.0 Btu/hr ft² °F. A value of 0.8 Btu/hr ft² °F is assumed to start. This can be changed if the analysis shows there is a large error.

$$R_2 = \frac{1}{h_c A} \quad \text{(ref. Eq. 5.28)}$$

Given $h_c = 0.8$ Btu/hr ft² °F (assumed to start)

$$A = \frac{8 \times 10 \text{ in}^2}{144} = 0.555 \text{ ft}^2$$

$$R_2 = \frac{1}{(0.8)(0.555)} = 2.25 \frac{\text{hr °F}}{\text{Btu}} \tag{5.34}$$

The temperature rise across each thermal resistor is determined from Eq. 5.35. For resistor R_1:

$$\Delta t_1 = Q R_1 \tag{5.35}$$

Given $Q = (5.5 \text{ watts})(3.413 \text{ Btu/hr/watt}) = 18.77$ Btu/hr

$R_1 = 2.62$ hr °F/Btu (ref. Eq. 5.33)

$$\Delta t_1 = (18.77)(2.62) = 49.2°F \tag{5.36}$$

For resistor R_2:

$$\Delta t_2 = Q R_2 = (18.77)(2.25) = 42.2°F \tag{5.37}$$

The natural convection coefficient was assumed to be 0.8 Btu/hr ft² °F. The actual value can now be determined from Eqs. 5.4 and 5.37 using a vertical chassis wall height of 8.0 in.

$$h_c = 0.29 \left(\frac{\Delta t}{L}\right)^{0.25} = 0.29 \left(\frac{42.2}{8/12}\right)^{0.25} = 0.81 \frac{\text{Btu}}{\text{hr ft}^2 \text{ °F}} \tag{5.38}$$

This compares well with the assumed value.

The surface temperature of the component on the end PCB can be determined as follows:

$$t_c = t_{\text{AMB}} + \Delta t_1 + \Delta t_2 \tag{5.39}$$

$$t_c = 110 + 49.2 + 42.2 = 201.4°F \ (94.1°C) \tag{5.40}$$

The average air temperature in the 0.20 in gap between the wall and the component can now be determined. A temperature of 175°F was assumed to obtain the air thermal conductivity.

The average air temperature in the gap is obtained from Eq. 5.40.

$$t_{av} = \frac{t_{comp} + t_{wall}}{2} = \frac{201.4 + (110 + 42.2)}{2} = 176.8°F$$

This compares well with the assumed value.

Therefore, since Eq. 5.40 shows that the component surface temperature is below the maximum allowable value of 212°F (100°C), the design is satisfactory.

If the inside and outside surfaces of the chassis are anodized or painted any color except silver, heat transfer by radiation will be increased and the surface of the PCB will be cooler.

5.14 HIGH ALTITUDE EFFECTS ON NATURAL CONVECTION

All of the previous natural convection equations were determined for sea level conditions, where the atmospheric pressure is 14.7 lb/in² (1034.4 g/cm²). Natural convection depends upon a reduction in the density of the cooling air as it picks up heat, which forces it to rise in a gravity field.

At high altitudes the air density is much lower than it is at sea level conditions, so that the ability of the air to pick up heat is sharply reduced. This, in turn, reduces the ability of the air to rise, which results in a lower convection coefficient.

An examination of Eq. 5.1 shows that the convection coefficient h is related to the air density as $(\rho^2)^m$, where m has a value of 0.25. This means that the convection coefficient is related to the square root of the air density, or $\sqrt{\rho}$. When this is combined with the basic gas law $PV = WRT$, it shows that the air density is directly related to the air pressure when the air temperature does not change. Therefore, the convection coefficient h is also proportional to the square root of the air pressure, as shown in Eq. 5.41.

$$h_{alt} = h_{sl}\sqrt{\frac{\rho_{alt}}{\rho_{sl}}} = h_{sl}\sqrt{\frac{P_{alt}}{P_{sl}}} \quad (5.41)$$

For example, when the natural convection coefficient at sea level is 0.75 Btu/hr ft² °F, the natural convection coefficient at an altitude of 30,000 ft can be determined from the atmospheric pressure, which is 4.37 psia, as shown in Table 6.5. When there is no change in the temperature,

$$h_{alt} = 0.75 \sqrt{\frac{4.37}{14.7}} = 0.41 \frac{\text{Btu}}{\text{hr ft}^2 \, °F} \qquad (5.42)$$

When there is a change in the temperature, the air density ratio should be used instead of the pressure ratio, for greater accuracy.

The thermal conductivity of air changes very little with a change in altitude or air pressure, up to an altitude of about 82,000 ft (see Figure 3.19). The thermal conductivity of air does change, however, with the air temperature, as shown in Figure 6.28. Therefore, at high altitudes the natural convection is reduced, but conduction heat transfer across small air gaps is not affected, as long as there is no change in the temperature. These altitude and temperature characteristics can be demonstrated with a sample problem.

5.15 SAMPLE PROBLEM—PCB COOLING AT HIGH ALTITUDES

For the previous sample problem (Section 5.13 and Figure 5.13) determine the component surface temperature that will result at an altitude of 30,000 ft. The aluminum chassis has an iridited finish, which has a low emissivity, so that heat lost by radiation is small and is ignored.

SOLUTION

The external convection coefficient from the outer surface of the box will be reduced at the high altitude. Its exact value is not known. Equation 5.41 might be used with the results from the previous sample problem, shown by Eq. 5.38, to obtain a rough approximation of the convection coefficient at 30,000 ft. However, this would not be accurate, because the outer surface of the box will be hotter at the high altitude condition than at the sea level condition, which will change the convection coefficient. A better approximation of the external convection coefficient at the 30,000 ft altitude condition can be obtained by combining the following equations:

$$h_{sl} = 0.29 \left(\frac{\Delta t}{L}\right)^{0.25} \qquad \text{ref. Eq. 5.4)}$$

$$h_{alt} = h_{sl} \sqrt{\frac{P_{alt}}{P_{sl}}} \qquad \text{(ref. Eq. 5.41)}$$

$$h_{alt} = 0.29 \left(\frac{\Delta t}{L}\right)^{0.25} \sqrt{\frac{P_{alt}}{P_{sl}}}$$

$$h_{alt} = 0.29 \left(\frac{\Delta t}{L}\right)^{0.25} \sqrt{\frac{4.37}{14.7}} = 0.158 \left(\frac{\Delta t}{L}\right)^{0.25}$$

5.15 Sample Problem—PCB Cooling at High Altitudes

Consider another equation:

$$Q = h_{alt} A \, \Delta t \quad \text{(at altitude, ref. Eq. 5.7)}$$

or

$$h_{alt} = \frac{Q}{A \, \Delta t}$$

Equate the convection coefficients h_{alt} and solve for Δt.

$$\frac{Q}{A \, \Delta t} = \frac{0.158 \, \Delta t^{0.25}}{L^{0.25}}$$

$$QL^{0.25} = 0.158 A \, \Delta t^{1.25}$$

$$\Delta t = \left(\frac{QL^{0.25}}{0.158 A} \right)^{0.8} \quad (5.43)$$

For a power dissipation of 5.5 watts (18.77 Btu/hr), a vertical height of 8.0 in (0.666 ft) and a surface area of $8 \times 10 = 80 \text{ in}^2$ (0.555 ft²), the resulting external temperature rise at an altitude of 30,000 ft becomes

$$\Delta t_2 = \left[\frac{(18.77)(0.666)^{0.25}}{(0.158)(0.555)} \right]^{0.8} = 67.4°F \quad (5.43a)$$

This compares to a temperature rise of only 42.2°F across the external convection film at sea level conditions, as shown by Eq. 5.37.

The resulting external convection coefficient at the 30,000 ft altitude can be obtained from Eq. 5.7.

$$h_{alt} = \frac{Q}{A \, \Delta t} = \frac{18.77 \text{ Btu/hr}}{(0.555 \text{ ft}^2)(67.4)}$$

$$h_{alt} = 0.50 \, \frac{\text{Btu}}{\text{hr ft}^2 \, °F}$$

The thermal conductivity of the air in the 0.20 in air gap will increase to about 0.018 Btu/hr ft °F because the average air temperature in the gap will increase to about 200°F. This average air gap temperature can be verified after the preliminary temperature rise calculations. Then, following Eq. 5.32.

$$h_{AG} = \frac{K}{L} = \frac{0.018 \text{ Btu/hr ft } °F}{0.0167 \text{ ft air gap}} = 1.078 \, \frac{\text{Btu}}{\text{hr ft}^2 \, °F} \quad (5.44)$$

Substitute into Eq. 5.28, using the area 0.375 ft², as shown in Eq. 5.33.

$$R = \frac{1}{(1.078)(0.375)} = 2.47 \frac{\text{hr °F}}{\text{Btu}}$$

From Eq. 5.35, using a heat value of $Q = 5.5$ watts = 18.77 Btu/hr, we obtain

$$\Delta t_1 = \left(18.77 \frac{\text{Btu}}{\text{hr}}\right)\left(2.47 \frac{\text{hr °F}}{\text{Btu}}\right) = 46.4°F \quad (5.45)$$

The component surface temperature at 30,000 ft is determined from Eq. 5.39. The ambient temperature is 110°F.

$$t_c = 110 + 46.4 + 67.4 = 223.8°F \ (106.5°C) \quad (5.46)$$

Since this exceeds the maximum allowable component surface temperature of 212°F (100°C), the design is not acceptable. Again, one more check must be made of the average temperature in the air gap to veryify the thermal conductivity used.

$$t_{av} = \frac{t_{comp} + t_{wall}}{2} = \frac{223.8 + (110 + 67.4)}{2} = 200.6°F$$

This is close to the assumed value of 200°F, so that the results are valid.

5.16 RADIATION COOLING OF ELECTRONICS

Hot bodies emit thermal radiation that is transferred by electromagnetic waves, which can travel through a vacuum or through a gas with relatively little absorption. The true nature of radiation and its transfer mechanism is not completely clear, but it is known that radiation in free space travels at the speed of light.

When radiation waves strike a second body, part of the radiated energy may be absorbed, part may be reflected, and part may be transmitted through the body, as in the case of glass and quartz. The relative amounts of energy transferred in each case depends upon the temperature, the surface characteristics, the body geometry, the material, and the wavelength.

The amount of heat radiated from a surface is related to a factor called the emittance e. This is dependent upon the condition of the surface, such as its temperature, roughness, finish, coating, and if it is a metal, the amount of oxidation. The emittance is a ratio of the amount of radient energy emitted by a body to that emitted by the ideal radiator, which is a black body. The emittance is always less than 1.0 because the perfect black body

5.16 Radiation Cooling of Electronics

does not really exist in nature. The black body is used as a standard for comparison with the radiation characteristics of other bodies.

A perfect black body is one that absorbs 100% of all the energy it receives, without reflecting any energy. The emissive power of a black body depends only upon its temperature. The black body radiates the maximum amount of energy at any wavelength, as shown in Figure 5.15.

The ability of a body to absorb part of the total radiated energy it receives is called the absorptivity α. Again, the amount of energy that the body receives is dependent upon the material and the temperature (or wavelength) of the emitter.

When a body is at thermal equilibrium, the emissivity and the absorptivity of that body will be equal. When this condition exists for the ideal radiator or black body, the emissivity and the absorptivity both equal unity.

The total area under any one of the indicated temperature curves in Figure 5.15 is used to determine the total energy radiated by the black body at that temperature.

Metals and plastics normally used in electronics are very opaque to radiation. Therefore, radiation can be considered as a surface phenomenon, where the effective surface depth for metals is only about 0.0001 in (0.000254 cm) and for nonmetals about 0.02 in (0.0508 cm).

The emissivities of clean metallic materials are generally very low while the emissivities of nonmetallic materials are much higher. Also, nonmetallic materials, in contrast to metals, show decreasing emissivities with an increase in temperature.

A spacecraft in close orbit around the earth or around the moon can receive radiation from the sun and the earth (or moon) at the same time. Most of the radiation is in the ultraviolet and visible light frequency range,

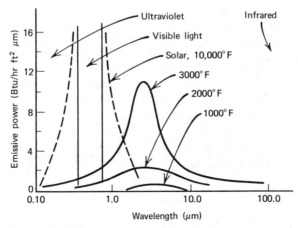

Figure 5.15 Change of emissive power with relation to the temperature of an ideal radiator or black body.

which has a wavelength up to about 1 micrometer (μm), as shown in Figure 5.15 [4, 5, 15].

In general, dull dark surfaces are good absorbers and good emitters of heat, so they have high emittance values. Polished metal surfaces have very low emissivities. However, as metals oxidize, their emissivities increase rapidly. Sandblasted metal surfaces also show emissivities that are much higher than those for polished surfaces.

That emissivity is a surface phenomenon can be demonstrated by placing a thin coat of paint on a highly polished metal surface. Before painting a polished copper surface, the emissivity will be about 0.03. However, with just a thin coat of lacquer, 0.0005 in (0.00127 cm) thick, the emissivity of the same surface will increase sharply to about 0.80.

For temperatures normally encountered in electronic equipment, the radiation wavelength is relatively long, about 7 μm, and is not perceived by the human eye. This represents the infrared region, where color is relatively unimportant, so that black paints have radiation characteristics that are very similar to white paints. In this region a good emitter is a good receiver of radient energy. Dark colors will not emit or absorb any more heat than

Table 5.3 Typical Emissivities 100°C [7, 8]

Material	Emissivity, e
Aluminum	
Polished	0.06
Commercial sheet	0.09
Rough polish	0.07
Iridited	0.10
Gold, highly polished	0.018–0.035
Steel, polished	0.06
Iron, polished	0.14–0.38
Cast iron, machine cut	0.44
Brass, polished	0.06
Copper, polished	0.023–0.052
Polished steel casting	0.52–0.56
Glass, smooth	0.85–0.95
Aluminum oxide	0.33
Anodized aluminum	0.81
White enamel on rough iron	0.9
Black shiny lacquer on iron	0.8
Black or white lacquer	0.80–0.95
Aluminum paint and lacquer	0.52
Rubber, hard or soft	0.86–0.94
Water (32–212°F)	0.95–0.96

5.16 Radiation Cooling of Electronics

light colors in the infrared region, because the emissivities and absorptivities are not related to the paint color.

Sometimes a surface will have characteristics similar to a black body, as shown in Figure 5.15. The shape or its emissive power, as a function of the temperature, will then be the same as the black body, except that the height is reduced by the numerical value of the emissivity. This type of surface is defined as being gray. In most electronic heat transfer calculations, the surfaces are assumed to be gray even though they do not meet the exact requirements for a gray surface.

As the temperature of a body increases, the radiation wavelength decreases until the visible spectrum is reached at about 0.7 μm. In the visible light spectrum, the amount of radiation energy absorbed by a body is sharply affected by the color. Dark colors in the visible spectrum absorb more radient energy than light colors. This can be dramatically shown by placing your hand on a black automobile roof and a white automobile roof at the same time, after both cars have been sitting in the hot sun for several hours. The black roof will be substantially hotter than the white roof.

Heat from both the black and white automobile surfaces will be radiated back to the surrounding areas in the long wavelength, or infrared region, where these paints act like gray surfaces.

Typical emissivity values for various materials are shown in Table 5.3.

Values of α/e for several different materials are shown in Table 5.4.

Relations for the emissivities for two different surfaces are shown in Table 5.5.

Radiation really involves an exchange of energy, since part of the energy received is always reflected back. For normal engineering purposes, the reflected energy is taken care of by the emissivity and the absorptivity of each body involved in the exchange.

The general radiation equation for the exchange of radiant heat energy between two non-black bodies is shown in Eq. 5.47 [7, 8, 15, 17, 24].

$$Q = \sigma f e A (T_1^4 - T_2^4) \tag{5.47}$$

Consistent sets of units must be used as defined for English units and metric units in Table 5.6.

The heat transmitted by radiation is directly related to the surface area A. If the surface is irregular, the projected area should be used. This is the area that would result when a tightly stretched string is held over various parts of the surface, as shown in Figure 5.16.

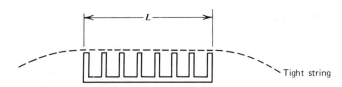

Figure 5.16 Projected area of an irregular surface.

Table 5.4 Some Representative Values of α/e

Material	Solar Absorptivity (α)	Total Normal Emissivity at 100°F (e)	α/e
Aluminum, freshly evaporated	0.10	0.025	4
Gold	0.16	0.02	8
Molybdenum	0.43	0.03	14
Palladium	0.41	0.03	14
Platinum	0.33	0.03	11
Rhodium	0.28	0.02	14
Silver	0.07	0.01	7
Tantalum	0.59	0.02	29
Aluminum, vapor deposit on 0.5 mil Mylar			
Aluminum side	0.13	0.04	3.25
Mylar side	0.18	0.90	0.20
White paint, 1 mil thick	0.15	0.94	0.16
Black paint, 1 mil thick	0.97	0.94	1.03
Clear varnish on aluminum 1 mil	0.20	0.80	0.25
Clear varnish on aluminum 0.05 mil coat	0.20	0.10	2.0
Black lacquer, 0.24 mil coat	0.96	0.48	2.0

Table 5.5 Emissivity Factors for Different Configurations

Configuration	e
Infinite parallel planes	$\dfrac{1}{1/e_1 + 1/e_2 - 1}$
One surface (1) completely surrounded by another (2)	$\dfrac{1}{1/e_1 + A_1/A_2 [1/e_2 - 1]}$
Completely enclosed body 1, small compared with enclosing body 2	e_1
General case for two surfaces	$e_1 \times e_2$

5.17 Radiation View Factor

Table 5.6 Units Used with Radiation Exchange Equation

Item	Symbol	English Units	Metric Units
Radiant heat exchange	Q	$\dfrac{\text{Btu}}{\text{hr}}$	$\dfrac{\text{cal}}{\text{sec}}$
Stefan–Boltzmann constant	σ	$\dfrac{0.1713 \times 10^{-8} \text{ Btu}}{\text{hr ft}^2 \, °R^4}$	$\dfrac{1.355 \times 10^{-12} \text{ cal}}{\text{sec cm}^2 \, °K^4}$
View factor	f	Dimensionless	Dimensionless
Emissivity	e	Dimensionless	Dimensionless
		$\dfrac{1}{1/e_1 + 1/e_2 - 1}$	$\dfrac{1}{1/e_1 + 1/e_2 - 1}$
Area	A	ft^2	cm^2
Absolute temperature	T	°F + 460 = °R	°C + 273 = °K

The finned surface increases the effective area for convection heat transfer but has very little affect on the effective area for radiation heat transfer. Radiating surfaces must be able to "see" each other in order to transfer heat. The fins block the line of sight so that only the projected area bounded by the tight string is effective for the transfer of heat.

5.17 RADIATION VIEW FACTOR

The view factor f is defined as the fraction of the radiation leaving surface 1 in all directions, which is intercepted by surface 2, as shown in Figure 5.17.

Heat transferred by radiation can be evaluated as taking place from surface 1 to surface 2, or from surface 2 to surface 1, if both are black bodies, whichever is more convenient. The product of the area times the

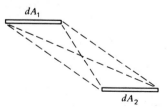

Figure 5.17 Areas involved in a radiation interchange.

Normally have to estimate f: what percentage of area of a body can I see from this part?

view factor works in both directions, as shown in Eq. 5.48 [15, 24].

$$A_1 f_{1-2} \equiv A_2 f_{2-1} \quad (5.48)$$

Many complex equations have been derived, and many computer programs are available, which relate the view factor between two surfaces as a function of the geometry. Sometimes it is possible to determine the view factor between two rectangular plates with a piece of string, using the cross string method.

For example, consider two long, narrow, rectangular parallel plates as in Figure 5.18, where only the narrow edges are shown. Using a piece of string, the dimensions across the free edges are shown by the dashed lines.

As shown by Eq. 5.48, the product of the area times the view factor, per unit length of depth, is the sum of the lengths of the crossed strings stretched from the edges of the plates, minus the sum of the lengths of the noncrossed strings similarly stretched from the edges of the plates, all divided by 2. This is shown as follows:

$$A_1 f_{1-2} = A_2 f_{2-1} = \frac{\text{sum cross strings} - \text{sum noncross strings}}{2} \quad (5.49)$$

$$A_1 f_{1-2} = A_2 f_{2-1} = \frac{(8.54 + 5.0) - (5.0 + 3.0)}{2} = 2.77 \, \frac{\text{in}^2}{\text{in of depth}}$$

The view factor from plates 1 to 2 (f_{1-2}) is

$$f_{1-2} = \frac{A_2 f_{2-1}}{A_1} = \frac{2.77 \, \text{in}^2/\text{in}}{(4)(40) \, \text{in}^2} (40 \, \text{in}) = 0.692 \, (\text{dimensionless}) \quad (5.50)$$

The view factor from plates 2 to 1 (f_{2-1}) is

$$f_{2-1} = \frac{A_1 f_{1-2}}{A_2} = \frac{2.77 \, \text{in}^2/\text{in}}{(8)(40) \, \text{in}^2} (40 \, \text{in}) = 0.346 \, (\text{dimensionless}) \quad (5.51)$$

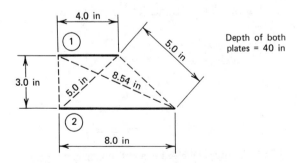

Figure 5.18 End view of two long rectangular parallel plates.

5.17 Radiation View Factor

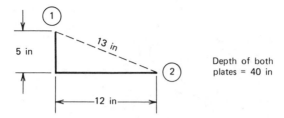

Figure 5.19 End view of two long rectangular perpendicular plates.

The view factor for two long rectangular plates that intersect in a right angle, as shown in Figure 5.19, can be determined with Eq. 5.49. Here the cross strings are the width of each plate, and the noncross string is the hypotenuse of the triangle.

$$A_1 f_{1-2} = A_2 f_{2-1} = \frac{5 + 12 - 13}{2} = 2 \frac{\text{in}^2}{\text{in depth}}$$

The view factor from plates 1 to 2 (f_{1-2}) is

$$f_{1-2} = \frac{A_2 f_{2-1}}{A_1} = \frac{2 \text{ in}^2/\text{in}}{(5)(40) \text{ in}^2} (40 \text{ in}) = 0.40 \text{ (dimensionless)} \quad (5.52)$$

The view factor from plates 2 to 1 (f_{2-1}) is

$$f_{2-1} = \frac{A_1 f_{1-2}}{A_2} = \frac{2}{(12)(40)} \times 40 = 0.167 \text{ (dimensionless)} \quad (5.53)$$

Figure 5.20 shows two infinite rectangular parallel plates of width b, directly above one another and separated by a distance h. The view factor

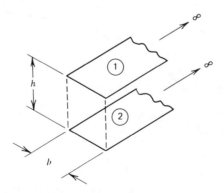

Figure 5.20 Two infinite parallel plates.

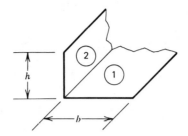

Figure 5.21 Two infinite perpendicular plates.

f_{1-2} is shown by Eq. 5.54.

$$f_{1-2} = \sqrt{1 + \left(\frac{h}{b}\right)^2} - \frac{h}{b} \tag{5.54}$$

When the distance between the plates (h) is equal to the width (b) of the plates, the view factor f_{1-2} is shown by Eq. 5.55.

$$f_{1-2} = \sqrt{1 + (1)^2} - 1 = 1.414 - 1.0 = 0.414 \tag{5.55}$$

Figure 5.21 shows two infinite rectangular plates of width b and h that intersect each other at an angle of 90°. The view factor f_{1-2} is shown by Eq. 5.56.

$$f_{1-2} = \frac{1}{2}\left[1 + \frac{h}{b} - \sqrt{1 + \left(\frac{h}{b}\right)^2}\right] \tag{5.56}$$

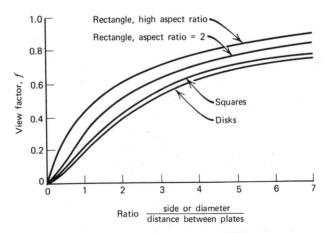

Figure 5.22 View factor for flat parallel plates directly opposed. (From Heat Transmission, by McAdams, McGraw Hill Book Co.)

5.17 Radiation View Factor

When the width of both plates are equal, the view factor f_{1-2} is shown by Eq. 5.57.

$$f_{1-2} = \frac{1}{2}[1 + 1 - \sqrt{1 + (1)^2}] = 0.293 \qquad (5.57)$$

Figure 5.22 shows view factors for two rectangular parallel plates of equal size but with different ratios of length to width, or aspect ratio [15].

Figure 5.23 shows view factors for two rectangular plates of different sizes that intersect at an angle of 90° [15].

The view factor f is not easily determined for radiation involving irregular shapes, unless the radiating surface (1) is completely enclosed by another surface (2), as shown in Figure 5.24. The view factor from surface 1 to surface 2 (f_{1-2}) is 1.0 for this condition.

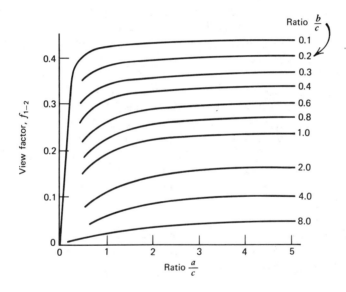

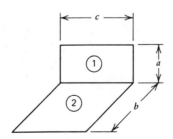

Figure 5.23 View factor for perpendicular and adjacent rectangles with a common side. (From Heat Transmission, by McAdams, McGraw Hill Book Co.)

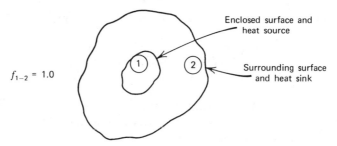

Figure 5.24 View factor for a completely enclosed irregular body.

When partial surfaces are involved in the radiation interchange, the view factor must be calculated or estimated. In electronic systems, the view factor from a transistor on a printed circuit board (PCB) to a cooler wall may be extremely complex and involve lengthy mathematics. If there are 50 transistors in different positions on different PCBs, the calculations may take several weeks. The amount of time required to perform these calculations may not be cost effective, so that the view factors for each transistor may simply be estimated.

Radiation heat transfer depends upon the line of sight. Therefore, it is important to determine approximately what fraction of the radiation leaving the surface of each transistor in all directions is intercepted by the cooler wall. This is what determines the view factor from the transistor to the wall.

The view factor from a hot transistor case on one PCB to the surface of an adjacent PCB is meaningless unless the adjacent PCB is much cooler than the transistor case. In general, there is very little heat lost by radiation between the hot components on adjacent PCBs. These hot components just "see" each other, so that there is a radiation interchange but no heat loss (see Figure 5.25).

Electronic components mounted on the end PCBs will usually have a good view of the end walls. Radiation heat transfer for these components will be relatively good if the end walls are cool.

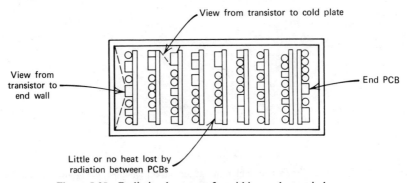

Figure 5.25 Radiation heat transfer within an electronic box.

5.18 Sample Problem—Radiation Heat Transfer from a Hybrid

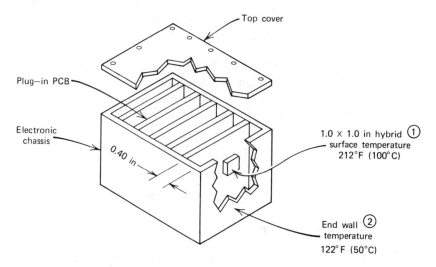

Figure 5.26 Hybrid mounted on a PCB that faces a chassis end wall.

5.18 SAMPLE PROBLEM—RADIATION HEAT TRANSFER FROM A HYBRID

A 1.0 in. square (2.54 cm) × 0.180 in. high (0.457 cm) flat pack hybrid is mounted on the end printed circuit board so that it faces the end wall of an electronic chassis, which has a temperature of 122°F (50°C), as shown in Figure 5.26. The hybrid is about 0.40 in (1.02 cm) from the wall, so that natural convection and conduction heat transfer will be negligible. The maximum allowable case temperature of the hybrid is 212°F (100°C). Determine the maximum allowable power dissipation for the hybrid with and without a conformal coating. The inside surfaces of the aluminum chassis are painted light blue.

SOLUTION

The top surface of the hybrid and its electrical lead wires have a good view of the chassis end wall. The view from the side walls of the hybrid to the side walls of the chassis is blocked by other electronic components around the hybrid. The bottom surface of the hybrid has no view of any wall. Ignoring radiation from the lead wires, the view factor from the top surface of the hybrid to the end wall will be very high, because the end wall is so close to the hybrid and the end wall is so much larger than the hybrid. The view factor f from the hybrid case to the chassis wall in this case would be about 0.95.

Equation 5.47 can be used to determine the amount of heat the hybrid is permitted to dissipate, in English and metric units.

Given $\sigma = 0.1713 \times 10^{-8}$ Btu/hr ft^2 °R^4 = 1.355×10^{-12} cal/sec cm^2 °K^4
$f = 0.95$ estimated, based on geometry
$e_1 = 0.066$ emissivity of bare hybrid steel case
$e_2 = 0.90$ emissivity of painted aluminum wall

$$e = \frac{1}{1/e_1 + 1/e_2 - 1} = \frac{1}{1/0.066 + 1/0.90 - 1} = 0.0655$$
(combined emissivity)

$A_1 = 1.0$ in^2 = 0.00694 ft^2 = 6.45 cm^2 (area)
$t_1 = 212$°F = 100°C (hybrid case temperature)
$t_2 = 122$°F = 50°C (chassis wall temperature)

Substitute into Eq. 5.47 for heat transferred from the hybrid by radiation, without a conformal coating, using English units.

$$Q = \left(0.1713 \times 10^{-8} \frac{\text{Btu}}{\text{hr ft}^2 \text{ °R}^4}\right)(0.95)(0.0655)(0.00694 \text{ ft}^2)$$

$$[(460 + 212)^4 - (460 + 122)^4] \text{ °R}^4$$

$$Q = (0.1713)(0.95)(0.0655)(0.00694)\left[\left(\frac{672}{100}\right)^4 - \left(\frac{582}{100}\right)^4\right]$$

$$Q = 0.065 \frac{\text{Btu}}{\text{hr}} \times 0.293 \frac{\text{watt}}{\text{Btu/hr}} = 0.019 \text{ watt} \qquad (5.58)$$

In metric units for the hybrid with no coating:

$$Q = \left(1.355 \times 10^{-12} \frac{\text{cal}}{\text{sec cm}^2 \text{ °K}^4}\right)(0.95)(0.0655)(6.45 \text{ cm}^2)$$

$$[(273 + 100)^4 - (273 + 50)^4] \text{ °K}^4$$

$$Q = (1.355)(0.95)(0.065)(6.45)\left[\left(\frac{373}{1000}\right)^4 - \left(\frac{323}{1000}\right)^4\right]$$

$$Q = 0.00457 \frac{\text{cal}}{\text{sec}} \times 4.187 \frac{\text{watts sec}}{\text{cal}} = 0.019 \text{ watt} \qquad (5.58a)$$

When a thin conformal coating is applied to the hybrid case, the emissivity will jump to about 0.80. The combined emissivity will jump to

$$e = \frac{1}{1/0.80 + 1/0.90 - 1} = 0.734$$

An examination of Eqs. 5.58 and 5.58a shows that the amount of heat transferred by radiation is directly proportional to the emissivity. Therefore,

5.19 Sample Problem—Junction Temperature of a Dual FET Switch

a direct ratio of the combined emissivities can be used to determine the heat transferred when a conformal coating is added to the hybrid case.

$$Q = 0.019 \text{ watt} \left(\frac{e_{\text{coat}}}{e_{\text{no coat}}} \right) = 0.019 \left(\frac{0.734}{0.0655} \right)$$

$$Q = 0.213 \text{ watt} \tag{5.59}$$

This shows that just adding a conformal coating to the hybrid will increase its heat transfer capability by a factor of 11.

5.19 SAMPLE PROBLEM—JUNCTION TEMPERATURE OF A DUAL FET SWITCH

Figure 5.27 shows a dual field effect transistor (FET) switch in a TO-8 size case that dissipates 0.150 watt. The switch is mounted on a PCB within an electronic chassis that has painted walls at a temperature of 185°F (85°C). The maximum allowable junction temperature is 302°F (150°C) and the thermal impedance (Θ_{jc}) from the junction of the switch to the case is 49.2°F/watt (27.3°C/watt) as rated by the manufacturer. The case of the switch is made of polished iron and the view factor f from the case to the chassis walls is estimated to be about 0.80. The chassis is enclosed, so that natural convection and conduction heat transfer from the switch are negligible. Determine if the design is satisfactory.

SOLUTION

The following information is required for Eq. 5.47.

$f = 0.80$ view factor, estimated from geometry
$e_1 = 0.24$ emissivity of switch with bare iron case
$e_2 = 0.84$ emissivity of painted chassis walls

$$e = \frac{1}{1/e_1 + 1/e_2 - 1} = \frac{1}{1/0.24 + 1/0.84 - 1} = 0.23$$

(combined emissivity)

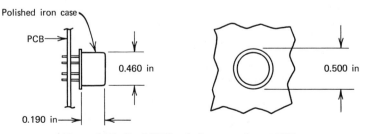

Figure 5.27 Dual FET switch mounted on a PCB.

The bottom surface of the switch faces the PCB, which is a poor heat conductor, so that its area is ignored. Also, radiation from the lead wires is ignored. The switch surface area is:

$$A = \frac{\pi(0.460)(0.190) + (\pi/4)(0.500)^2 \text{ in}^2}{144 \text{ in}^2/\text{ft}^2} = 0.00327 \text{ ft}^2$$

$$Q = 0.150 \text{ watt} \times 3.413 \frac{\text{Btu/hr}}{\text{watt}} = 0.512 \frac{\text{Btu}}{\text{hr}} \text{ (heat dissipation)}$$

Substitute into Eq. 5.47 for the case temperature.

$$0.512 = 0.1713 \times 10^{-8} (0.8)(0.23)(0.00327)[(460 + t_1)^4 - (460 + 185)^4]$$

$$0.512 = 1.031 \times 10^{-12} (460 + t_1)^4 - 0.1784$$

$$460 + t_1 = \left(\frac{0.512 + 0.1784}{1.031 \times 10^{-12}}\right)^{1/4}$$

$$t_1 = 444.6°\text{F} \ (229.2°\text{C}) \ \text{(case temperature)} \qquad (5.60)$$

The temperature rise from the switch case to the junction is determined from the manufacturer's thermal impedance from junction to case (Θ_{jc}), 49.2°F/watt.

$$\Delta t_{jc} = 49.2 \frac{°\text{F}}{\text{watt}} \times 0.150 \text{ watt} = 7.4°\text{F} \qquad (5.61)$$

The junction temperature becomes

$$t_j = 444.6 + 7.4 = 452°\text{F} \ (233.3°\text{C}) \qquad (5.62)$$

This is substantially above the maximum allowable junction temperature of 302°F (150°C), so that the design is not satisfactory.

To improve the radiation heat transfer, a black oxide finish is added to the surface of the switch, which raises its emissivity to a value of 0.95. The combined emissivity is now

$$e = \frac{1}{1/0.95 + 1/0.84 - 1} = 0.804 \text{ (combined emissivity)}$$

Substitute back into Eq. 5.47 for the case temperature.

$$0.512 = 0.1713 \times 10^{-8}(0.8)(0.804)(0.00327)[(460 + t_1)^4 - (460 + 185)^4]$$

$$460 + t_1 = \left(\frac{0.512 + 0.623}{3.60 \times 10^{-12}}\right)^{1/4}$$

$$t_1 = 289°\text{F} \ (142.7°\text{C}) \ \text{(case temperature)} \qquad (5.63)$$

5.20 Radiation Heat Transfer in Space

The thermal impedance Θ_{jc} from the junction to the case is still the same. The junction temperature of the switch with the black case is shown as follows:

$$t_j = 289 + 7.4 = 296.4°F \ (146.9°C) \tag{5.64}$$

Since the junction temperature is less than 302°F (150°C), the design is satisfactory.

NOTE: Could have raised e with conformal coat, but it might crack off

5.20 RADIATION HEAT TRANSFER IN SPACE

In the hard vacuum environment of space, the amount of radient energy that a body receives will depend upon its distance from the sun, the view the body has of the sun, and its proximity to other planets or satellites. A spacecraft in close orbit around the earth or around the moon can receive radiation from the sun, the earth, and the moon at the same time. Most of this radiation is in the ultraviolet and visible light frequency range, which has a wavelength up to about 1 μm, which is 1×10^{-6} m, as shown in Figure 5.15. In a near earth orbit the direct radiation intensity is about 444 Btu/hr ft² (130 watts/ft²) (0.0334 cal/sec cm²) [4, 5].

In addition to the direct solar radiation energy, there is reflected radiation energy, or albedo. Albedo is defined as the fraction of solar radiation that is returned to space due to reflections of solar energy from the atmosphere, clouds, and surface of a planet. For the earth, the albedo is about 0.40, depending upon the amount of cloud cover in the sky. For the moon, the albedo is very low, about 0.07, because the moon has virtually no atmosphere.

When the earth is viewed from a distance, its average temperature appears to be about −20°F (−29°C). The mean surface temperature appears to have a value of about 57°F (14°C), and the mean temperature of the upper atmosphere, at about 150,000 ft, is about −85°F (−65°C). The sun, when viewed from a distance, appears to have a temperature of about 10,400°F (5760°C).

The amount of radiant energy that a body in space will receive will depend upon the proximity and the view the body has of the sun and, or the moon, or perhaps some other planet. Table 5.7 shows how much radiant energy various planets receive from the sun. This is the energy that reaches the planet's upper atmosphere, not the energy that reaches the surface (except for the moon).

The upper atmosphere of the earth attenuates the radiation energy received from the sun, and this reduces the amount of solar radiation that actually reaches the surface of the earth. The amount of radiation energy that actually reaches the surface of the earth depends upon the month, hour, parallel of latitude, and cloud cover. Some typical maximum values

Table 5.7 Average Radiation Characteristics of the Planets

Planet	Incident Radiation Intensity (Btu/hr ft^2)	Reflected Radiation (Albedo)	Equivalent Black Body Temperature (°F)	
Mercury	2922	0.06	342	
Venus	860	0.61	127	
Earth	444	0.40	−5	Night sky = −40°F
(Moon)	434	0.072	51	
Mars	192	0.15	−56	
Jupiter	16.4	0.41	−242	
Saturn	4.9	0.42	−297	
Uranus	1.2	0.45	−346	
Neptune	0.5	0.54	−366	
Pluto	0.3	0.15	−380	

on a clear day are as follows: [37, 38]:

40° north latitude USA: 300 Btu/hr ft^2

32° north latitude USA: 310 Btu/hr ft^2

24° north latitude USA: 320 Btu/hr ft^2

Deep space has the ability to absorb virtually an infinite amount of heat that makes it a very useful heat sink with a temperature of −460°F (−273°C), absolute zero. The view factor of a body in deep space is 1.0, because the body can radiate to deep space in all directions.

If two aluminum spheres are placed in deep space and if one sphere has a polished aluminum surface while the other sphere has a surface that is painted white, the white sphere will show a steady state surface temperature of about 80°F (26.6°C) and the polished sphere will show a surface temperature of about 450°F (232°C), as shown in Figure 5.28.

5.21 EFFECTS OF α/e ON TEMPERATURES IN SPACE

When thermal radiation energy strikes a body, the fraction of the incident radiation energy that is absorbed is called the absorptivity α. In the infrared, or long wavelength range, the color has no effect on the absorptivity, and a black surface will absorb about the same amount of thermal radiation energy as a white surface. In the visible, or short wavelength range (see Figure 5.15), the amount of radiation energy absorbed by a body is sharply

5.21 Effects of α/e on Temperatures in Space

Figure 5.28 Temperatures of painted and unpainted aluminum spheres in space.

affected by its color. The stabilized temperature a body reaches in space will therefore be related to the ratio of the solar absorptivity (α_s) to the infrared emissivity (e), which is shown as the ratio α_s/e. For the polished aluminum sphere this ratio is shown in Eq. 5.65 and for the white-painted aluminum sphere, this ratio is shown in Eq. 5.66.

Polished aluminum sphere in space:

$$\frac{\alpha_s}{e} = \frac{\text{solar absorptivity}}{\text{infrared emissivity}} = \frac{0.10}{0.025} = 4 \qquad (5.65)$$

White-painted aluminum sphere in space:

$$\frac{\alpha_s}{e} = \frac{\text{solar absorptivity}}{\text{infrared emissivity}} = \frac{0.15}{0.94} = 0.16 \qquad (5.66)$$

A high ratio of α/e means that more solar radiation energy will be absorbed, so that the resulting temperature of a body in space will be higher. Conversely, a low α/e means that less solar radiation energy will be absorbed, so that the resulting temperature will be lower. The equilibrium temperature of a body in space can be controlled, to a great extent, by controlling the ratio of α_s/e.

A body in space will receive radiation from the sun and emit radiation to space at $-460°F$, which is absolute zero. If the amount of solar energy absorbed by a body (including albedo) is shown as $\alpha_s Q_s A_s$, the general radiation relation shown by Eq. 5.47 can be modified as shown by Eq. 5.67.

$$Q_i + \alpha_s Q_s A_s = \sigma e A T^4 \qquad (5.67)$$

where α_s = absorptivity of surface receiving solar radiation
  Q_s = solar irradiation on absorbing surface
  Q_i = internal power dissipation
  A_s = projected area of body normal to sun
  σ = Stefan-Boltzmann constant
  e = emissivity of surface radiating to space
  A = surface area of body radiating to space
  T = absolute temperature of body in space

When the body in space is away from all the planets, with no internal power dissipation and no heat transferred into or out of the back side of the body, and when the projected area of the body is normal to the sun. The absolute temperature of the body in space is shown by Eq. 5.68.

$$T = \left(\frac{\alpha_s Q_s}{\sigma e}\right)^{0.25} \tag{5.68}$$

In a deep space orbit similar to that of Venus, but away from all planets, the direct solar radiation intensity Q_s is about 860 Btu/hr ft^2 (252 watts) (0.0648 cal/sec cm^2). Substitute this value into Eq. 5.68 to obtain the relation for the absolute temperature of a body in a deep space orbit similar to that of Venus.

$$T = \left(\frac{860 \alpha_s}{0.1713 \times 10^{-8} e}\right)^{0.25} = 842 \left(\frac{\alpha_s}{e}\right)^{0.25} \tag{5.69}$$

Since the solar absorptivity (α_s) and the infrared emissivity (e) are both characteristics of the body surface, the control of these values will help to control the body temperature.

5.22 SAMPLE PROBLEM—TEMPERATURES OF AN ELECTRONIC BOX IN SPACE

An electronic box has an insulated panel with a surface area of 1.0 ft^2. It is in deep space approximately the same distance from the sun as the earth. The panel is painted white and always faces the sun. Determine the stabilized temperature of the panel with no internal power dissipation and with an internal power dissipation of 50 watts.

SOLUTION

Given Q_s = 444 Btu/hr ft^2 solar radiation on panel in an orbit similar to that of Earth
α_s = 0.15 solar absorptivity of white paint (ref. Table 5.4)
σ = 0.1713 × 10^{-8} Btu/hr ft^2 °R^4 constant
e = 0.9 emissivity of white paint (ref. Table 5.4)

Substitute into Eq. 5.68 for the stabilized temperature of the panel with no internal power dissipation.

$$T = \left[\frac{(0.15)(444)}{(0.1713 \times 10^{-8})(0.9)}\right]^{0.25} = 455.9°R$$

$$t = 455.9 - 460 = -4.1°F \tag{5.70}$$

The stabilized temperature of the panel with power on can be obtained with Eqs. 5.67 and 5.68, which requires adding the internal power of 50 watts (170.6 Btu/hr).

$$T = \left[\frac{(0.15)(444) + 170.6}{(0.1713 \times 10^{-8})(0.9)}\right]^{0.25} = 626.3°R$$

$$t = 626.3 - 460 = 166.3°F \tag{5.71}$$

If the electronic box was in an orbit around the earth, where it was possible for the panel to receive reflected albedo radiation (40%) from the earth, the panel would receive an additional incident radiation of 444 × 0.40 or 177.6 Btu/hr ft. Substituting into Eq. 5.68 for the panel temperature, we obtain

$$T = \left[\frac{(0.15)(444 + 177.6) + 170.6}{(0.1713 \times 10^{-8})(0.9)}\right]^{0.25} = 643.2°R$$

$$t = 643.2 - 460 = 183°F \tag{5.72}$$

There would also be an interchange of radiation energy between the electronic box and the earth, which would depend upon the orbit of the electronic box around the planet. This energy interchange was ignored in the problem.

5.23 SIMPLIFIED RADIATION HEAT TRANSFER EQUATION

It is often convenient to express the radiation heat loss in a form similar to the convection heat loss, as shown by Eq. 5.7. This can be accomplished by setting Eq. 5.7 equal to Eq. 5.47. The radiation heat transfer equation is then simplified as shown by Eq. (5.73) [15, 17, 24].

$$Q = h_r A \, \Delta t \tag{5.73}$$

In this expression, h_r represents the radiation heat transfer coefficient in terms of Btu/hr ft² °F (cal/sec cm² °C). The radiation coefficient is then defined as shown in Eq. 5.74 for English units and in Eq. 5.74a for metric units.

$$h_r = \frac{0.1713\,fe\,\{[(t_1 + 460)/100]^4 - [(t_2 + 460)/100]^4\}}{t_1 - t_2}$$

$$= \frac{\text{Btu}}{\text{hr ft}^2\,°F} \tag{5.74}$$

$$h_r = \frac{1.355 fe \{[(t_1 + 273)/1000]^4 - [(t_2 + 273)/1000]^4\}}{t_1 - t_2}$$

$$= \frac{\text{cal}}{\text{sec cm}^2 \, ^\circ\text{C}} \tag{5.74a}$$

A plot of Eq. 5.74 is shown in Figure 5.29 for various heat receiver temperatures and various temperature differences between the emitting and receiving surfaces [70].

5.24 SAMPLE PROBLEM—RADIATION HEAT LOSS FROM AN ELECTRONIC BOX

An electronic box, for a space application, has hybrid circuits mounted on the cover facing the inside of the box. The backside of the cover can "see" the back side of a space radiator panel within the spacecraft, as shown in Figure 5.30. The maximum allowable hybrid case temperature is 212°F (100°C). The temperature rise from the component case to the cover is 12°F (6.7°C), which means that the maximum allowable cover temperature is 200°F (93.3°C). The radiator panel has a temperature of 20°F (−6.7°C) and an emissivity e_2 of 0.75. The electronic box cover has a painted finish with an emissivity e_1 of 0.85. Determine the maximum allowable power the electronic components on the cover can dissipate.

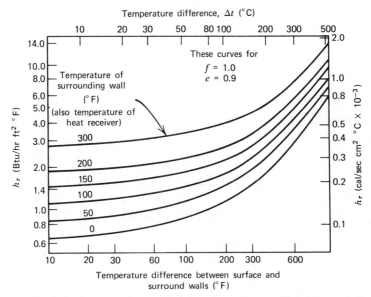

Figure 5.29 Radiation heat transfer as a function of temperature. (From General Electric Co. Heat Transfer/Fluid Flow Data Book)

5.24 Sample Problem—Radiation Heat Loss from an Electronic Box

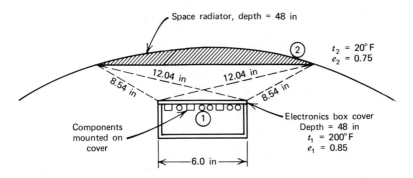

Figure 5.30 Electronic box mounted in a spacecraft.

SOLUTION

The cross string method shown in Eq. 5.49 is used to determine the view factors between the radiating surfaces.

$$A_1 f_{1-2} = A_2 f_{2-1} = \frac{2(12.04) - 2(8.54)}{2} = 3.5 \frac{\text{in}^2}{\text{in depth}}$$

$$f_{1-2} = \frac{A_2 f_{2-1}}{A_1} = \frac{3.5 \text{ in}^2/\text{in}}{(6)(48) \text{ in}^2} \times 48 \text{ in} = 0.583$$

$$f_{2-1} = \frac{A_1 f_{1-2}}{A_2} = \frac{3.5}{(12)(48)} \times 48 \text{ in} = 0.292$$

The combined emissivity is obtained from Table 5.5.

$$e = \frac{1}{1/e_1 + 1/e_2 - 1} = \frac{1}{1/0.85 + 1/0.75 - 1} = 0.662$$

The view factor from area 1 to area 2 ($f_{1-2} = 0.583$) will be used for the radiation heat transfer, so area 1 must be computed.

$$A_1 = \frac{(6)(48) \text{ in}^2}{144 \text{ in}^2/\text{ft}^2} = 2.0 \text{ ft}^2 = 1858 \text{ cm}^2 \tag{5.75}$$

Substitute into Eq. 5.74 to obtain the radiation coefficient in English Units.

$$h_r = \frac{0.1713(0.583)(0.662) \{[(200 + 460)/100]^4 - [(20 + 460)/100]^4\}}{200 - 20}$$

$$h_r = 0.502 \frac{\text{Btu}}{\text{hr ft}^2 \, °\text{F}} \tag{5.76}$$

Substitute into Eq. 5.74a to obtain the radiation coefficient in metric units.

$$h_r = \frac{1.355(0.583)(0.662)\{[(93.33 + 273)/1000]^4 - [(-6.7 + 273)/1000]^4\}}{93.3 - (-6.7)}$$

$$h_r = 0.0000679 \frac{\text{cal}}{\text{sec cm}^2 \,°\text{C}} \tag{5.76a}$$

Substitute Eqs. 5.75 and 5.76 into Eq. 5.73 for the maximum heat that can be transferred by radiation, using English units.

$$Q = \left(0.502 \frac{\text{Btu}}{\text{hr ft}^2 \,°\text{F}}\right)(2.0 \text{ ft}^2)(200 - 20)°\text{F}$$

$$Q = 180.7 \frac{\text{Btu}}{\text{hr}} \times 0.293 \frac{\text{watt}}{\text{Btu/hr}} = 52.9 \text{ watts} \tag{5.77}$$

Substitute into Eq. 5.73 for the maximum heat that can be transferred by radiation using metric units.

$$Q = \left(0.0000679 \frac{\text{cal}}{\text{sec cm}^2 \,°\text{C}}\right)(1858 \text{ cm}^2)[(93.3 - (-6.7)]$$

$$Q = 12.61 \frac{\text{cal}}{\text{sec}} \times 4.187 \frac{\text{watts}}{\text{cal/sec}} = 52.8 \text{ watts} \tag{5.77a}$$

The heat transfer coefficient h_r could also have been obtained from Figure 5.29 for English units and metric units. Notice that Figure 5.29 is based upon a view factor of 1 and a combined emissivity of 0.9. If the geometry under consideration has a different view factor and a different emissivity, corrections must be applied to obtain the true radiation coefficient.

For the previous problem, using a heat sink temperature of 20°F and a ΔT of $200 - 20 = 180$°F, $h_r = 1.18$ Btu/hr ft² °F for English units. Corrections must be applied for view factors and emissivity.

$$h_r = 1.18 \times 0.583 \times \left(\frac{0.662}{0.9}\right) = 0.506 \frac{\text{Btu}}{\text{hr ft}^2 \,°\text{F}} \tag{5.78}$$

Equation 5.78 compares well with Eq. 5.76.

5.25 COMBINING CONVECTION AND RADIATION HEAT TRANSFER

In many electronic boxes, radiation heat transfer and convection (both natural and forced) occur simultaneously. Therefore, it is convenient to write the combined heat transfer equation in a form similar to the general

5.26 Sample Problem—Electronic Box in an Airplane Cockpit Area

relation shown by Eqs. 5.7 and 5.73. The combined relation is shown in Eq. 5.79 [15, 17, 24].

$$Q = (h_c + h_r) A \, \Delta t \qquad (5.79)$$

where h_c = Btu/hr ft² °F or cal/sec cm² °C (convection coefficient)
h_r = Btu/hr ft² °F or cal/sec cm² °C (radiation coefficient)
A = ft² or cm² (area)
Δt = °F or °C (temperature difference, surface to ambient)

For most small electronic boxes, typical values of h_c are about 0.8 Btu/hr ft² °F (0.000108 cal/sec cm² °C) and typical values of h_r are about 1.2 Btu/hr ft² °F (0.000162 cal/sec cm² °C), which results in a combined heat transfer coefficient, $h_c + h_r$, of about 2.0 Btu/hr ft² °F (0.00027 cal/sec cm² °C).

5.26 SAMPLE PROBLEM—ELECTRONIC BOX IN AN AIRPLANE COCKPIT AREA

An electronic box is mounted in the cockpit area of an airplane that must operate at 50,000 ft without a pressurized cabin. Plug-in PCBs within the box are cooled by conducting their heat to the side walls of the electronic box. Natural convection and radiation are used to cool the external surfaces of the box. The maximum expected cockpit temperature is 100°F and the maximum allowable box surface temperature is 140°F. Because the top and bottom covers are fastened to the box with a few small screws, the covers make poor contact with the box and will not be effective heat transfer surfaces (see Figure 5.31). Determine the maximum amount of heat that can

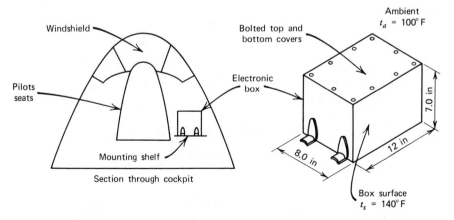

Figure 5.31 Electronic box mounted on a shelf in an airplane.

be transferred from the box to the surrounding ambient at sea level and at 50,000 ft. The box has painted external surfaces.

SOLUTION

Only the vertical side and end walls will be effective for heat transfer by natural convection and radiation. The top and bottom covers are ignored here because the covers make poor contact with the chassis. The natural convection coefficient for the vertical side and end walls is determined first for the sea level condition.

$$h_c = 0.29 \left(\frac{\Delta t}{L}\right)^{0.25} \quad \text{(ref. Eq. 5.4)}$$

Given $\Delta t = 140 - 100 = 40°F$ (temperature difference)

$$L = \frac{7.0 \text{ in}}{12 \text{ in/ft}} = 0.583 \text{ ft (vertical height)}$$

$$h_c = 0.29 \left(\frac{40}{0.583}\right)^{0.25} = 0.83 \frac{\text{Btu}}{\text{hr ft}^2 \text{ °F}} \text{ (convection coefficient)} \quad (5.80)$$

The radiation heat transfer coefficient is determined next, using an emissivity of 0.9 for the painted finish and a view factor of 1.0 for the electronic box. Using Figure 5.29 with an ambient heat sink temperature of 100°F and a Δt of 40°F,

$$h_r = 1.20 \frac{\text{Btu}}{\text{hr ft}^2 \text{ °F}} \text{ (radiation coefficient)} \quad (5.81)$$

The surface area available for heat transfer is the sides and ends of the box only.

$$A = \frac{(2 \times 12 + 2 \times 8)(7) \text{ in}^2}{144 \text{ in}^2/\text{ft}^2} = 1.94 \text{ ft}^2 \quad (5.82)$$

Substitute Eqs. 5.80 through 5.82 into Eq. 5.79 for the heat transferred at sea level conditions.

$$Q = (0.83 + 1.20)(1.94)(40)$$

$$Q = 157.5 \frac{\text{Btu}}{\text{hr}} \times 0.293 \frac{\text{watt}}{\text{Btu/hr}} = 46.1 \text{ watts} \quad (5.83)$$

At an altitude of 50,000 ft, there will be a small change in the radiation characteristics of the box, but there will be a big change in the natural convection characteristics. (The surface temperature of the box will increase slightly at altitude.)

$$h_{alt} = h_c \sqrt{\frac{\text{altitude pressure}}{14.7 \text{ psia}}} \quad \text{(ref. Eq. 5.41)}$$

Using the pressure altitude from Table 6.5, the pressure at 50,000 ft is 1.692 lb/in².

$$h_{alt} = 0.83 \sqrt{\frac{1.69 \text{ psia}}{14.7 \text{ psia}}} = 0.28 \frac{\text{Btu}}{\text{hr ft}^2 \text{ °F}} \quad (5.84)$$

Equation 5.79 is used once again to determine the amount of heat that can be transferred at the 50,000 ft altitude.

$$Q = (h_{alt} + h_r)\dot{A}\,\Delta T \quad \text{(ref. Eq. 5.79)}$$

$$Q = (0.28 + 1.20)(1.94)(40)$$

$$Q = 114.8 \frac{\text{Btu}}{\text{hr}} \times 0.293 \frac{\text{watt}}{\text{Btu/hr}} = 33.6 \text{ watts} \quad (5.85)$$

This shows that the amount of heat that can be removed at the 50,000 ft altitude has been reduced about 27%.

5.27 EQUIVALENT AMBIENT TEMPERATURE FOR RELIABILITY PREDICTIONS

Reliability failure rates for electronic component parts are usually shown to be related to the *ambient* temperature in which the component is operating. MIL-HDBK-217B, for example, shows failure rates for various components as a function of the ambient temperature [39].

In many electronic systems, the electronic components are mounted on thermally conductive circuit boards, which spread the heat to eliminate hot spots and also carry the heat to cold plates and heat exchangers. A low thermal resistance path is provided from the component surface directly to the chassis heat sink. This technique is very effective in providing a low component *surface* temperature.

The reliability engineer is not interested in component surface temperatures, because there are very few data relating surface temperatures of components to failure rates. Most data will show that component failure rates are related to the ambient temperature. Therefore, reliability engineers will usually request information relating to ambient temperatures when they are computing failure rates for a system.

When components are cooled by conduction, there are no real ambient operating temperatures that can be used to predict failure rates. However, an equivalent ambient temperature can be calculated based upon the surface temperature of the component. Thus, for a given set of conditions and for specific components, the equivalent ambient temperature can be determined from the component surface temperature. This permits failure rates to be determined for components when only the surface temperatures are known.

Since heat travels from high temperature areas to low temperature areas, it follows that a heat dissipating component will be hotter than the ambient temperature surrounding the component.

When an electronic component is suspended in a controlled environmental chamber, heat is transferred from the component to the surrounding ambient by means of radiation and natural convection. Component manufacturers and component testing engineers often use this technique to determine how much power a component can dissipate and what the failure rates are for various ambient temperature conditions.

The surface temperature of a component on a PCB is usually known from test data, or it can be calculated. The equivalent ambient temperature for the same component (the ambient temperature that will result in the same component surface temperature) is determined by obtaining the temperature rise from the component to the surrounding ambient using Eq. 5.86.

$$t_s = t_a + \Delta t_{s-A} \qquad (5.86)$$

where t_s = component surface temperature
t_a = ambient temperature
Δt_{s-A} = temperature rise from surface to ambient

The temperature rise from the component surface to the ambient Δt_{s-A} is determined from Eq. 5.79.

$$\Delta t_{s-A} = \frac{Q}{(h_c + h_r)A} \qquad \text{(ref. Eq. 5.79)}$$

The physical properties of the component are used to determine the surface area, the radiation coefficient, and the natural convection coefficient. The electrical lead wires are ignored in these calculations because they are usually very short in most applications, so that they will transfer very little heat by radiation and natural convection.

Radiation depends upon the actual surface temperature of the component, which is unknown. Therefore, the method of solution is to assume a temperature rise (Δt) that permits the radiation and natural convection coefficients to be determined. A Δt value is then calculated from the coefficients and compared with the assumed value. If the Δt's do not agree, a correction is made and the process is repeated until the assumed Δt agrees relatively well with the calculated Δt. A sample problem is used to demonstrate the technique.

5.28 SAMPLE PROBLEM—EQUIVALENT AMBIENT TEMPERATURE OF AN RC07 RESISTOR

A MIL style RC07 ¼ watt carbon composition resistor, as shown in Figure 5.32, dissipates 0.125 watt when it is mounted on a PCB that is cooled by conduction. A small thermocouple on the resistor shows that the surface temperature is 212°F (100°C) in the operating environment. The reliability engineer must determine the mean time between failures (MTBF) for this resistor. The only failure rate data are based on ambient temperatures. Determine the equivalent *ambient* temperature that will produce the same failure rate as a component *surface* temperature of 212°F (100°C).

SOLUTION

The lead wires are small compared to the body of the resistor, so that they are ignored. Assume that the resistor body is suspended in a temperature-controlled chamber. The natural convection coefficient can then be determined from Eq. 5.3 and Table 5.1, using a C value of 1.39 for small components in free air.

$$h_c = 0.52C \left(\frac{\Delta t}{L}\right)^{0.25} \quad \text{(ref. Eq. 5.3)}$$

$$h_c = 0.52(1.39)\left(\frac{\Delta t}{L}\right)^{0.25} = 0.72 \left(\frac{\Delta t}{L}\right)^{0.25} \quad (5.87)$$

Since the Δt is not known yet, assume a value of 67°F to start, and calculate the natural convection coefficient from Eq. 5.87.

Given $\Delta t = 67°F$ (assumed to start)

$$L = \frac{0.098 \text{ in}}{12 \text{ in/ft}} = 0.0082 \text{ ft diameter}$$

$$h_c = 0.72 \left(\frac{67}{0.0082}\right)^{0.25} = 6.84 \frac{\text{Btu}}{\text{hr ft}^2 \text{ °F}} \quad (5.88)$$

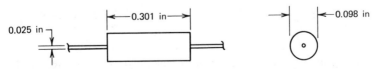

Figure 5.32 MIL style RC 07 carbon composition ¼ watt resistor.

The radiation coefficient h_r is determined from Figure 5.29, which is based upon an emissivity of 0.9 and a view factor of 1.0. The view factor of the resistor in the chamber will be 1.0 and the emissivity e of the resistor surface is about 0.9, so that the curves in Figure 5.29 can be used directly with no correction factors. The temperature of the surrounding walls is the same as the ambient temperature t_a, which is determined from Eq. 5.86.

Given $t_s = 212°F$ (surface temperature)
$\Delta t_{s-A} = 67°F$ (assumed to start)
$t_a = t_s - \Delta t_{s-A} = 212 - 67 = 145°F$ (ambient)

From Figure 5.29:

$$h_r = 1.61 \frac{\text{Btu}}{\text{hr ft}^2 \, °F} \tag{5.89}$$

Also,

$$Q = 0.125 \text{ watts} \times \frac{3.413 \text{ Btu/hr}}{\text{watt}} = 0.427 \frac{\text{Btu}}{\text{hr}} \text{ (heat)}$$

$$A = \frac{(2)(\pi/4)(0.098)^2 + \pi(0.098)(0.301) \text{ in}^2}{144 \text{ in}^2/\text{ft}^2} = 0.000748 \text{ ft}^2 \text{ (area)}$$

Substitute into Eq. 5.79 and solve for Δt.

$$\Delta t = \frac{Q}{(h_c + h_r) A} = \frac{0.427}{(6.84 + 1.61)(0.000748)} = 67.5°F \tag{5.90}$$

Since the calculated Δt of 67.5°F is close to the assumed value of 67°F, the assumed Δt is valid.

The equivalent ambient temperature that can be used for reliability predictions is determined from Eq. 5.86.

$$t_a = t_s - \Delta t_{s-A} = 212 - 67 = 145°F \tag{5.91}$$

This means that the RC07 resistor will have a surface temperature of 212°F in an ambient temperature of 145°F when the resistor dissipates 0.125 watt. Reliability failure rate predictions will be based upon operation in a 145°F ambient environment.

5.29 INCREASE IN EFFECTIVE EMITTANCE ON EXTENDED SURFACES

Extended surfaces, such as fins, are often used to improve the heat transfer characteristics of electronic equipment. A substantial improvement is ob-

5.29 Increase in Effective Emittance on Extended Surfaces

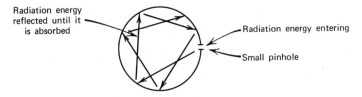

Figure 5.33 Black body represented by a hollow sphere with a pin hole.

tained with convection heat transfer, because the heat removed is directly related to the fin surface area. With radiation, the heat transfer is related to the projected fin area and the emittance of the surface.

A rough surface has a higher emittance than the same surface when it is smooth. Also, a finned surface has a higher emittance than does the same surface without fins. In both cases the emittance increases because the surfaces have many small cavities. These cavities act somewhat like many small partial black bodies [24].

A perfect black body is one that absorbs 100% of all the energy it receives without reflecting any energy. A small hole in a hollow sphere is often used to represent a black body. The energy enters the small opening and strikes the opposite wall, where part of the energy is absorbed and part is reflected. The reflected energy again strikes the opposite wall, where part of the energy is absorbed and again part is reflected. This process continues until all of the energy is absorbed, as shown in Figure 5.33.

Since finned surfaces act somewhat like cavities, a deeper fin should have a higher emittance than a shallow fin. When the emittance of the flat surface is known, the effective emittance of the finned surface can be approximated from Figure 5.34.

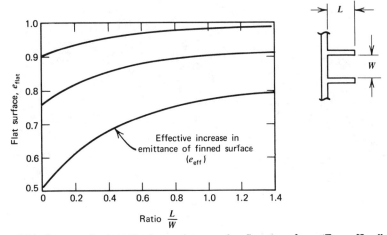

Figure 5.34 Increase in the effective emittance of a finned surface. (From Handbook of Electronic Packaging, by Harper, McGraw Hill Book Co.)

The height of the fin is shown by L and the distance between the fins is shown by W. Dimensions can be in inches or centimeters.

Figure 5.34 shows that the effective emittance of the finned surface can be substantially increased for low emittance surfaces, when the ratio of L/W is greater than 1.0, as shown by Eq. 5.92.

Figure 5.35 Tall single bay cabinet, with a blower at the bottom, for cooling panel and drawer mounted electronics. (Courtesy Litton Systems, Inc.)

5.29 Increase in Effective Emittance on Extended Surfaces

$$e_{\text{eff}} = e_{\text{flat}} + \left(\frac{1 - e_{\text{flat}}}{2}\right) \qquad (5.92)$$

When e_{flat} is 0.5, then for a L/W ratio of 1.0 the effective emissivity e_{eff} becomes

$$e_{\text{eff}} = 0.5 + \frac{1 - 0.5}{2} = 0.5 + 0.25 = 0.75 \text{ (approximately)}$$

Figure 5.35 shows a tall cabinet that is cooled by means of a blower in the base.

6

Forced Air Cooling for Electronics

6.1 FORCED COOLING METHODS

Heat transfer by forced convection generally makes use of a fan, blower, or pump to provide high velocity fluid (air or liquid) past a heated surface. The high velocity fluid results in a decreased thermal resistance across the boundary layer from the fluid to the heated surface. This, in turn, increases the amount of heat that is carried away by the fluid [40-42].

Forced air systems can provide heat transfer rates in electronic systems that are 10 times greater than those available with natural convection and radiation. Forced liquid cooling systems can provide heat transfer rates that are 10 times greater than those for forced air cooling systems. The penalty for increased cooling is increased costs, power, noise, and complexity. For liquid cooled systems the size and weight may also increase. Since a simple system is generally a more reliable system, heat transfer by natural convection and radiation should be used wherever possible.

The rewards of forced convection cooling techniques are generally reduced size for air cooled systems and higher component densities, with lower hot spot temperatures. This increases the electronic component reliability but requires added maintenance for the extra fans or pumps.

A liquid cooled system is usually larger and heavier than an air cooled system, because a reservoir is generally required for the liquid. A fan cooled system, on the other hand, normally has a large supply of air readily available with no storage requirements, which reduces the size and weight of the system.

Laminar flow conditions as well as turbulent flow conditions can exist with forced convection in liquids and air. Turbulent flow conditions are much more desirable, because they permit more heat to be removed. However, turbulent flow will usually result in a higher pressure drop through the system, which requires that larger pumps and fans and more power be used to overcome the added resistance.

6.2 COOLING AIR FLOW DIRECTION FOR FANS

When a fan is used for cooling electronic equipment, the air flow direction can be quite important. The fan can be used to draw air through a box or to blow air through a box. A blowing fan system will raise the internal air pressure within the box, which will help to keep dust and dirt out of a box that is not well sealed. A blowing system will also produce slightly more turbulence, which will improve the heat transfer characteristics within the box. However, when an axial flow fan is used in a blowing system, the air may be forced to pass over the hot fan motor, which will tend to heat the air as it enters the electronic box, as shown in Figure 6.1.

An exhaust fan system, which draws air through an electronic box, will reduce the internal air pressure within the box. If the box is located in a dusty or dirty area, the dust and dirt will be pulled into the box through all of the small air gaps if the box is not sealed. In an exhaust system, the cooling air passes through an axial flow fan as the air exits from the box, as shown in Figure 6.2. The cooling air entering the electronic box is therefore cooler.

The position of the fan blades within an axial flow fan housing can be a critical factor in determining how well a fan will perform. This is especially important in high speed fans that have speeds greater than about 8000 rpm.

An examination of most high speed axial flow fans will show that the fan blades are not located at the center of the tubular housing but near one end. When the fan is located adjacent to a restricted area, such as a 90° bend, the fan blades should be positioned so that they are at the downstream end of the housing, for the best performance. Air has weight and kinetic energy, so that the air velocity must be allowed to develop to effectively overcome the flow resistance. When the fan blades are located at the downstream end of the fan housing, the air has a slightly longer flow path. This improves the velocity profile, as shown in Figures 6.3 and 6.4.

It makes no difference if a blowing or an exhaust fan system is used; the velocity profile must have a chance to develop to provide an efficient air delivery system.

The reduced flow efficiency for the fan shown in Figure 6.4 will not be obvious to a casual observer. When the fan is in operation, you can place your hand over the exhaust and feel a large volume of air flowing through the fan. However, if a thin strip of paper is slowly passed across the fan exhaust, it will show that some of the air is being short-circuited. The air

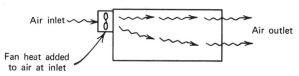

Figure 6.1 Axial flow fan blowing cooling air through a box.

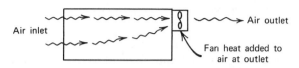

Figure 6.2 Axial flow fan drawing cooling air through a box.

flow at the outer perimeter of the fan will be moving away from the fan, but the air flow at the center will be moving toward the fan, resulting in a short circuit, as shown in Figure 6.5.

Some good fan installations and some poor ones are shown in Figure 6.6. Test data on these types of installations have shown that the cooling air flow rate can be more than doubled just by properly orienting the position of the fan blade within the fan housing when the fan is located adjacent to an area that restricts the free flow of cooling air.

6.3 STATIC PRESSURE AND VELOCITY PRESSURE

Air flow through an electronic box is due to a pressure difference between two points in the box, with the air flowing from the high pressure area to the low pressure area. The flow of air will result in a static pressure and a velocity pressure.

Static pressure is the pressure that is exerted on the walls of the container or electronic box, even when there is no flow of air; it is independent of the air velocity. Static pressure can be positive or negative, depending upon whether it is greater or less than the outside ambient pressure.

Velocity pressure is the pressure that forces the air to move through the electronic box at a certain velocity. The velocity pressure depends upon the velocity of the air and always acts in the direction of the air flow.

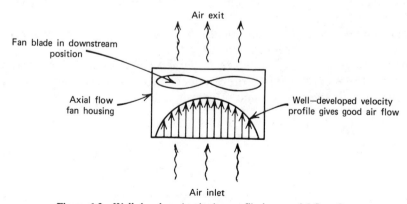

Figure 6.3 Well-developed velocity profile in an axial flow fan.

6.3 Static Pressure and Velocity Pressure

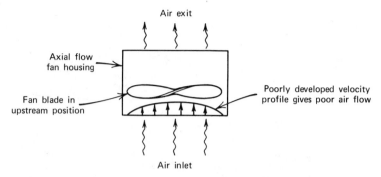

Figure 6.4 Poorly developed velocity profile in an axial flow fan.

The amount of cooling air flowing through an electronic box will usually determine the amount of heat removed from the box. If more air flows through the box, more heat will be removed. As the air flow through the box increases, however, it requires an even greater pressure to force the air through the box.

Static and velocity pressures can be expressed in lb/in^2 and g/cm^2. However, these values are usually very small, so that it is often more convenient to express these pressures in terms of the height of a column of water. For example, 1.0 lb/in^2 (70.37 g/cm^2) equals 27.7 in of water (70.37 cm of water), or 1.0 in of water (2.54 cm of water) equals 0.036 lb/in^2 (2.54 g/cm^2). The water column height is called the "head" of water and is shown as H_s for the static head and as H_v for the velocity head.

The velocity head (H_v) is a convenient reference that is often used to determine pressure drops through electronic boxes. The velocity head can be related to the air flow velocity with the use of Eq. 6.1 [10, 18, 69].

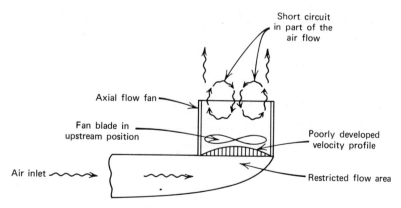

Figure 6.5 Short circuit in part of the cooling air flow due to improper positioning of the fan blades adjacent to a restricted flow area.

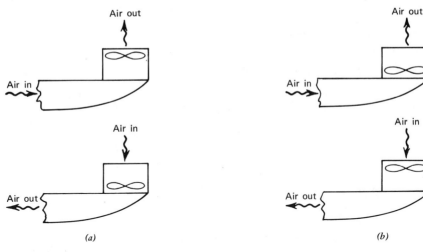

Figure 6.6 Examples of good fan installations and poor fan installations on axial flow fans located adjacent to restricted areas. (*a*) Good fan installations; (*b*) poor fan installations.

$$V = \sqrt{2gH} \tag{6.1}$$

where V = velocity = ft/sec
 g = acceleration of gravity = ft/sec^2
 H = height = ft

When the air velocity is expressed in feet per minute and the height is expressed as the velocity head H_v in inches of water, Eq. 6.1 can be modified using standard air with a density of 0.0750 lb/ft^3 at 69°F and 14.7 psia. For English units, this is shown in Eq. 6.2.

$$V = 60\,\frac{\text{sec}}{\text{min}}\sqrt{\frac{2(32.14\ \text{ft/sec}^2)(62.4\ \text{lb/ft}^3\ \text{water})(H_v\ \text{in water})}{(12\ \text{in/ft})(0.0750\ \text{lb/ft}^3\ \text{standard air})}}$$

$$V = 4005\,\sqrt{H_v\ (\text{in water})} = \frac{\text{ft}}{\text{min}} \tag{6.2}$$

For metric units, this is shown in Eq. 6.2a.

$$V = \sqrt{\frac{2(979.6\ \text{cm/sec}^2)(1\ \text{g/cm}^3\ \text{water})(H_v\ \text{cm water})}{0.0012\ \text{g/cm}^3\ \text{air at 20.5°C}}}$$

$$V = 1277\,\sqrt{H_v\ (\text{cm water})} = \frac{\text{cm}}{\text{sec}} \tag{6.2a}$$

A sealed electronic box with a gauge pressure of 0.05 lb/in^2 (3.518 g/cm^2)

6.3 Static Pressure and Velocity Pressure

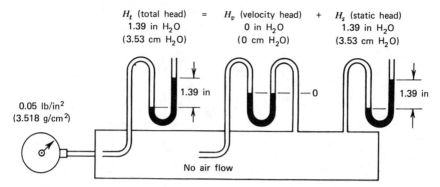

Figure 6.7 A pressurized electronic box with no air flow.

with no air flow will have a pressure equal to the static pressure. If small holes are drilled through the walls and if water filled manometers are used to measure the various pressure heads, the results will appear as shown in Figure 6.7 [40, 41, 69].

When a fan blows air through the electronic box, the pressure within the box will be slightly higher than the outside air pressure. A velocity head will now be developed, as shown in Figure 6.8 [40, 41, 69].

The total head will be the sum of the velocity head and the static head, as shown by Eq. 6.3 [40, 41, 69].

$$H_t = H_v + H_s \tag{6.3}$$

When an exhaust fan is used, and the air is drawn through the box, the pressure within the box will be slightly lower than the outside air pressure, and the pressure head characteristics will appear as shown in Figure 6.9. The total head is still shown by Eq. 6.3 [40, 41, 69].

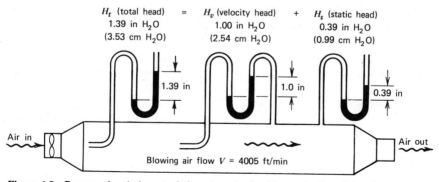

Figure 6.8 Pressure head characteristics when the fan blows air through an electronic box.

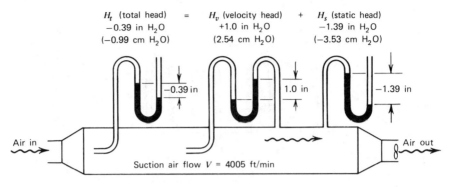

Figure 6.9 Pressure head characteristics when fan draws air through an electronic box.

6.4 STATIC LOSSES EXPRESSED IN TERMS OF VELOCITY HEADS

As the cooling air flows through the electronic box, it will experience friction and dynamic losses. There are many ways of expressing these losses. One convenient method is to express the *static* pressure losses in terms of the *velocity* head for the air at some specific point. For example, there will be friction and dynamic losses at the inlets to the fans shown in Figures 6.1 through 6.6. These losses can be expressed as some ratio of the velocity head at the fan inlet. If the loss is judged to be equal to one velocity head, or perhaps two velocity heads for a severely restricted flow area, Eq. 6.2 can be used to determine the actual pressure loss in inches of water, based upon the velocity of the air flow at that section [40, 41, 69].

The loss in a 90° elbow can be expressed as some ratio of the velocity head at that elbow, which is related to the velocity of the air flowing through the elbow. A loss in a sudden expansion is expressed as a ratio of the velocity head just before the expansion. A typical sudden expansion will result in a loss of one velocity head, because the air velocity at the expansion is sharply reduced. Most of the energy required to provide that velocity at the point just before the expansion is lost. That energy was originally supplied by the fan. If more energy is lost, a larger fan is required to supply that energy. If the losses in an air flow system can be reduced, by avoiding such things as sharp turns and sudden expansions, a smaller fan can be used.

Air velocities must be kept low to avoid large losses. High air velocities result in high losses. A system with high losses will require a fan or blower that can supply the required cooling air at a higher pressure. This usually results in a larger and heavier fan or blower, with greater power requirements and higher costs [42].

6.5 SAMPLE PROBLEM—AIR FLOW LOSS AT A FAN ENTRANCE

The axial flow fan shown in Figure 6.1 has an air inlet velocity of 3500 ft/min (1778 cm/sec). It is estimated that the turbulence and contraction in the air at the entrance to the fan will result in a static loss of about 1.5 velocity heads. Determine the static head loss in terms of the height of a column of water.

SOLUTION

The static loss represented by one velocity head can be determined by rewriting Eq. 6.2 slightly, as shown in Eq. 6.4 for English units.

$$H_v = \left(\frac{V}{4005}\right)^2 \tag{6.4}$$

where $V = 3500$ ft/min

$$H_v = \left(\frac{3500}{4005}\right)^2 = 0.764 \text{ in } H_2O \text{ for one velocity head} \tag{6.5}$$

Using Eq. 6.2a for metric units yields

$$H_v = \left(\frac{V}{1277}\right)^2$$

$$H_v = \left(\frac{1778}{1277}\right)^2 = 1.94 \text{ cm } H_2O \text{ for one velocity head} \tag{6.5a}$$

The static head loss at the fan entrance is expected to be 1.5 velocity heads. The static loss expected in English units is shown in Eq. 6.6.

$$H_{\text{fan loss}} = 1.5(0.764 \text{ in } H_2O) = 1.146 \text{ in } H_2O \tag{6.6}$$

The static loss expected, in metric units, is shown in Eq. 6.6a.

$$H_{\text{fan loss}} = 1.5(1.94 \text{ cm}H_2O) = 2.91 \text{ cm } H_2O \tag{6.6a}$$

The air exit velocity should also be kept low, if possible, to reduce the velocity head lost at the exit. Sometimes a gradual enlargement can be incorporated at the exit. This will help to convert some of the velocity head to a higher static head, which will help to increase the flow of air through the box.

6.6 ESTABLISHING THE FLOW IMPEDANCE CURVE FOR AN ELECTRONIC BOX

Electronic boxes that are cooled with the use of fans must be carefully evaluated to make sure the fan will provide the proper cooling. If the fan is too small for the box, the electronic system may overheat and fail. If the fan is too big for the box, the cooling will be adequate, but the larger fan will be more expensive, heavier, and will draw more power.

Air flowing through the electronic box will encounter resistance as it enters different chambers and is forced to make many turns. This flow resistance is approximately proportional to the square of the velocity, so that it is approximately proportional to the square of the flow rate in cubic foot per minute (cfm). When the static pressure of the air flow through a box is plotted against the cfm air flow rate, the result will be a parabolic curve. This curve can be generated by considering the various flow resistances the moving air will encounter as it flows through the box. The method of analysis is to assume several different cfm flow rates through the box and then to calculate a static pressure drop through the box for each flow rate. The result will be a curve similar to Figure 6.10.

Once the box flow impedance curve has been developed, it is necessary to examine different fan impedance curves to see how well the fans will match the box. A typical fan impedance curve is shown in Figure 6.11.

If the impedance curve for the box is superimposed on the impedance curve for the fan, they will intersect. The point of intersection represents the actual operating point for the system, as shown in Figure 6.12 [42].

The method of pressure drop analysis just outlined can be demonstrated by considering a typical fan cooled electronic box.

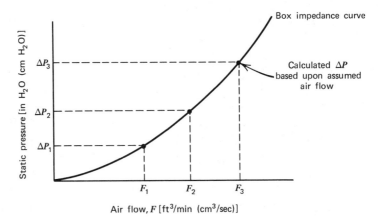

Figure 6.10 Typical air flow impedance curve for an electronic box.

6.7 Sample Problem—Fan Cooled Electronic Box

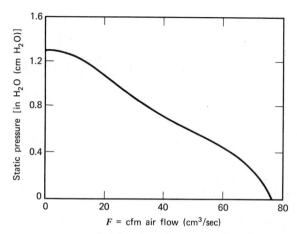

Figure 6.11 Typical flow impedance curve for a fan.

6.7 SAMPLE PROBLEM—FAN COOLED ELECTRONIC BOX

Determine the pressure drop characteristics of the electronic box in Figure 6.13, and determine a fan size for the system. The system must be capable of continuous operation in a 131°F (55°C) ambient at sea level conditions. The maximum allowable hot spot component surface temperature is limited to 212°F (100°C). The system contains seven PCBs, each dissipating 20 watts, for a total power dissipation of 140 watts. This does not include the fan power dissipation.

SOLUTION

The box must be examined in two phases to ensure the integrity of the complete design. In phase 1, the thermal design of the box is examined,

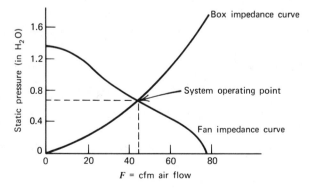

Figure 6.12 Intersection of fan and box impedance curves.

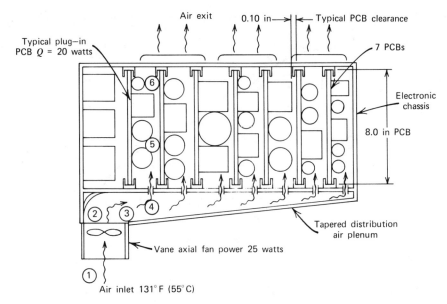

Figure 6.13 Plan view of fan cooled electronic box.

with the proposed fan, to make sure the component hot spot temperature of 212°F (100°C) is not exceeded. In phase 2, the electronic chassis air flow impedance curve is developed and matched with several fans, to make sure there is sufficient cooling air available for the system.

Phase 1. Electronic Box Thermal Design

A preliminary thermal analysis must be made of the system to determine approximately how much cooling air is required. The preliminary analysis can also be used to establish an approximate fan size.

The maximum allowable temperature rise from the air inlet to the component surface hot spot temperature is 212°F − 131°F = 81°F (100°C − 55°C = 45°C). There are two major contributors to this temperature rise.

1. Δt due to the heat input from the fan and the electronics to the cooling air.
2. Δt due to the thermal resistance across the convection film from the cooling air to the surface of the component.

These temperature rise sources can be examined in more detail to see how much each contributes to the total value.

1. Δt due to the heat input from the fan motor to the cooling air.

6.7 Sample Problem—Fan Cooled Electronic Box

Heat generated by the fan motor is usually removed by convection from the cooling air. If the fan is bolted to a large heat sink away from the electronic box, a large part of the heat from the fan motor may be removed by direct conduction and radiation from the fan housing to the heat sink.

When the fan is bolted directly to the chassis that houses the electronics, the chassis walls may be quite hot, so relatively little fan heat may be transferred to the chassis. Therefore, it is a good practice to be slightly conservative and to assume that all of the fan power will be picked up by the cooling air as it passes through the fan and into the chassis housing.

Several types of fans can be used for cooling the electronics. Centrifugal fans (sometimes called radial or squirrel cage blowers) can deliver a large quantity of air against a high static pressure. Also, there are probably more centrifugal fans in general use for cooling than any other type of fan. However, centrifugal fans are generally larger and heavier than tube axial and vane axial fans for the same performance in small fan sizes. Axial flow fans will be required to run at higher speeds to get the same performance, so that they will generate more noise than will centrifugal fans.

If a small, high speed axial flow fan is selected, the noise generated may be objectionable to personnel working in areas adjacent to the fan. Simple mufflers can be made of plywood and foam packing material to reduce the noise levels when these fans are operating with room ambient cooling air.

Past experience with air cooled electronic systems has shown that satisfactory thermal performance can be obtained if the cooling air exit temperature from the electronic chassis does not exceed 160°F (71°C). The allowable cooling air temperature rise then becomes

$$\Delta t_A = 160°F - 131°F = 29°F$$

or

$$\Delta t_A = 71°C - 55°C = 16°C$$

The required cooling air weight flow for the system can then be determined from the general weight flow equation [15, 17, 24].

$$W = \frac{Q}{\Delta t C_p} \tag{6.7}$$

Given

Item	English Units	Metric Units
Q = heat	= Btu/hr	= cal/sec
Δt = temperature rise	= °F	= °C
C_p = specific heat	= Btu/lb °F	= cal/g °C
W = air flow	= lb/hr	= g/sec

Considering a lightweight vane axial flow fan similar to the Rotron Aximax 2, which weighs only about 0.25 lb, the power dissipation for a 15,000 rpm fan will be about 25 watts, The total box power dissipation in English units then becomes:

$$Q = 140 + 25 = 165 \text{ watts} \times 3.413 \frac{\text{Btu/hr}}{\text{watts}} = 563 \frac{\text{Btu}}{\text{hr}} \text{ (heat)}$$

$$\Delta t = 160 - 131 = 29°F \text{ (cooling air temperature rise)}$$

$$C_p = 0.24 \frac{\text{Btu}}{\text{lb °F}} \text{ (specific heat)}$$

The cooling air flow through the box must be

$$W = \frac{563 \text{ Btu/hr}}{(29°F)(60 \text{ min/hr})(0.24 \text{ Btu/lb °F})} = 1.35 \frac{\text{lb}}{\text{min}} \quad (6.8)$$

In metric units:

$$Q = 140 + 25 = 165 \text{ watts} \times 0.239 \frac{\text{cal/sec}}{\text{watt}} = 39.4 \text{ cal/sec (heat)}$$

$$\Delta t = 71 - 55 = 16°C$$

$$C_p = 0.24 \frac{\text{cal}}{\text{g °C}} \text{ (specific heat)}$$

$$W = \frac{39.4 \text{ cal/sec}}{(16°C)(0.24 \text{ cal/g °C})} = 10.3 \frac{\text{g}}{\text{sec}} \quad (6.8a)$$

2 Δt due to the thermal resistance across the convection film from the cooling air to the surface of the component.

This temperature rise can be determined from the convection relation shown by Eq. 5.7.

$$\Delta t = \frac{Q}{h_c A} \quad \text{(ref. Eq. 5.7)}$$

The convection coefficient h_c is determined from the geometry of the air duct that is formed by the air space between the plug-in PCBs. Its value is determined from Eq. 6.9 [3, 15, 43].

$$h_c = JC_p G \left(\frac{C_p \mu}{K}\right)^{-2/3} \quad (6.9)$$

6.7 Sample Problem—Fan Cooled Electronic Box

Given J = Colburn factor
C_p = specific heat of fluid (ref. Figure 6.28)
G = weight velocity flow through duct
μ = viscosity of fluid (ref. Figure 6.28)
K = thermal conductivity of fluid (ref. Figure 6.28)

The Colburn J factor can be determined from the Reynolds number (N_R). For laminar flow conditions with N_R between 200 and 1800, the approximate J value for rectangular ducts with aspect ratios greater than about 8 is shown in Eq. 6.10 (compiled from data shown in Kays and London, *Compact Heat Exchangers*) [43].

$$J = \frac{6}{(N_R)^{0.98}} \qquad (6.10)$$

For square ducts (aspect ratio = 1.0) the approximate J value for the same laminar flow conditions is as follows:

$$J = \frac{2.7}{(N_R)^{0.95}} \qquad (6.10a)$$

For turbulent conditions, where N_R is between 10,000 and 120,000 McAdams [15] shows that

$$J = \frac{0.023}{(N_R)^{0.2}} \qquad (6.10b)$$

The air duct formed by the clearance between the components and the PCBs is shown in Figure 6.14.

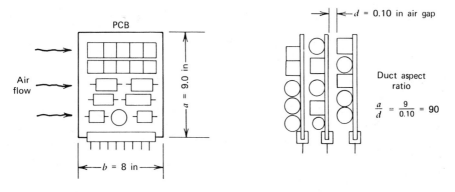

Figure 6.14 Air flow duct formed between printed circuit boards.

The Reynolds number for the air duct formed by the PCBs is shown in Eq. 6.11.

$$N_R = \frac{VD\rho}{\mu} = \frac{GD}{\mu} \tag{6.11}$$

Given $\quad D = \dfrac{4 \times \text{area}}{\text{perimeter}} = \dfrac{4ad}{2(a+d)} = \dfrac{2ad}{a+d}$ (hydraulic diameter)

In English units:

$$D = \frac{2(9.0)(0.10)}{9.0 + 0.10} = 0.198 \text{ in} = 0.0165 \text{ ft}$$

In metric units:

$$D = \frac{2(22.86)(0.254)}{22.86 + 0.254} = 0.502 \text{ cm}$$

The weight velocity flow G through the duct formed by the PCBs is determined from Eq. 6.12.

$$G = \frac{W}{A} \tag{6.12}$$

Given $\quad W = 1.35$ lb/min (10.3 g/sec) (ref. Eq. 6.8)

$$A = \frac{(9.0)(0.10) \text{ in}^2}{144 \text{ in}^2/\text{ft}^2} = 0.00625 \text{ ft}^2 \text{ (area one duct)}$$

In English units:

$$G = \frac{(1.35 \text{ lb/min})(60 \text{ min/hr})}{0.00625 \text{ ft}^2 \times 7 \text{ ducts}} = 1851 \frac{\text{lb}}{\text{hr ft}^2} \tag{6.13}$$

In metric units:

$$G = \frac{10.3 \text{ g/sec}}{(22.86)(0.254) \text{ cm}^2 \times 7 \text{ ducts}} = 0.253 \frac{\text{g}}{\text{sec cm}^2} \tag{6.13a}$$

$\mu = 0.050 \dfrac{\text{lb}}{\text{ft hr}} = 0.000207 \dfrac{\text{g}}{\text{cm sec}}$ [ref. Fig. 6.28; viscosity at 150°F (65.5°C)]

6.7 Sample Problem—Fan Cooled Electronic Box

Substitute into Eq. 6.11 for the Reynolds number, in English units:

$$N_R = \frac{(1851 \text{ lb/hr ft}^2)(0.0165 \text{ ft})}{0.050 \text{ lb/ft hr}} = 611 \text{ (dimensionless)} \quad (6.14)$$

In metric units:

$$N_R = \frac{(0.253 \text{ g/sec cm}^2)(0.502 \text{ cm})}{0.000207 \text{ g/cm sec}} = 613 \text{ (dimensionless)} \quad (6.14a)$$

$$K = 0.017 \frac{\text{Btu}}{\text{hr ft °F}} = 0.000070 \frac{\text{cal}}{\text{sec cm °C}} \text{ (ref. Fig. 6.28; thermal conductivity at 150°F)}$$

Substitute into Eq. 6.10 for the J factor.

$$J = \frac{6}{(N_R)^{0.98}} = \frac{6}{(611)^{0.98}} = 0.0111 \text{ (dimensionless)}$$

Substitute into Eq. 6.9 for the forced air convection coefficient, in English units.

$$h_c = \frac{(0.0111)(0.24)(1851 \text{ lb/hr ft}^2)}{[(0.24 \times 0.05)/0.017]^{0.666}} = 6.22 \frac{\text{Btu}}{\text{hr ft}^2 \text{ °F}} \quad (6.15)$$

In metric units:

$$h_c = \frac{(0.0111)(0.24)(0.253 \text{ g/sec cm}^2)}{[(0.24 \times 0.000207)/0.000070]^{0.666}} = 0.000846 \frac{\text{cal}}{\text{sec cm}^2 \text{ °C}} \quad (6.15a)$$

Section 5.11 points out that some of the surface area on the backside of flow soldered PCBs is available for heat dissipation when the component lead wires extend through the PCB. When there is copper on the back side of such a PCB, the heat can flow through the PCB to the copper, which acts as a heat spreader. As much as 30% of the back side of a PCB may be available for effective heat transfer. The total heat transfer surface area may then be 1.3 times the area of one face on the PCB.

The forced convection coefficient h_c can be increased by using a smaller air gap between the PCBs. For example, if the air gap is reduced from 0.10 in to 0.08 in, the forced convection coefficient will increase to a value of about 7.8 Btu/hr ft² °F. However, if the electronic system must operate in a severe vibration environment, there must be sufficient sway space between the PCBs, to prevent tall electronic components from striking the sharp lead wires on the back surface of the adjacent PCB.

The temperature rise across the convection film from the air stream to the component surface can now be determined using Eq. 5.7. The heat from the fan is not included with the heat generated by the PCBs. Each PCB dissipates 20 watts = 68.3 Btu/hr.

Given $Q = 68.3$ Btu/hr = 4.78 cal/sec (heat)
$h_c = 6.22$ Btu/hr ft² °F = 0.000846 cal/sec cm² °C (convection)

In English units:

$$A = \frac{(1.3)(8)(9) \text{ in}^2}{144 \text{ in}^2/\text{ft}^2}$$

$$= 0.65 \text{ ft}^2 \text{ (PCB area, including the back side of the PCB)}$$

In metric units:

$$A = (1.3)(20.32)(22.86) = 603.8 \text{ cm}^2 \text{ (PCB area)}$$

The temperature rise, in English units, is

$$\Delta t_B = \frac{68.3}{(6.22)(0.65)} = 16.9°F \tag{6.16}$$

In metric units:

$$\Delta t_B = \frac{4.78}{(0.000846)(603.8)} = 9.3°C \tag{6.16a}$$

The component hot spot surface temperature t_s is determined from the cooling air inlet temperature, plus the temperature rises to the component surface, as shown in Eq. 6.17.

$$t_s = t_{\text{in}} + \Delta t_A + \Delta t_B \tag{6.17}$$

In English units, the component surface temperature is

$$t_s = 131 + 29 + 16.9 = 176.9°F \tag{6.18}$$

In metric units:

$$t_s = 55 + 16 + 9.3 = 80.3°C \tag{6.18a}$$

Since the component hot spot surface temperature is less than 212°F (100°C), the initial design concept with the vane axial flow fan appears to be satisfactory.

6.7 Sample Problem—Fan Cooled Electronic Box

Phase 2. Electronic Chassis Air Flow Impedance Curve

The air flow conditions are examined at six different points in the chassis, where the maximum static pressure losses are expected to occur, as shown in Figure 6.13. These static pressure losses are itemized as follows:

1. Air inlet to fan
2. 90° turn and transition to an oval section
3. Contraction and transition to a rectangular section
4. Plenum entrance to PCB duct
5. Flow through PCB channel duct
6. Exhaust from PCB duct and chassis

The static pressure losses are examined in greater detail using the velocity head loss technique discussed in Section 6.4.

1. *Air Inlet to Fan.* The air inlet to the axial flow fan is very similar to —page 196 the plain end duct, as shown in Table 6.2 (Section 6.12). Ducts of this type can be expected to show a static pressure loss of 0.93 velocity head. The loss is conservatively rounded off to 1.0 velocity head, as shown in Eq. 6.19.

$$H_1 = 1.0 H_{v1} \qquad (6.19)$$

2. *90° Turn and Transition from a Round Section at the Fan to an Oval Section.* The radius of the turn is rather sharp, probably less than 1 diameter. Also, the section is changing from round to oval (the section will eventually become rectangular).

$$\begin{aligned} 90° \text{ turn} &= 0.7 H_v \\ \text{Transition} &= 0.2 H_v \end{aligned} \biggr\} \text{ estimated}$$

$$\overline{0.9 H_v}$$

Conservatively round off this loss to unity.

$$H_2 = 1.0 H_{v2} \qquad (6.20)$$

3. *Contraction and Transition to Rectangular Section*

$$\begin{aligned} \text{Contraction} &= 0.2 H_v \\ \text{Transition} &= 0.2 H_v \end{aligned} \biggr\} \text{ estimated}$$

$$\overline{0.4 H_v}$$

Conservatively round off to $0.5H_v$.

$$H_3 = 0.5H_{v3} \qquad (6.21)$$

4 Plenum Entrance to Printed Circuit Board (PCB) Duct. The air must make an abrupt change at this point. It must make a 90° turn and enter the PCB duct area. This is a critical area with respect to pressure drop. If the inlet extends slightly into the air stream, as shown in Figure 6.15a, the pressure loss can exceed 3 velocity heads as a result of the turbulence generated ($2.7H_v$ entry + $1.0H_v$ for 90° turn = $3.7H_v$ total loss) [41, 69].

An attempt should be made to keep the entrance flush. It is much better to bevel the edges as shown in Figure 6.15b, where the pressure loss is only about 1.5 velocity heads ($0.5H_v$ entry + $1.0H_v$ turn = $1.5H_v$ total loss).

Conservatively assume a flush edge.

$$\begin{aligned}\text{Entrance} &= 1.0H_v \\ 90° \text{ turn} &= 1.0H_v \\ \hline H_4 &= 2.0H_{v4}\end{aligned} \qquad (6.22)$$

5 Air Flow through PCB Channel Duct. In this case the duct is really the space between the backside of one board and the electronic components on the adjacent board. The back side of the PCB has small exposed tips of the electrical lead wires, and the front side of the PCB is populated with electronic components. The loss can be expressed in terms of air flowing through a rectangular duct as shown in Section 6.24 (Eq. 6.82 or 6.83). Using velocity heads for convenience, a high pressure loss is expected for this condition because of the narrow (0.10 in) air gap.

$$H_5 = 1.0H_{v5} \qquad (6.23)$$

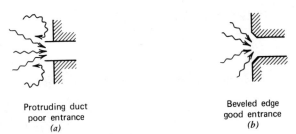

Protruding duct
poor entrance
(a)

Beveled edge
good entrance
(b)

Figure 6.15 Protruding duct creates turbulence that increases pressure loss at air inlet (a); beveled edge reduces pressure loss at air inlet (b).

6.7 Sample Problem—Fan Cooled Electronic Box

6 *Exhaust from PCB Duct and Chassis.* This calculation is normally not included in the static pressure drop loss through the box. However, it is always a good idea to see how much energy is being thrown away at the exit. Remember that the fan must supply all of the air-moving energy. If too much is thrown away at the exhaust, the fan has to work harder. When the exhaust loss is too high in a critical design, some attempt should be made to recover part of the loss by slightly enlarging the exhaust area to reduce the exhaust velocity.

As the cooling air exhausts from the box, it will expand suddenly, so that it will lose its velocity completely.

$$H_6 = 1.0 H_{v6} \qquad (6.24)$$

Note that the air flows across all of the PCBs in a parallel flow path. The pressure drop across this section will all be about the same for all of the PCBs. Therefore, only one air flow path across the circuit board stack has to be examined to obtain the pressure drop.

All of the information is now available to calculate the approximate pressure drop through the box based upon the geometry of the air flow path. Three different cfm air flow conditions will be assumed: 10 cfm (4719 cm³/sec), 20 cfm (9438 cm³/sec), and 30 cfm (14157 cm³/sec). The pressure drops at each of the six points in Figure 6.13 will be determined for each of the three assumed flow conditions, starting with the 10 cfm (4719 cm³/sec) flow condition.

1 *Air Inlet to Fan.* Equation 6.4 shows that the velocity head H_v is a function of the air flow velocity in feet per minute. Starting with an assumed air flow of 10 cfm, the cross section flow area through the axial flow fan is required, and is shown in Figure 6.16.

$$A_1 = \frac{(\pi/4)[(1.88)^2 - (1.10)^2] \text{ in}^2}{144 \text{ in}^2/\text{ft}^2} = 0.0127 \text{ ft}^2 = 11.8 \text{ cm}^2$$

$$V_1 = \frac{F}{A_1} = \frac{10 \text{ ft}^3/\text{min}}{0.0127 \text{ ft}^2} = 787 \frac{\text{ft}}{\text{min}} = 400 \frac{\text{cm}}{\text{sec}}$$

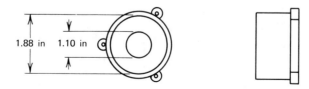

Figure 6.16 Cross section flow area for a small lightweight vane axial flow fan (similar to Rotron Aximax 2).

Substitute into Eq. 6.4 for the velocity head at point 1 in English units.

$$H_{v1} = \left(\frac{V}{4005}\right)^2 = \left(\frac{787}{4005}\right)^2 = 0.0386 \text{ in } H_2O \qquad (6.25)$$

In metric units:

$$H_{v1} = \left(\frac{V}{1277}\right)^2 = \left(\frac{400}{1277}\right)^2 = 0.098 \text{ cm } H_2O \qquad (6.25a)$$

The losses for each flow rate of 10 cfm, 20 cfm, and 30 cfm can be determined from Eq. 6.19 by using a ratio of the 10 cfm flow rate.

$$\left.\begin{array}{l}
\text{10 cfm:} \quad H_1 = (1.0)(0.0386) = 0.0386 \text{ in } H_2O = 0.098 \text{ cm } H_2O \\
\text{20 cfm:} \quad H_1 = \left(\frac{20}{10}\right)^2 (0.0386) = 0.1544 \text{ in } H_2O = 0.392 \text{ cm } H_2O \\
\text{30 cfm:} \quad H_1 = \left(\frac{30}{10}\right)^2 (0.0386) = 0.3474 \text{ in } H_2O = 0.882 \text{ cm } H_2O
\end{array}\right\} (6.26)$$

2 *90° Turn and Transition from a Round Section at the Fan to an Oval Section.* The cross section area at section 2 is shown in Figure 6.17.

$$A_2 = \frac{(\pi/4)(0.5)^2 + (3.5)(0.5) \text{ in}^2}{144 \text{ in}^2/\text{ft}^2} = 0.0135 \text{ ft}^2 = 12.56 \text{ cm}^2$$

$$V_2 = \frac{10 \text{ ft}^3/\text{min}}{0.0135 \text{ ft}^2} = 741 \frac{\text{ft}}{\text{min}} = 376 \frac{\text{cm}}{\text{sec}}$$

Substitute into Eq. 6.4 for the velocity head at point 2.

$$H_{v2} = \left(\frac{741}{4005}\right)^2 = 0.034 \text{ in } H_2O = 0.087 \text{ cm } H_2O$$

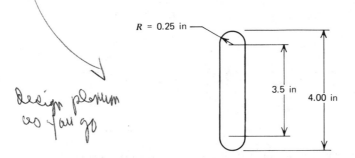

Figure 6.17 Cross section of air flow path at point 2.

6.7 Sample Problem—Fan Cooled Electronic Box

The losses for each flow rate at point 2 are determined from Eq. 6.20.

$$
\begin{aligned}
&10 \text{ cfm:} \quad H_2 = (1.0)(0.034) = 0.034 \text{ in } H_2O = 0.086 \text{ cm } H_2O \\
&20 \text{ cfm:} \quad H_2 = \left(\frac{20}{10}\right)^2 (0.034) = 0.136 \text{ in } H_2O = 0.345 \text{ cm } H_2O \\
&30 \text{ cfm:} \quad H_2 = \left(\frac{30}{10}\right)^2 (0.034) = 0.306 \text{ in } H_2O = 0.777 \text{ cm } H_2O
\end{aligned} \quad (6.27)
$$

3 *Contraction and Transition to Rectangular Section.* The cross section area at section 3 is shown in Figure 6.18.

$$V_3 = \frac{10 \text{ cfm}}{0.0104} = 961 \frac{\text{ft}}{\text{min}} = 488 \frac{\text{cm}}{\text{sec}}$$

$$H_{v3} = \left(\frac{961}{4005}\right)^2 = 0.0575 \text{ in } H_2O = 0.146 \text{ cm } H_2O$$

The losses for each flow rate at point 3 are determined from Eq. 6.21.

$$
\begin{aligned}
&10 \text{ cfm:} \quad H_3 = (0.5)(0.0575) = 0.0287 \text{ in } H_2O = 0.0729 \text{ cm } H_2O \\
&20 \text{ cfm:} \quad H_3 = \left(\frac{20}{10}\right)^2 (0.0287) = 0.115 \text{ in } H_2O = 0.292 \text{ cm } H_2O \\
&30 \text{ cfm:} \quad H_3 = \left(\frac{30}{10}\right)^2 (0.0287) = 0.259 \text{ in } H_2O = 0.657 \text{ cm } H_2O
\end{aligned} \quad (6.28)
$$

4 *Plenum Entrance to PCB Duct.* The cross section at section 4 is shown in Figure 6.19. Since there are seven card slots, approximately ⅐ of the total flow will pass through each slot if the tapered distribution duct shown in Figure 6.13 does its job properly. When the power dissipation is not the same on each PCB, more air should be distributed to the PCBs that have a higher power dissipation. A mockup of the

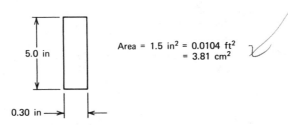

Figure 6.18 Cross section of air flow path at point 3.

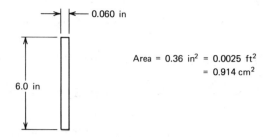

Figure 6.19 Inlet slot for distributing cooling air to PCB duct.

air flow system is usually built and tested, to make sure the air distribution to each PCB is adequate.

$$V_4 = \frac{10 \text{ cfm}}{(0.0025 \text{ ft}^2)(7 \text{ slots})} = 571 \frac{\text{ft}}{\text{min}} = 290 \frac{\text{cm}}{\text{sec}}$$

$$H_{v4} = \left(\frac{571}{4005}\right)^2 = 0.0203 \text{ in } H_2O = 0.0516 \text{ cm } H_2O$$

The losses for each flow rate at point 4 are determined from Eq. 6.22.

$$\left. \begin{array}{l} 10 \text{ cfm:} \quad H_4 = (2)(0.0203) = 0.0406 \text{ in } H_2O = 0.103 \text{ cm } H_2O \\ 20 \text{ cfm:} \quad H_4 = \left(\frac{20}{10}\right)^2 (0.0406) = 0.162 \text{ in } H_2O = 0.411 \text{ cm } H_2O \\ 30 \text{ cfm:} \quad H_4 = \left(\frac{30}{10}\right)^2 (0.0406) = 0.365 \text{ in } H_2O = 0.927 \text{ cm } H_2O \end{array} \right\} \quad (6.29)$$

5 Air Flow through PCB Channel Duct. The cross section of the duct formed by the PCBs is shown in Figure 6.14.

$$A_5 = (0.10)(9.0) = 0.90 \text{ in}^2 = 0.00625 \text{ ft}^2 = 5.80 \text{ cm}^2$$

$$V_5 = \frac{10 \text{ cfm}}{(0.00625 \text{ ft}^2)(7 \text{ slots})} = 228 \frac{\text{ft}}{\text{min}} = 116 \frac{\text{cm}}{\text{sec}}$$

$$H_{v5} = \left(\frac{228}{4005}\right)^2 = 0.0032 \text{ in } H_2O = 0.0082 \text{ cm } H_2O$$

The losses for each flow rate at point 5 are determined from Eq. 6.23.

6.7 Sample Problem—Fan Cooled Electronic Box

$$
\begin{aligned}
&10 \text{ cfm:} \quad H_5 = (1.0)(0.0032) = 0.0032 \text{ in H}_2\text{O} = 0.0081 \text{ cm H}_2\text{O} \\
&20 \text{ cfm:} \quad H_5 = \left(\frac{20}{10}\right)^2 (0.0032) = 0.0128 \text{ in H}_2\text{O} = 0.0325 \text{ cm H}_2\text{O} \\
&30 \text{ cfm:} \quad H_5 = \left(\frac{30}{10}\right)^2 (0.0032) = 0.0288 \text{ in H}_2\text{O} = 0.0731 \text{ cm H}_2\text{O}
\end{aligned}
\quad (6.30)
$$

6 *Exhaust from PCB Duct and Chassis.* The velocity of the air at the PCB exit is the same as the velocity at point 5. However, at the exit, one full velocity head is lost, as shown by Eq. 6.24. The losses for each flow rate are as follows:

$$
\begin{aligned}
&10 \text{ cfm:} \quad H_6 = (1.0)(0.0032) = 0.0032 \text{ in H}_2\text{O} = 0.0081 \text{ cm H}_2\text{O} \\
&20 \text{ cfm:} \quad H_6 = \left(\frac{20}{10}\right)^2 (0.0032) = 0.0128 \text{ in H}_2\text{O} = 0.0325 \text{ cm H}_2\text{O} \\
&30 \text{ cfm:} \quad H_6 = \left(\frac{30}{10}\right)^2 (0.0032) = 0.0288 \text{ in H}_2\text{O} = 0.0731 \text{ cm H}_2\text{O}
\end{aligned}
\quad (6.31)
$$

The exhaust losses are relatively small, so that very little energy is wasted at the exit.

The static pressure loss at each flow is the sum of the individual pressure losses, which are shown in Tables 6.1 and 6.1A.

Table 6.1 is used to plot the chassis air flow impedance curve, which is shown in Figure 6.20. The static pressure curve for two different vane axial flow fans are superimposed on the same curve, using the manufacturers data. Both fans have 400 cycle motors in similar 2 in diameter housings.

Table 6.1 Static Pressure Loss for Different Air Flow Rates through Electronic Box in English Units

Station Point	Pressure Loss (in H$_2$O)		
	10 cfm	20 cfm	30 cfm
1	0.0386	0.1544	0.3470
2	0.0340	0.1360	0.3060
3	0.0287	0.1150	0.2590
4	0.0406	0.1620	0.3650
5	0.0032	0.0128	0.0288
6	0.0032	0.0128	0.0288 ← Exhaust
Total in H$_2$O	0.1483	0.5930	1.3346

Table 6.1A Static Pressure Loss for Different Air Flow Rates through Electronic Box in Metric Units

Station Point	Pressure Loss (cm H₂O)		
	$4719 \frac{cm^3}{sec}$	$9438 \frac{cm^3}{sec}$	$14{,}157 \frac{cm^3}{sec}$
1	0.0980	0.3920	0.8820
2	0.0860	0.3450	0.7770
3	0.0729	0.2920	0.6570
4	0.1030	0.4110	0.9270
5	0.0081	0.0325	0.0731
6	0.0081	0.0325	0.0731 ← Exhaust
Total cm H₂O	0.3761	1.5050	3.3892

Fan A is a 3 phase 15,500 rpm, 25 watt motor. Fan B is a single phase 11,000 rpm 18 watt motor which speeds up at high altitudes [44].

If fan A is used, the air flow rate through the chassis will be about 30 cfm with a constant cfm flow rate at all altitudes. If fan B is used, the sea level air flow rate will be 23 cfm. Fan B, however, is a variable speed fan which increases in speed as the altitude is increased (Rotron trade name is Altivar). This type of fan is usually very desirable for cooling electronic boxes at high altitudes where the air density is reduced [44].

The air weight flow through the electronic box can be determined from the average air density (ρ) in the box, using Eq. 6.32, which is obtained from the gas laws [45–47].

$$\rho = \frac{P}{RT} \quad (6.32)$$

where $P = 14.7$ lb/in² sea level pressure (English units)
$P = 1034.4$ g/cm² sea level pressure (metric units)
$R = 53.3$ ft/°R gas constant for air (English units)
$R = 2924$ cm/°K gas constant for air (metric units)

$$t_{av} = \frac{131 + 160}{2} = 145.5°F = 63°C \text{ (average temperature)}$$

$T = 460 + 145.5 = 605.5°R$ (absolute air temperature)

$$\rho = \frac{(14.7 \text{ lb/in}^2)(144 \text{ in}^2/\text{ft}^2)}{(53.3 \text{ ft/°R})(605.5°R)} = 0.0656 \frac{\text{lb}}{\text{ft}^3} \text{ (English units)} \quad (6.33)$$

The weight flow is determined from Eq. 6.34.

$$W = \rho F \quad (6.34)$$

6.8 Hollow Core PCBs

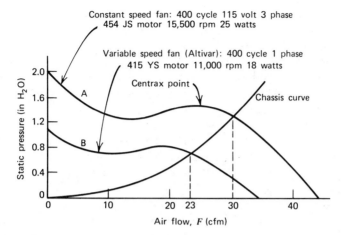

Figure 6.20 Chassis air flow impedance curve superimposed on two different vane axial fan curves with different speeds (both fans have same 2 in diameter outer housing).

Using fan A, where the air flow $F = 30$ cfm, the weight flow is

$$W = \left(0.0656 \frac{\text{lb}}{\text{ft}^3}\right)\left(30 \frac{\text{ft}^3}{\text{min}}\right) = 1.97 \frac{\text{lb}}{\text{min}} \quad (6.35)$$

Using fan B, where the air flow is 23 cfm, the weight flow is:

$$W = \left(0.0656 \frac{\text{lb}}{\text{ft}^3}\right)\left(23 \frac{\text{ft}^3}{\text{min}}\right) = 1.51 \frac{\text{lb}}{\text{min}} \quad (6.36)$$

Equation 6.8 showed that the minimum required cooling air flow rate is 1.35 lb/min. Since fans A and B can both supply more than the minimum required flow rate, either fan will be acceptable.

6.8 HOLLOW CORE PCBs

Cooling air is often contaminated with moisture, which can bridge printed circuits and electrical connector contacts. This can result in a high resistance path across two conductors, which can affect the electrical operation of high impedance circuits. Many of the new specifications for advanced electronic systems will, therefore, not permit cooling air to come into direct contact with the electronic components or circuits. Under these circumstances, the fan cooling air is usually ducted through the walls of the chassis, which have been made into heat exchangers. Plug-in types of printed circuit boards (PCBs) with metal cores of strips, similar to those shown in Figures 4.1 and 4.2, and then used to conduct away the heat.

The length of the heat flow path from the PCB components to the cooling air in the chassis walls, as shown in Figure 4.2, is quite long. This may result in a large temperature rise. If a hollow core can be provided for the cooling air between each circuit board, with the components on the outside, the length of the heat flow path from the component to the cooling air will be sharply reduced. This will reduce the temperature rise to the components, so that they will run cooler and last longer.

Electrical lead wires must not extend into the cooling air ducts in the hollow cores, so that components must be lap soldered to the circuit boards. Existing flow soldering and wave soldering equipment, normally used for high volume production of printed circuits, could not be used to assemble lap soldered components.

The most difficult task associated with plug-in hollow core PCBs is the air seal. A low resistance air flow path must be provided to the PCB, with a seal that can be engaged and disengaged many times with very little air leakage.

Figure 6.21a, b, and c shows some different types of air seals which have been used successfully for plug-in types of hollow PCBs. Figure 6.21a relies on tapered PCB sides to provide a seal with a soft gasket interface. Figure 6.21b uses an overlap interface, with a gasket that seals the outer surfaces at the extreme edges of the PCB. In the foregoing two designs, the inlet cooling air plenum and the exhaust air plenum are part of the chassis. In Figure 6.21c, the inlet cooling air slot and the exhaust air slot are part of the PCB itself. The PCBs are stacked together and clamped using four axial clamp bars, located at the four corners. This stacking arrangement forms a

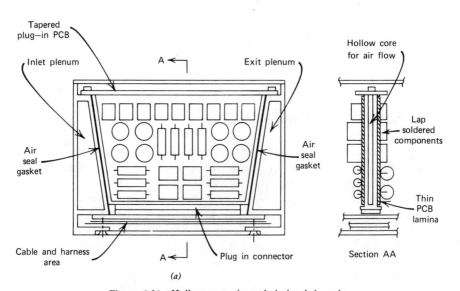

Figure 6.21 Hollow core air cooled circuit boards.

6.8 Hollow Core PCBs

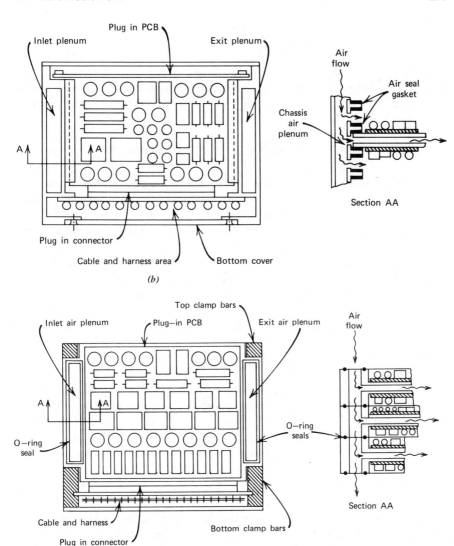

Figure 6.21 (Continued)

cooling air plenum out of the slots on each side of the PCB. An O-ring seal is used around the perimeter of each side slot to reduce the air leakage [48].

Equation 6.9 can still be used to determine the forced convection coefficient h_c for the hollow core PCB. If a slot width of 0.050 in (0.127 cm) is used for the hollow core PCB, the average forced convection coefficient shown in Eq. 6.15 can be doubled to about 12 Btu/hr ft² °F.

Extreme care must be exercised if the hollow core PCBs will be operating in a vibration environment. The thin epoxy fiberglass circuit boards will have a low resonant frequency, which will result in large vibration displacement amplitudes and rapid fatigue failures [1].

To reduce the failure rate and to provide a long fatigue life, stiffening ribs should be provided in the hollow core air space between the circuit boards. If more stiffening is required, external stiffening ribs can be added to the outside surface of the circuit boards between the electronic components.

6.9 COOLING AIR FANS FOR ELECTRONIC EQUIPMENT

Many different types of fans are available for cooling electronic equipment. These can generally be broken down into four major types: propeller, tube axial, vane axial, and centrifugal blowers. These fans can be single phase, three phase, 60 cycles, 400 cycles, 800 cycle ac/dc, constant speed, variable speed, and flow rates from 1 cfm to several thousand cfm [44].

Propeller fans can be obtained in a large variety of sizes, with different blade shapes and arrangements. Large propeller fans may be belt driven from an electric motor, while small fans may have the propeller fan attached directly to the electric motor, as shown in Figure 6.22. This type of fan can normally provide a static pressure of about 0.30 in (0.76 cm) of water at 0 cfm, or no delivery. Propeller fans are well suited for flushing operations through large racks or cabinets, which require large volumes or air at relatively low static pressures.

Tube axial fans have their motors mounted in the fan hub. This results in a fan with an axial depth that is much shorter than a propeller fan but which has a greater static pressure capability than a propeller fan of the same diameter. This type of fan, which is shown in Figure 6.23, is normally used where low noise, long life, and low cost are important. Typical applications are in areas that might use propeller fans, but where the depth is the limiting factor and the air flow requirements exceed those of a flushing operation [44].

Figure 6.24 shows a small vane axial fan which operates at a high rotational speed and provides static pressures of more than 4 in (10.16 cm) of water at zero delivery. Fans of this type are usually used for cooling electronic assemblies with high heat densities, which must operate in severe thermal and vibration environments. They generally operate at a high frequency of 400 Hz, and are available with variable speed [44] motors, which

Figure 6.22 Typical small propeller fan.

6.9 Cooling Air Fans for Electronic Equipment

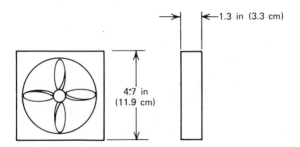

Figure 6.23 Typical flat tube axial fan that has a rated life of more than 10 years of continuous duty at 20°C.

speed up at high altitudes to provide extra cooling with low density air. These high speed fans generate a high level of acoustic noise, so that mufflers are recommended for fans that will be used in areas adjacent to working personnel.

The pressure flow characteristics of these axial flow fans, with high blade angles, produces a region of discontinuity which corresponds to a stalling condition on the blade airfoil. It is generally not advisable to operate fans in this area (which is sometimes called the centrax point) or at lower flow volumes, as shown in Figure 6.25. Operation should take place in the area to the right of the centrax point.

When the system operates to the left of the centrax point, pressure surges can occur. For example, the static pressure at points A and B in Figure 6.25 are the same, but the cfm flow conditions are not. Therefore, the flow conditions through the box can switch back and forth between points A and B. If the higher flow at point B is required, the reduced flow at point A may produce overheating in the electronics.

If the box impedance curve falls to the left of the centrax point, it is usually possible to bleed air from the fan with shims or holes. This will move the box impedance curve to the right of the centrax point.

The fan will then be delivering an excess cfm flow rate, so that some air will be thrown away. Since the fan will be operating to the right side of the

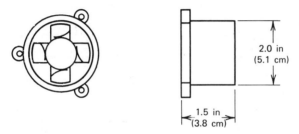

Figure 6.24 High speed, small size vane axial fan.

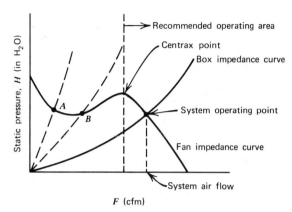

Figure 6.25 Recommended operating area for axial flow fans.

centrax point, its operation will be stable and not be subjected to pressure surges. Tests should be run to ensure that the proper amount of air is still flowing through the box for cooling.

High static pressures and high air flow rates can be obtained with low rotational speeds using centrifugal blowers (sometimes called squirrel cage blowers). Three types of blades are commonly used: backward curved, radial or paddle blade, and forward curved. A typical centrifugal blower is shown in Figure 6.26.

The highest pressures for a given propeller diameter and speed are usually developed by the forward curved fans. However, a relatively small increase in volume flow may cause a considerable increase in the power required.

Backward curved fans made with air foil section blades have produced efficiencies well above 80%. These fans are often referred to as nonoverloading fans, because they work well over a large range of volume flow.

6.10 AIR FILTERS

Electronic equipment may be required to operate around heavy machinery or in dusty areas where the air can pick up oil, lint, and other impurities.

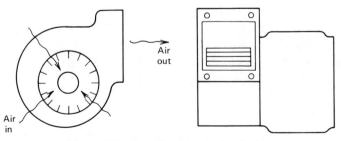

Figure 6.26 Centrifugal (squirrel cage) fan.

If these contaminates are permitted to enter the cooling air passages, they might plug up the openings and reduce the air flow, causing overheating of the electronics. To prevent this from happening, air filters are often used.

The pressure drop across an air filter can be very large if the air velocity through the filter is too high. To keep the pressure drop to a minimum, the air velocity through the filter should be less than about 250 ft/min (127 cm/sec). A lower air velocity will reduce the pressure drop through the filter even more, but will require a large filter surface area. This is desirable if there is sufficient room available for the larger filter.

Some filter manufacturers recommend a minimum air flow velocity of about 200 ft/min (102 cm/sec) through their filters. This velocity is supposed to make the suspended particles strike the filter with sufficient force to capture the particle. A lower velocity through their filter might permit some particles to change their direction so that they pass through larger openings in a filter that is not too dense.

6.11 CUTOFF SWITCHES

When cooling air is required to provide a safe operating temperature for electronic equipment, some means should be provided to turn the equipment off when the cooling air flow is interrupted, to prevent damage to the equipment. The required air flow may not be supplied because of a fan failure, power failure, clogged ducts, or clogged filters.

An air flow interlock switch is convenient for turning off the power in electronic systems that are not required to operate in severe vibration, shock, or acceleration environments. This device has an air flow sensor that extends into the air stream. If the air flow falls below a certain critical level, the sensor shuts off the main power to prevent the electronics from overheating.

Electronic equipment that must operate in severe vibration, shock, or acceleration environments can use miniature thermal cutoff switches. These switches should be mounted adjacent to the hottest components to prevent them from overheating. They can be purchased to open at virtually any temperature, and the cutoff point is accurate to about ±2°C.

6.12 STATIC PRESSURE LOSS TABLES AND CHARTS

Static pressure losses for various flow relations and geometry changes are shown in terms of the velocity head for different structures and conditions in Tables 6.2, 6.3, and 6.4 [40, 41, 69].

Table 6.2 Entry Losses for Different Shapes

Type of Opening	Description	Number of Velocity Heads (H_v) Lost
	Plain duct end	0.93
	Flanged duct end	0.49
	Well-rounded entry	0.04
	Protruding entry	2.70

Similar data available from industrial ventilation handbooks

6.13 HIGH ALTITUDE CONDITIONS

Electronic equipment is often required to operate at high altitudes, without conditioned cooling air. For fan cooled systems, the ambient air within the electronic equipment bay is normally used as the air source.

At high altitudes, the air density is very low. The weight flow of cooling air may be so sharply reduced that the equipment can be damaged as a result of overheating. If the system is protected with a thermal switch, it will simply shut the equipment off.

Many types of fans are available for cooling electronic systems. Some fans will speed up as the air density is reduced, so that they will deliver more cfm and more weight flow at the higher altitudes [44, 49].

Constant speed fans, either single phase or three phase, will not change their speeds very much. They will run at approximately the same speed, and deliver approximately the same cfm, regardless of the altitude. The weight flow the fan delivers will change as the density or altitude changes. A constant speed fan will, therefore, deliver a different weight flow for different altitudes. As the weight flow through the electronic box changes, the temperatures within the box will also change.

As the altitude (or pressure) changes, the cfm flow through the box will remain the same if a constant speed fan is used. When the altitude is increased, the air density is decreased and therefore the weight flow is decreased. There is less resistance to the flow through the box, so the box flow impedance curve goes down, as shown in Figure 6.27 [42, 44, 49].

With the reduced density at high altitudes, the ability of the fan to draw against a high static pressure is reduced. The static pressure curve for the

6.13 High Altitude Conditions

Table 6.3 Turn Losses in Elbows.

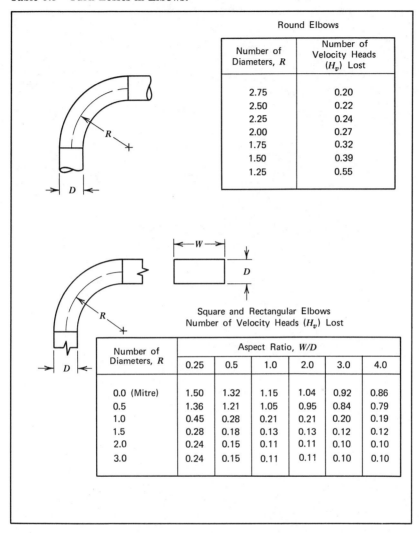

Round Elbows

Number of Diameters, R	Number of Velocity Heads (H_v) Lost
2.75	0.20
2.50	0.22
2.25	0.24
2.00	0.27
1.75	0.32
1.50	0.39
1.25	0.55

Square and Rectangular Elbows
Number of Velocity Heads (H_v) Lost

Number of Diameters, R	Aspect Ratio, W/D					
	0.25	0.5	1.0	2.0	3.0	4.0
0.0 (Mitre)	1.50	1.32	1.15	1.04	0.92	0.86
0.5	1.36	1.21	1.05	0.95	0.84	0.79
1.0	0.45	0.28	0.21	0.21	0.20	0.19
1.5	0.28	0.18	0.13	0.13	0.12	0.12
2.0	0.24	0.15	0.11	0.11	0.10	0.10
3.0	0.24	0.15	0.11	0.11	0.10	0.10

fan therefore goes down, as shown in Figure 6.27. The rate of reduction for the box impedance curve is the same as the rate of reduction for the fan static pressure curve, so the two curves cross at the same cfm flow point. As a result, the cfm flow through the box remains approximately constant as the altitude changes.

The static pressure drop through the box at the altitude condition can be determined from the static pressure drop through the box at sea level

Table 6.4 Expansion and Contraction Losses.

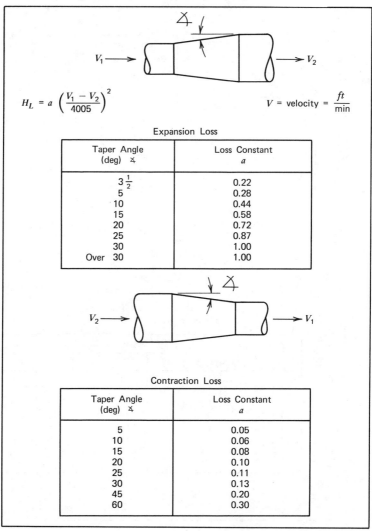

Taper Angle (deg) ⊀	Loss Constant a
$3\frac{1}{2}$	0.22
5	0.28
10	0.44
15	0.58
20	0.72
25	0.87
30	1.00
Over 30	1.00

Contraction Loss

Taper Angle (deg) ⊀	Loss Constant a
5	0.05
10	0.06
15	0.08
20	0.10
25	0.11
30	0.13
45	0.20
60	0.30

conditions if the altitude and sea level densities are known. The relation is linear as shown in Eq. 6.37 [42, 44, 49].

$$\Delta P_{\text{alt}} = \Delta P_{\text{sl}} \left(\frac{\rho_{\text{alt}}}{\rho_{\text{sl}}} \right) \tag{6.37}$$

6.14 Sample Problem—Fan Cooled Box At 30,000 Feet

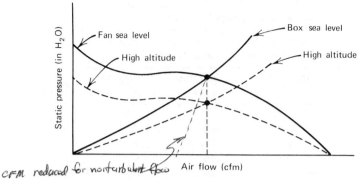

Figure 6.27 Effects of density (or altitude) on fan performance for turbulent flow

(handwritten annotations: "CFM reduced for nonturbulent flow"; "for turbulent flow")

6.14 SAMPLE PROBLEM—FAN COOLED BOX AT 30,000 FEET

Determine if the fan cooled box shown in Figure 6.13 is capable of operating at an altitude of 30,000 ft, using a constant speed fan delivering 30 cfm, as shown in Figure 6.20. The total power dissipation of the box is 165 watts, or 563 Btu/hr. This includes the power dissipation of 25 watts for the fan. The ambient air temperature in the equipment bay is expected to be 104°F (40.0°C) according to MIL-E-5400, the electronic equipment specification for class 1X equipment.

SOLUTION

The average density of the cooling air passing through the box is not known because the exit air temperature is not known. Therefore, assume a temperature rise and determine the average air density. Compute the temperature rise based upon the power dissipation and compare it with the assumed value. Correct it if necessary, and repeat the process until reasonably good correlation is obtained. Start by assuming a temperature rise of 66°F (36.6°C). The average air temperature is then

$$t_{av} = \frac{t_{in} + t_{out}}{2} = \frac{104 + 170}{2} = 137°F$$

The air pressure at 30,000 ft is obtained from Table 6.5 as 629.6 lb/ft². Now the air density is obtained from Eq. 6.32.

$$\rho = \frac{P}{RT} = \frac{629.6}{(53.3)(460 + 137)} = 0.0198 \; \frac{\text{lb}}{\text{ft}^3} \qquad (6.38)$$

Table 6.5 Changes of Temperature and Pressure with Altitude [37, 38]

Temperature and Pressure as Functions of Geometric Altitude: English Units

Altitude (ft)	Temperature (°R)	Pressure (P, lbf ft^{-2})
−16,500	577.58	3.6588 + 3
−15,000	572.22	3.5462
−10,000	554.37	3.0020
−5,000	536.52	2.5277
0	518.69	2.1162
5,000	500.86	1.7609
10,000	483.04	1.4556
15,000	465.23	1.1948
20,000	447.43	9.7327 + 2
25,000	429.64	7.8633
30,000	411.86	6.2966 + 2
35,000	394.08	4.9934
40,000	389.99	3.9312
45,000	389.99	3.0945
50,000	389.99	2.4361
55,000	389.99	1.9180
60,000	389.99	1.5103
65,000	389.99	1.1893 + 2
70,000	389.99	9.3672 + 1
75,000	389.99	7.3784
80,000	389.99	5.8125 + 1
85,000	394.32	4.5827
90,000	402.48	3.6292
95,000	410.64	2.8878
100,000	418.79	2.3085
105,000	426.94	1.8536
110,000	435.09	1.4947
115,000	443.23	1.2102 + 1
120,000	451.37	9.8372 + 0
125,000	459.50	8.0267
130,000	467.63	6.5735 + 0
135,000	475.76	5.4025
140,000	483.88	4.4552
145,000	492.00	3.6862
150,000	500.11	3.0597
155,000	508.22	2.5475
160,000	508.79	2.1247
165,000	508.79	1.7723
170,000	508.79	1.4784
175,000	508.79	1.2334
180,000	497.49	1.0272 + 0

6.14 Sample Problem—Fan Cooled Box At 30,000 Feet

Table 6.5 Continued

Temperature and Pressure as Functions of
Geometric Altitude: English Units

Altitude (ft)	Temperature (°R)	Pressure (P, lbf ft^{-2}) ← NOT PSI
185,000	485.36	8.5160 − 1
190,000	473.24	7.0278
195,000	461.12	5.7713
200,000	449.00	4.7151
210,000	424.79	3.0955
220,000	400.60	1.9835
230,000	376.44	1.2368
240,000	352.30	7.4774 − 2
250,000	328.20	4.3640

The average air weight flow through the box is determined from Eq. 6.34.

$$W = \rho F = \left(0.0198 \, \frac{\text{lb}}{\text{ft}^3}\right)\left(30 \, \frac{\text{ft}^3}{\text{min}}\right) = 0.594 \, \frac{\text{lb}}{\text{min}} \quad (6.39)$$

The temperature rise of the cooling air as it passes through the box is determined from Eq. 6.7.

$$\Delta t_A = \frac{Q}{W C_p} = \frac{563 \text{ Btu/hr}}{(0.594 \text{ lb/min})(60 \text{ min/hr})(0.24 \text{ Btu/lb °F})} = 65.8°F \quad (6.40)$$

The assumed temperature rise of 66°F agrees well with the calculated value of 65.8°F, so that the assumed value is accurate.

The temperature rise across the convection air film to the surface of the electronic components for the altitude condition is still determined from Eq. 5.7, using Eqs. 6.9 through 6.18 for the required information.

Substitute Eq. 6.39 into Eq. 6.12.

$$G = \frac{W}{A} = \frac{(0.594 \text{ lb/min})(60 \text{ min/hr})}{(0.00625 \text{ ft}^2)(7 \text{ ducts})} = 814 \, \frac{\text{lb}}{\text{hr ft}^2} \quad (6.41)$$

Substitute into Eq. 6.11 for the Reynolds number:

$$N_R = \frac{(814 \text{ lb/hr ft}^2)(0.0165 \text{ ft})}{0.050 \text{ lb/ft hr}} = 269 \text{ (dimensionless)} \quad (6.42)$$

Substitute into Eq. 6.10 for the J factor:

$$J = \frac{6}{(N_R)^{0.98}} = \frac{6}{(269)^{0.98}} = 0.0249 \qquad (6.43)$$

Substitute into Eq. 6.9 for the forced air convection coefficient at 30,000 ft (physical properties of air are obtained from Figure 6.28 for an average air temperature of 137°F).

$$h_c = \frac{(0.0249)(0.24)(814)}{[(0.24)(0.05)/0.017]^{0.666}} = 6.13 \frac{\text{Btu}}{\text{hr ft}^2 \, °\text{F}} \qquad (6.44)$$

Notice that although there has been a relatively large reduction in the Reynolds number, the forced convection coefficient at the altitude condition, Eq. 6.44, has not changed much from the forced convection coefficient at sea level, as shown in Eq. 6.15. This is quite typical for laminar flow conditions. The forced convection coefficient there depends much more on the hydraulic diameter than on the Reynolds number. A smaller hydraulic diameter increases the value of the coefficient (see Figure 6.38).

The temperature rise across the convection film to the components is determined from Eq. 5.7.

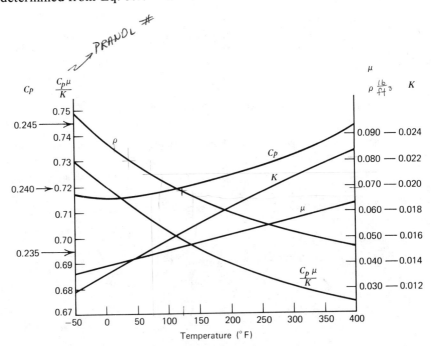

Figure 6.28 Physical properties of dry air at 1 atm (English units).

Given $Q = 20$ watts $= 68.3$ Btu/hr (PCB heat)

$h_c = 6.13$ Btu/hr ft² °F (ref. Eq. 6.44)

$$A = \frac{(1.3)(8)(9)}{144} = 0.65 \text{ ft}^2 \text{ [effective area 1 PCB (including back side)]}$$

$$\Delta t_B = \frac{Q}{h_c A} = \frac{68.3}{(6.13)(0.65)} = 17.1°F \tag{6.45}$$

The component hot spot temperature is determined from Eq. 6.17.

Given $t_{in} = 104°F$

$\Delta t_A = 65.8°F$ (ref. Eq. 6.40)

$\Delta t_B = 17.1°F$ (ref. Eq. 6.45)

$$t_S = 104 + 65.8 + 17.1 = 186.9°F \tag{6.46}$$

Since this temperature is less than 212°F (100°C), the design is satisfactory for the altitude condition.

If a slightly cooler operating box is desired, the cooling air fan can be used as an exhaust fan. The 25 watts of heat from the fan will not be added to the inlet cooling air. This will reduce the cooling air temperature about 10°F and will also reduce the component hot spot temperature about 10°F, as shown in Eq. 6.47.

$$\Delta T = \frac{Q}{WC_p} = \frac{(25 \text{ watts})(3.413 \text{ Btu/hr/watt})}{(0.594 \text{ lb/min})(60 \text{ min/hr})(0.24 \text{ Btu/lb °F})} = 10°F \tag{6.47}$$

The pressure drop through the box at the 30,000 ft altitude condition can be determined from Eq. 6.37.

Given $\Delta P_{sl} = 1.30$ in H₂O (ref. Figure 6.20 sea level for constant speed fan)

$\rho_{alt} = 0.0198$ lb/ft³ 30,000 ft (ref. Eq. 6.38)

$\rho_{sl} = 0.0704$ lb/ft³ sea level and 104°F

$$\Delta P_{alt} = 1.30 \left(\frac{0.0198}{0.0704}\right) = 0.366 \text{ in H}_2\text{O} \tag{6.48}$$

6.15 OTHER CONVECTION COEFFICIENTS

Convection coefficients can become quite high for turbulent air flow conditions through ducts. These ducts may be air passages that are formed between plug-in PCBs or ducts that deliver air between two electronic boxes. The convection coefficient is shown in Eq. 6.49.

$$h = 0.0198 \left(\frac{K}{D}\right)(N_R)^{0.8} \qquad (6.49)$$

Equation 6.49 can be used to find the forced convection coefficient for a pressurized enclosed duct. For example, consider a flat rectangular duct with a cross section $a = 3.0$ in, $b = 0.50$ in, and many feet long. Air at 20 psig and 150°F flows at 200 cfm. Find the convection heat transfer coefficient. The physical properties of air are needed. These are shown in Figure 6.28.

$$D = \frac{2ab}{a+b} \text{ (hydraulic diameter)}$$

$$D = \frac{2(3.0)(0.50)}{3.0 + 0.50} = 0.857 \text{ in} = 0.0714 \text{ ft}$$

$$\rho = \frac{P}{RT} \text{ (air density; ref. Eq. 6.32)}$$

$$\rho = \frac{(20.0 + 14.7)(144 \text{ in}^2/\text{ft}^2)}{(53.3)(460 + 150)°R} = 0.154 \frac{\text{lb}}{\text{ft}^3}$$

$$A = \frac{(3.0)(0.5) \text{ in}^2}{144 \text{ in}^2/\text{ft}^2} = 0.0104 \text{ ft}^2 \text{ (cross section area)}$$

$$v = \frac{\text{cfm}}{A} = \frac{(200 \text{ ft}^3/\text{min})(60 \text{ min/hr})}{0.0104 \text{ ft}^2} = 1.15 \times 10^6 \frac{\text{ft}}{\text{hr}} \text{ (velocity)}$$

$\mu = 0.055$ lb/ft hr (viscosity at 34.7 psia, 150°F; ref. Figure 6.28)

$K = 0.0175$ Btu/hr ft °F (conductivity at 150°F; ref. Figure 6.28)

The Reynolds number is

$$N_R = \frac{VD\rho}{\mu} = \frac{(1.15 \times 10^6)(0.0714)(0.154)}{0.055} = 230,000 \text{ (dimensionless)}$$

Since N_R is well above 2300, the flow is obviously turbulent. Substitute into Eq. 6.49.

$$h = (0.0198)\left(\frac{0.0175}{0.0714}\right)(230,000)^{0.8} = 94.5 \frac{\text{Btu}}{\text{hr ft}^2 °F} \qquad (6.50)$$

For the flow of gasses parallel to plane surfaces, the general heat transfer convection coefficient can be determined from Eq. 6.51.

$$h = 0.055 \left(\frac{K}{L}\right) (N_R)^{0.75} \qquad (6.51)$$

The length of the surface is shown by L. This is to be limited to 2 ft, even if the length is greater than 2 ft.

When the air is blowing across a cylinder or wire, the convection coefficient can take the form shown in Eq. 6.52 [7, 8].

$$h = b \left(\frac{K}{D}\right) (N_R)^m \qquad (6.52)$$

The constants b and m, which are a function of the Reynolds number, are shown in Table 6.6 for different conditions [7, 8].

The characteristic dimension D to be used in the equation for the square and diamond cross section is that of a circular cross section that has the same perimeter dimension.

For air flow over spheres, the convection heat transfer coefficient can be determined from Eq. 6.53.

$$h = 0.33 \left(\frac{K}{D}\right) (N_R)^{0.6} \qquad (6.53)$$

6.16 SAMPLE PROBLEM—COOLING A TO-5 TRANSISTOR

A 2N2905 transistor (TO-5 case size) dissipates 0.25 watt under steady state power conditions in a 160°F (71°C) ambient environment. The transistor is

Table 6.6 Values to Be Used with Eq. 6.52.

Shape	Reynolds Number	b	m
○ (cylinder)	0.4–4.0	0.891	0.330
	4.0–40	0.821	0.385
	40–4,000	0.615	0.466
	4,000–40,000	0.174	0.618
	40,000–400,000	0.0239	0.805
□ (square)	5,000–10,000	0.0921	0.675
◇ (diamond)	5,000–100,000	0.222	0.585

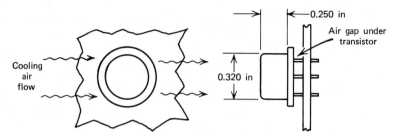

Figure 6.29 Cooling air flow over a transistor.

mounted above the circuit board as shown in Figure 6.29. Determine the case temperature when the air velocity over the transistor is 250 ft/min.

SOLUTION

The Reynolds number (N_R) must be computed first, using the following information. (Physical properties of air are determined from Figure 6.28.)

$$V = 250 \text{ ft/min} \times 60 \text{ min/hr} = 15{,}000 \text{ ft/hr (air velocity)}$$

$$D = \frac{0.320 \text{ in}}{12 \text{ in/ft}} = 0.0267 \text{ ft (hydraulic diameter)}$$

$$\rho = \frac{P}{RT} = \frac{(14.7 \text{ lb/in}^2)(144 \text{ in}^2/\text{ft}^2)}{(53.3 \text{ ft/}°R)(460 + 160)°R} = 0.064 \frac{\text{lb}}{\text{ft}^3} \text{ (air density)}$$

$$\mu = 0.050 \text{ lb/ft hr (viscosity at 160°F)}$$

$$K = 0.017 \text{ Btu/hr ft °F (thermal conductivity at 160°F)}$$

Substitute into Eq. 6.11 for the Reynolds number:

$$N_R = \frac{(15000)(0.0267)(0.064)}{0.050} = 512 \text{ (dimensionless)} \qquad (6.54)$$

The Reynolds number is between 40 and 4000, so that the values shown in Table 6.6 are $b = 0.615$ and $m = 0.466$. Substituting these values into Eq. 6.52 results in Eq. 6.55.

$$h = 0.615 \left(\frac{K}{D}\right)(N_R)^{0.466} \qquad (6.55)$$

$$h = 0.615 \left(\frac{0.017}{0.0267}\right)(512)^{0.466} = 7.17 \frac{\text{Btu}}{\text{hr ft}^2 \text{ °F}} \qquad (6.56)$$

The temperature rise across the boundary layer, from the ambient air to the transistor surface, is determined from Eq. 5.7.

$$\Delta t = \frac{Q}{hA} \text{ (ref. Eq. 5.7)}$$

Given $Q = 0.25$ watt $\times$ 3.413 Btu/hr/watt = 0.85 Btu/hr

$h = 7.16$ Btu/hr ft² °F (ref. Eq. 6.56)

A = surface area of transistor

$$= \frac{2(\pi/4)(0.32)^2 + \pi(0.32)(0.25) \text{ in}^2}{144 \text{ in}^2/\text{ft}^2} = 0.00286 \text{ ft}^2$$

$$\Delta t = \frac{0.85 \text{ Btu/hr}}{(7.16 \text{ Btu/hr ft}^2 \text{ °F})(0.00286 \text{ ft}^2)} = 41.5\text{°F} \quad (6.57)$$

The case temperature of the transistor becomes

$$t_{\text{case}} = 160 + 41.5 = 201.5\text{°F} \quad (6.58)$$

The junction temperature of the transistor is determined using the thermal impedance from the junction to the case (Θ_{jc}) which is usually obtained from the component manufacturer. When the value of $\Theta_{jc} = 59.4$°F/watt (33°C/watt) the temperature rise from the case to the junction is:

$$\Delta t_{jc} = 59.4 \frac{\text{°F}}{\text{watt}} \times 0.25 \text{ watt} = 14.8\text{°F} \quad (6.59)$$

The junction temperature of the transistor can then be obtained from Eq. 6.60.

$$t_j = t_{\text{case}} + \Delta t_{jc} \quad (6.60)$$

$$t_j = 201.5 + 14.8 = 216.3\text{°F} \quad (6.61)$$

6.17 CONDITIONED COOLING AIR FROM AN EXTERNAL SOURCE

Conditioned cooling air is often supplied to the electronic equipment from an external source, such as the compressor section of an aircraft engine. For example, in the F-16 airplane, the electronic equipment cooling air is obtained from the seventeenth stage of the engine compressor. The air temperature and pressure are both very high, and the air may have a high moisture content. This air is therefore throttled, cooled, and dried before it is passed through the electronic equipment.

The amount of conditioned air required to cool an electronic system will vary with the temperature of the cooling air. Less air is required when the cooling air temperature is low, and more air is required when the cooling air temperature is high. Extensive test data and analysis have shown that an efficient thermal design can be achieved if the temperature of the cooling air is not allowed to exceed 160°F (71°C) as it exits from the electronic chassis. This condition should result in a maximum component surface temperature of about 212°F (100°C) when the component mounting techniques shown in Chapter 4 are used.

Equation 6.7 can be used to establish the cooling air flow curve for an electronic box when the power dissipation of the box is known. The technique is demonstrated with a sample problem.

6.18 SAMPLE PROBLEM—GENERATING A COOLING AIR FLOW CURVE

An electronic box, with a power dissipation of 330 watts, has conditioned cooling air supplied by the airplane engine. Generate the flow curve required to maintain a box cooling air exit temperature of 160°F (71°C) when the cooling air inlet temperature varies from −65°F (−54°C) to +131°F (+55°C). The conversion tables in Section 1.11 are used to convert to English and metric units.

SOLUTION

$Q = 330$ watts $= 1126.3$ Btu/hr $= 78.87$ cal/sec (heat dissipation)

$C_p = 0.24$ Btu/lb °F $= 0.24$ cal/g °C (specific heat)

$\Delta t = 160°F - t_{in}$ (English) $(71°C - t_{in}$ metric)

Start with an air inlet temperature of −65°F for English units, which corresponds to −54°C for metric units.

$\Delta t = 160°F - (-65)°F = 225°F$ (English units)

$\Delta t = 71°C - (-54)°C = 125°C$ (metric units)

Substituting into Eq. 6.7, the cooling air weight flow is determined using English units.

$$W = \frac{1126.3 \text{ Btu/hr}}{(225°F)(0.24 \text{ Btu/lb °F})(60 \text{ min/hr})} = 0.347 \frac{\text{lb}}{\text{min}} \quad (6.62)$$

6.19 Static Pressure Losses for Various Altitude Conditions

Using Eq. 6.7, the weight flow for metric units is

$$W = \frac{78.87 \text{ cal/sec}}{(125°C)(0.24 \text{ cal/g }°C)} = 2.63 \frac{g}{\text{sec}} \quad (6.62a)$$

At an air inlet temperature of 80°F (26.6°C)

$$\Delta t = 160°F - 80°F = 80°F \text{ (English units)}$$
$$\Delta t = 71°C - 26.6°C = 44.4°C \text{ (metric units)}$$

Substituting into Eq. 6.7 for the weight flow, we obtain

$$W = \frac{1126.3 \text{ Btu/hr}}{(80°F)(0.24)(60 \text{ min/hr})} = 0.978 \frac{\text{lb}}{\text{min}} \text{ (English units)} \quad (6.63)$$

$$W = \frac{78.87 \text{ cal/sec}}{(44.4°C)(0.24)} = 7.40 \frac{g}{\text{sec}} \text{ (metric units)} \quad (6.63a)$$

At an air inlet temperature of 131°F (55°C):

$$\Delta t = 160°F - 131°F = 29°F \text{ (English units)}$$
$$\Delta t = 71°C - 55°C = 16°C \text{ (metric units)}$$

Substituting into Eq. 6.7 yields

$$W = \frac{1126.3 \text{ Btu/hr}}{(29°F)(0.24)(60 \text{ min/hr})} = 2.70 \frac{\text{lb}}{\text{min}} \text{ (English units)} \quad (6.64)$$

$$W = \frac{78.87 \text{ cal/sec}}{(16°C)(0.24)} = 20.54 \frac{g}{\text{sec}} \text{ (metric units)} \quad (6.64a)$$

When the cooling air flow rate is plotted against the cooling air temperature, the cooling air flow curve will be produced, as shown in Figure 6.30.

6.19 STATIC PRESSURE LOSSES FOR VARIOUS ALTITUDE CONDITIONS

Electronic systems are often required to operate at different altitude conditions. As the altitude varies, it is desirable to maintain a constant cooling air weight flow rate, similar to the curve shown in Figure 6.30. At high altitudes the air density is lower, so that the cooling air flow velocity must be increased to provide the same air flow weight. As the air velocity increases, however, the static pressure drop through the electronic system

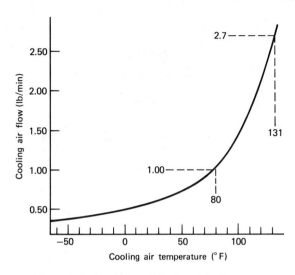

Figure 6.30 Typical cooling air weight flow curve.

also increases. It is therefore important to know how these pressure losses will change with the altitude, to ensure an adequate flow of cooling air to the electronics under all conditions.

The complexity of the static pressure drop relations for various altitude conditions can be simplified by considering the density of the cooling air at the altitude condition and comparing it with the density of air at some standard condition. The standard selected is usually based on a temperature of 59°F (15°C) and a pressure of 14.7 lb/in² (1034.4 g/cm²), where the air density is 0.0765 lb/ft³ (0.00122 g/cm³).

It is convenient to use a density ratio that is called sigma and is normally shown by the Greek letter sigma (σ), which is defined as follows:

$$\sigma = \frac{\text{air density at any condition}}{\text{air density at standard conditions}} = \frac{\rho}{\rho_{\text{std}}}$$

$$\sigma = \frac{\rho}{0.0765 \text{ lb/ft}^3} \text{ (for English units)} \qquad (6.65)$$

$$\sigma = \frac{\rho}{0.00122 \text{ g/cm}^3} \text{ (for metric units)} \qquad (6.65a)$$

It is desirable to work with the average density of the cooling air for high altitude conditions. At high altitudes, the outside air pressure and density are very low, so that the cooling air velocity must be very high to provide a reasonable weight flow. High air velocities mean high flow losses, so that a high pressure must be used to push the cooling air through the box. Under these circumstances, the inlet air density can be many times greater than

6.19 Static Pressure Losses for Various Altitude Conditions

the outlet air density. If these density changes are ignored at high altitude conditions, it can lead to large pressure drop errors. The average air density within the box is therefore used, to improve the accuracy of the calculations.

The density ratio (σ) can be written in another way by using the density relation obtained from the gas law relation shown in Eq. 6.32. Then substituting into Eq. 6.65, we obtain

$$\sigma = \frac{P_1/R_1 T_1}{P_2/R_2 T_2} = \frac{\dfrac{P_1}{(53.3 \text{ ft/°R}) T_1}}{\dfrac{14.7 \text{ lb/in}^2}{(53.3 \text{ ft/°R})(460 + 59)\text{°R}}} = 35.3 \frac{P_1}{T_1} \text{ (English units)}$$

$$\sigma = \frac{\dfrac{P_1}{(2924 \text{ cm/°K}) T_1}}{\dfrac{1034.4 \text{ g/cm}^2}{(2924 \text{ cm/°K})(273 + 15)\text{°K}}} = 0.279 \frac{P_1}{T_1} \text{ (metric units)}$$

The inlet and the outlet air pressures should be averaged to obtain the average value for the electronic box in English units and metric units. The resulting equation is convenient for working with electronic boxes at high altitude conditions.

$$\sigma = \frac{35.3}{2}\left(\frac{P_{in}}{T_{in}} + \frac{P_{out}}{T_{out}}\right) \text{ (English units)} \qquad (6.66)$$

where P = pressure = lb/in²
  T = absolute temperature (460 + °F) = °R
  $P_{in} = P_{out} + \Delta P$ = lb/in² (inlet pressure)
  P_{out} = outside pressure = lb/in²

$$\sigma = \frac{0.279}{2}\left(\frac{P_{in}}{T_{in}} + \frac{P_{out}}{T_{out}}\right) \text{ (metric units)} \qquad (6.66a)$$

where P = pressure = g/cm²
  T = absolute temperature (273 + °C) = °K
  $P_{in} = P_{out} + \Delta P$ = g/cm² (inlet pressure)
  P_{out} = outside pressure = g/cm²

solve by iteration

The value of σ is usually multiplied by the static pressure drop ΔP and plotted against the cooling air weight flow on a log-log plot. This results in a straight line, as shown in Figure 6.31.

For fully turbulent flow conditions, the slope of the $\sigma \Delta P$ line on a log-log plot is 2.0. For laminar flow conditions, the slope will usually vary from about 1.2 to about 1.7, depending upon the severity of the restrictions in the cooling air flow path.

$$\Delta P_{ALT} = \frac{\sigma \Delta P}{\sigma_{ALT}} \quad \text{— constant at all altitudes for given flow rate}$$

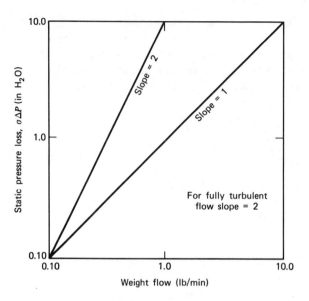

Figure 6.31 Typical $\sigma \Delta P$ pressure loss curves for electronic boxes.

(log-log plot)

Once the $\sigma \Delta P$ curve has been established for a particular electronic box, it is very easy to determine the pressure drop through the box for any altitude and for any cooling air flow condition.

6.20 SAMPLE PROBLEM—STATIC PRESSURE DROP AT 65,000 FEET

An electronic box with a power dissipation of 330 watts has conditioned cooling air supplied (by the airplane turbine) with a weight flow rate as shown in Figure 6.30. This flow rate will provide a cooling air exit temperature of 160°F (71°C) for any inlet air temperature shown in the curve. Test data show that the static pressure drop through the box is 2.0 in of water (5.08 cm of water) when the air inlet temperature is 80°F (26.6°C) and the cooling air flow rate is 1.0 lb/min (7.56 g/sec).

1. Determine the static pressure drop through the box at an altitude of 65,000 ft (19,812 m) with an air flow rate of 1.0 lb/min at 80°F. The outside air pressure at this altitude is 0.83 lb/in² (58.4 g/cm²).
2. Determine the static pressure drop through the box at sea level and at an altitude of 65,000 ft when the air inlet temperature is 131°F (55°C) and the cooling air flow rate is 2.7 lb/min (20.43 g/sec).

→ can't be done: $V_{air} > M1$

6.20 Sample Problem—Static Pressure Drop at 65,000 Feet

SOLUTION

For the solution of part 1 of the problem, the value of σ is first computed for the sea level condition. Equations 6.65 and 6.65a will give about the same results as Eqs. 6.66 and 6.66a because the 2.0 in pressure drop only represents a $2 \times 0.036 = 0.072$ lb/in² pressure drop. This is small compared to the atmospheric pressure of 14.7 lb/in², so that P_{in} is virtually the same as P_{out} in Eq. 6.66.

The average air density for the 80°F (26.6°C) sea level condition is computed from Eq. 6.32, using English units first. The average temperature in the box must be used.

$$t_{av} = \frac{t_{in} + t_{out}}{2} = \frac{80 + 160}{2} = 120°F \text{ (English units)}$$

$$t_{av} = \frac{26.6 + 71}{2} = 48.8°C \text{ (metric units)}$$

In English units:

$$\rho = \frac{(14.7 \text{ lb/in}^2)(144 \text{ in}^2/\text{ft}^2)}{(53.3 \text{ ft/°R})(460 + 120)°R} = 0.0685 \frac{\text{lb}}{\text{ft}^3} \qquad (6.67)$$

In metric units:

$$\rho = \frac{1034.4 \text{ g/cm}^2}{(2924 \text{ cm/°K})(273 + 48.8)°K} = 0.001099 \frac{\text{g}}{\text{cm}^3} \qquad (6.67a)$$

Substitute into Eq. 6.65 for the sea level value of σ using English units.

$$\sigma = \frac{\rho}{0.0765} = \frac{0.0685}{0.0765} = 0.895 \text{ (dimensionless)} \qquad (6.68)$$

Substitute into Eq. 6.65a for the sea level value of σ using metric units.

$$\sigma = \frac{\rho}{0.001224} = \frac{0.001099}{0.001224} = 0.897 \text{ (dimensionless)} \qquad (6.68a)$$

The value of σ is multiplied by the static pressure drop (ΔP) head of water, shown previously as 2.0 in H₂O (5.08 cm H₂O), to obtain the $\sigma \Delta P$ values of static pressure drop:

$$\sigma \Delta P = (0.895)(2.0 \text{ in H}_2\text{O}) = 1.79 \text{ in H}_2\text{O (English units)} \qquad (6.69)$$

$$\sigma \Delta P = (0.897)(5.08 \text{ cm H}_2\text{O}) = 4.55 \text{ cm H}_2\text{O (metric units)} \qquad (6.69a)$$

The values shown above are valid only for a cooling air flow rate of 1.0 lb/min (7.56 g/sec) and an air inlet temperature of 80°F (26.6°C). However, this $\sigma \Delta P$ value is exactly the same *at any altitude* condition, *as long as the air flow rate and temperature stay the same*.

The static pressure drop at an altitude of 65,000 ft, with the 1.0 lb/min flow rate, can be determined if the value of σ can be obtained for the high altitude condition. Using Eqs. 6.66 and 6.66a, the following information is required.

Item	English Units	Metric Units
P_{out}	0.83 lb/in²	58.4 g/cm²
T_{out}	460 + 160 = 620°R	273 + 71 = 344°K
T_{in}	460 + 80 = 540°R	273 + 26.6 = 299.6°K
P_{in}	0.83 lb/in² + ΔP	58.4 g/cm² + ΔP

The static pressure drop ΔP through the electronic box is not known for the high altitude condition, but its value is required. Therefore, a value is assumed and the calculation is made. The assumed value is then compared with the calculated value. If they are not close, another guess is made and the calculation is repeated. Close agreement can usually be obtained with two or three iterations. Considering English units first, assume a static pressure drop, ΔP, of 22 in H$_2$O.

$$\Delta P = 22 \text{ in H}_2\text{O} \times 0.036 \frac{\text{lb/in}^2}{\text{in H}_2\text{O}} = 0.792 \text{ lb/in}^2$$

$$P_{in} = 0.83 \text{ lb/in}^2 + 0.792 \text{ lb/in}^2 = 1.622 \text{ lb/in}^2$$

Substitute into Eq. 6.66 for the σ value at 65,000 ft with a flow rate of 1.0 lb/min at 80°F.

$$\sigma_{alt} = \frac{35.3}{2}\left(\frac{1.622}{540} + \frac{0.83}{620}\right) = 0.0766 \qquad (6.70)$$

For metric units, assume a static pressure drop ΔP of 56 cm H$_2$O.

$$\Delta P = 56 \text{ cm H}_2\text{O} = 56 \text{ g/cm}^2$$

$$P_{in} = 58.4 \text{ g/cm}^2 + 56 \text{ g/cm}^2 = 114.4 \text{ g/cm}^2$$

Substitute into Eq. 6.66a for the σ value at 19,812 m with a flow rate of 7.56 g/sec at 26.6°C.

$$\sigma_{alt} = \frac{0.279}{2}\left(\frac{114.4}{299.6} + \frac{58.4}{344}\right) = 0.0769 \qquad (6.70a)$$

6.20 Sample Problem—Static Pressure Drop at 65,000 Feet

The static pressure drop at the altitude condition is determined from Eq. 6.71.

$$\Delta P_{alt} = \frac{\sigma \Delta P}{\sigma_{alt}} \tag{6.71}$$

For English units, substitute Eqs. 6.69 and 6.70 into Eq. 6.71.

$$\Delta P_{alt} = \frac{1.79}{0.0766} = 23.4 \text{ in } H_2O \tag{6.72}$$

For metric units, substitute Eqs. 6.69a and 6.70a into Eq. 6.71.

$$\Delta P_{alt} = \frac{4.556}{0.0769} = 59.2 \text{ cm } H_2O \tag{6.72a}$$

A pressure drop of 22 in H_2O (and 56 cm H_2O) was assumed for the box at an altitude of 65,000 ft (19,812 m), with a flow rate of 1.0 lb/min (7.56 g/sec) at 80°F (26.6°C). The calculated value turned out to be 23.4 in H_2O (and 59.2 cm H_2O) on the first try. A second iteration would show that the correct value is closer to 23 in H_2O (58.4 cm H_2O).

For the solution to part 2 of the problem, the slope of the $\sigma \Delta P$ curve on a log-log plot similar to Figure 6.31 must be known. For fully turbulent flow systems the $\sigma \Delta P$ slope is 2.0. This is the theoretical maximum that can be obtained with an air flow system. In actual systems, however, this slope is seldom obtained. The actual slope for a typical box will generally have a range from about 1.2 for a box with very few widely spaced PCBs to about 1.8 for a box with heat exchanger side walls containing 20 wavy fins per inch.

When the electronic box has a $\sigma \Delta P$ slope of 1.4, for example, the pressure drop at any cooling air weight flow condition can be determined from the general definition of the slope, as shown in Figure 6.32 and Eq. 6.73.

$$\text{Slope } S = \frac{\Delta Y}{\Delta X} = \frac{\log Y_2 - \log Y_1}{\log X_2 - \log X_1} \tag{6.73}$$

Common logarithms to the base 10 or natural logarithms to the base e can be used for Eq. 6.73. Considering common logarithms to the base 10 with English units, the $\sigma \Delta P$ static pressure drop through the box for a flow rate of 2.7 lb/min is computed.

$$1.4 = \frac{\log \sigma \Delta P_2 - \log 1.79}{\log 2.7 - \log 1.0} = \frac{\log \sigma \Delta P_2 - 0.2528}{0.4313 - 0}$$

$$\log \sigma \Delta P_2 = 1.4(0.4313) + 0.2528 = 0.8566$$

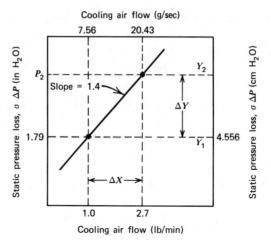

Figure 6.32 $\sigma \Delta P$ cooling curve with a slope of 1.4.

Take the antilog of both sides:

$$\sigma \Delta P_2 = 7.188 \text{ in } H_2O \quad (6.74)$$

Considering common logarithms to the base 10 with metric units, the $\sigma \Delta P$ static pressure drop through the box for a flow rate of 20.43 g/sec is computed.

$$1.4 = \frac{\log \sigma \Delta P_2 - \log 4.556}{\log 20.43 - \log 7.56} = \frac{\log \sigma \Delta P_2 - 0.6586}{1.3102 - 0.8785}$$

$$\log \sigma \Delta P_2 = 1.4 \,(0.4317) + 0.6586 = 1.2629$$

Take the antilog of both sides:

$$\sigma \Delta P_2 = 18.32 \text{ cm } H_2O \quad (6.74a)$$

Before the static pressure drop through the box at sea level conditions, with a flow rate of 2.7 lb/min, can be computed from the $\sigma \Delta P$ values, the average air temperature and average air density must be computed. In English units:

$$t_{av} = \frac{t_{in} + t_{out}}{2} = \frac{131 + 160}{2} = 145.5°F$$

$$\rho = \frac{(14.7)(144) \text{ lb/ft}^2}{(53.3 \text{ ft/°R})(460 + 145.5)°R} = 0.0656 \frac{\text{lb}}{\text{ft}^3} \quad (6.75)$$

6.20 Sample Problem—Static Pressure Drop at 65,000 Feet

In metric units:

$$t_{av} = \frac{55 + 71}{2} = 63\,°C$$

$$\rho = \frac{1034.4 \text{ g/cm}^2}{(2924 \text{ cm/°K})(273 + 63)\,°K} = 0.001053 \frac{g}{cm^3} \qquad (6.75a)$$

Substitute Eq. 6.75 into Eq. 6.65 for English units:

$$\sigma = \frac{\rho}{0.0765} = \frac{0.0656}{0.0765} = 0.857 \text{ (dimensionless)} \qquad (6.76)$$

Substitute Eq. 6.75a into Eq. 6.65a for metric units:

$$\sigma = \frac{\rho}{0.001224} = \frac{0.00105}{0.001224} = 0.857 \text{ (dimensionless)} \qquad (6.76a)$$

The sea level static pressure drop through the box with a flow rate of 2.7 lb/min (20.43 g/sec) is now computed from Eqs. 6.74 and 6.75 in English units.

$$\Delta P_{sl} = \frac{\sigma \Delta P_2}{\sigma} = \frac{7.188}{0.857} = 8.39 \text{ in } H_2O \qquad (6.77)$$

In metric units, with Eqs. 6.74a and 6.75a:

$$\Delta P_{sl} = \frac{\sigma \Delta P_2}{\sigma} = \frac{18.32}{0.857} = 21.38 \text{ cm } H_2O \qquad (6.77a)$$

The static pressure drop through the box at an altitude of 65,000 ft, with a flow rate of 2.7 lb/min, can be calculated as before. Assume a pressure drop of 62 in H_2O to start.

$\Delta P = 62$ in $H_2O \times 0.036$ (lb/in²)/in $H_2O = 2.232$ lb/in²

$P_{in} = 0.83$ lb/in² + 2.232 lb/in² = 3.062 lb/in²

$P_{out} = 0.83$ lb/in²

$T_{in} = 460 + 131 = 591\,°R$

$T_{out} = 460 + 160 = 620\,°R$

$$\sigma_{alt} = \frac{35.3}{2}\left(\frac{3.062}{591} + \frac{0.83}{620}\right) = 0.115$$

The $\sigma \Delta P$ value at 65,000 ft must be the same as the value shown by Eq.

6.74 for the 2.7 lb/min flow rate. The static pressure drop through the box at 65,000 ft is, therefore,

$$\Delta P_{\text{alt}} = \frac{\sigma \, \Delta P}{\sigma_{\text{alt}}} = \frac{7.188}{0.115} = 62.5 \text{ in } H_2O \tag{6.78}$$

A static pressure drop of 62 in H_2O was assumed for the 65,000 ft altitude condition for a flow rate of 2.7 lb/min. The calculated value is 62.5 in H_2O, which is close enough for these calculations and shows that the assumed value is accurate.

The force that pushes the cooling air through the electronic box is the pressure differential across the box. As this pressure difference increases, the velocity of the cooling air through the box will increase until the critical velocity is reached. Once the critical velocity is reached, any further increase in the pressure difference across the box will not produce any further increase in cooling air velocity.

Critical flow velocity will be reached when the ratio of the downstream pressure (or outside pressure) to the upstream pressure (or inside pressure) falls below the critical pressure ratio (P_{cr}) of 0.53, as shown below.

$$P_{cr} = \frac{P_{\text{out}}}{P_{\text{in}}} = 0.53 \text{ (critical pressure ratio)}$$

In the previous sample problem, this pressure ratio (R_p) at the 65,000 ft altitude is:

$$R_p = \frac{P_{\text{out}}}{P_{\text{in}}} = \frac{0.83}{3.062} = 0.27$$

Since this value is less than 0.53, it shows that the critical velocity has been reached. Therefore, the 2.7 lb/min cooling air flow rate at the 65,000 ft altitude condition can never be obtained.

6.21 TOTAL PRESSURE DROP FOR VARIOUS ALTITUDE CONDITIONS

Section 6.3 and Figures 6.7 through 6.9 show the basic relations between static pressures and velocity pressures for electronic boxes. Equation 6.3 shows that when the cooling air is forced through an electronic system, the total pressure is simply the sum of the velocity pressure and the static pressure.

$$H_t = H_v + H_s \quad \text{(ref. Eq. 6.3)}$$

6.23 Finned Cold Plates and Heat Exchangers

Cooling requirements for electronic equipment will sometimes specify the allowable pressure drop through the system in terms of the *static* pressure and sometimes in terms of the *total* pressure. Since the two pressures are different, it is necessary to understand the method for converting from one to another.

The static pressure drop through an electronic box is usually determined first. Section 6.4 shows how the static pressure losses are related to the velocity of the cooling air as it is being forced through the box.

After the static pressure losses have been determined for a box, it is necessary to compute the velocity pressure (or velocity head) for that box if the total pressure (or total head H_t) losses are required.

The velocity head H_v should be based upon the velocity of the cooling air in the most restricted flow section through the box, where the air velocity is the highest. These values can be computed from Eq. 6.2 or 6.4 for English units and Eq. 6.2a for metric units.

6.22 SAMPLE PROBLEM—TOTAL PRESSURE LOSS THROUGH AN ELECTRONIC BOX

An air flow analysis of an electronic box shows that the static pressure drop (H_s) through the box is 1.75 in of water. The most severe flow restriction through the box is at the inlet, where the cooling air velocity is 2200 ft/min. Determine the total pressure drop through the box.

SOLUTION

The velocity head at the air inlet is determined from Eq. 6.4.

$$H_v = \left(\frac{V}{4005}\right)^2 = \left(\frac{2200}{4005}\right)^2 = 0.302 \text{ in } H_2O \qquad (6.79)$$

The total pressure drop through the box is determined from Eq. 6.3.

$$H_t = H_v + H_s = 0.302 + 1.75 = 2.052 \text{ in } H_2O \qquad (6.80)$$

6.23 FINNED COLD PLATES AND HEAT EXCHANGERS

Cold plates and heat exchangers are often fabricated of aluminum, with many thin plate fins spaced close together. When the fins are spaced close together, there is a large increase in the forced convection heat transfer coefficient (h) and a large increase in the surface area (A) available for heat transfer. The product of $h_c \times A$ therefore increases very rapidly, which improves the thermal efficiency of the design. This results in a lower component hot spot temperature and improves the reliability of the electronic equipment [50].

There is a practical limit to the number of thin plate fins that can be used. In a dip brazed box, the maximum number of straight fins is about 22 fins per inch. More fins will often result in plugged passages because it is difficult to clean out the dip brazing salts, which may become trapped between the fins. If eutectic bonding or fluxless brazing is used, as many as 30 fins per inch can be used without plugging passages. Increasing the number of fins, however, will increase the pressure drop through the box, which can sharply reduce the cooling air flow rate and may result in overheating of the electronics.

Extensive test data [3] and analysis have shown that the forced convection heat transfer coefficient for multiple plate fin types of cold plates and heat exchangers can be determined with Eq. 6.9 [15, 43].

$$h_c = JC_p G \left(\frac{C_p \mu}{K}\right)^{-2/3} \quad \text{(ref. Eq. 6.9)}$$

The heat transfer coefficient J can be determined from the Reynolds number (N_R). For laminar flow conditions, with N_R between about 400 and 1500, the approximate J value for finned cold plates and heat exchangers is shown in Eq. 6.81, compiled from data given in [43].

$$J = \frac{0.72}{(N_R)^{0.7}} \quad (6.81)$$

The typical cross section through a multiple fin cold plate or heat exchangers is shown in Figure 6.33.

The cold plate or heat exchanger structure is often used for the side walls of electronic boxes. Printed circuit boards (PCBs) are then supported between the side walls. Copper or aluminum sheets are often laminated to the PCBs to provide a good thermal conduction heat flow path along the PCBs, as shown in Figure 4.1. The heat can then flow along the PCBs to the air cooled side walls of the chassis, which contain multiple fin heat exchangers, as shown in Figure 4.2.

Since the cooling medium used is air, it is important to bring the cooling air as close to the electronic components as possible. This will reduce the temperature rise and improve the thermal efficiency.

In many new packaging applications the cooling air is ducted directly to the circuit boards, which have hollow cores, as shown in Figure 6.21. For improved heat transfer efficiency, multiple fin heat exchangers are often added to the hollow core PCBs. The convection heat transfer coefficient, shown by Eqs. 6.9 and 6.81, can be used to determine the thermal charac-

6.24 Pressure Losses in Multiple Fin Heat Exchangers

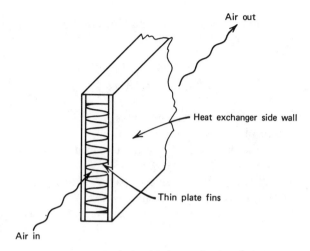

Figure 6.33 Typical cold plate or heat exchanger.

teristics of systems which have hollow core air passages through the center of the PCBs.

6.24 PRESSURE LOSSES IN MULTIPLE FIN HEAT EXCHANGERS

Multiple fin heat exchangers and cold plates require a substantial amount of energy to overcome the friction that resists the flow of cooling air through these units. When the fin density is high or when the surface finish on the fins is very rough, even more energy is required to force the air through the long, narrow tubes that are formed by the heat exchanger fins.

It is convenient to express the friction losses in terms of the height of a column of water H_L in inches (or centimeters) because the magnitudes of the numbers expressed in pounds per square inch (or grams per square centimeter) are so small.

Friction losses increase very rapidly as the velocity of the cooling air flow increases. However, the losses appear to be linear with respect to the length of the flow path through the heat exchanger.

The static head loss due to friction resulting from the flow through the multiple fin heat exchanger tubes only is shown in Eq. 6.82. Losses due to entrance resistance, turns, expansions, and contractions are not included in this equation. These other losses must be evaluated separately using techniques similar to those shown in Section 6.7, and then added to Eq. 6.82 to obtain the full static head loss for the electronic box. This is known as the Darcy Equation, and it is valid for English or metric units as long as a consistent set of units is used. This equation can be used for laminar and turbulent flow conditions [3, 15, 17, 43, 51].

$$H_L = 4f\left(\frac{L}{D}\right)\left(\frac{V^2}{2g}\right) \qquad (6.82)$$

where V = flow velocity = ft/sec or cm/sec
L = length of air flow path = in, ft, cm
D = hydraulic diameter of one duct = in, ft, cm
g = acceleration of gravity
g = 32.2 ft/sec^2 = 386 in/sec^2 = 980 cm/sec^2
f = friction factors, dimensionless
$f = 16/N_R$ = Fanning friction factor for laminar flow in round tubes (ref. Table 6.7)*

Equation 6.82 will result in the static pressure loss in terms of the height of a column of air. Sometimes this equation is written with the density (ρ) included, as shown in Eq. 6.83. When the density is included, it changes the units of the losses from a height dimension to a pressure dimension, lb/in^2 or lb/ft^2 for English units and g/cm^2 for metric units.

$$\Delta P_L = 4f\left(\frac{L}{D}\right)\left(\frac{\rho V^2}{2g}\right) \qquad (6.83)$$

Sometimes it is more convenient to obtain the static head loss in terms of $\sigma \Delta P$ directly in inches of water, as shown by Eq. 6.84. English units must be used with the constant of 0.226; then the weight flow W = lb/min and the cross section area A = in^2.

$$\sigma \Delta P = 0.226(4f)\left(\frac{L}{D}\right)\left(\frac{W}{A}\right)^2 \qquad (6.84)$$

where $\sigma \Delta P$ = static pressure loss = in H$_2$O
f = Fanning friction factor = dimensionless (ref. Table 6.7)
L = length of heat exchanger = in
D = hydraulic diameter, one duct = in
W = cooling air weight flow, one duct = lb/min
A = cross section flow area, one duct = in^2

The friction factor f is usually determined from test data for laminar and turbulent flow conditions and for different surface roughness. In the laminar flow region, for smooth wall surfaces, the friction factor can generally be

---

* Do not become confused with the Hagen–Poiseulle equation for laminar flow in round tubes, shown in Eqs. 10.3 and 10.4. In the Hagen–Poiseulle equation, the laminar flow friction factor $f = 64/N_R$. ↳ used primarily with liquids

6.25 Fin Efficiency Factor

Table 6.7 Laminar Flow Friction Factors (f) for Fully Developed Velocity and Temperature Profiles in Smooth Pipes with Different Cross Sections.

Geometry	Friction Factor
triangle	$f = \dfrac{13.3}{N_R}$
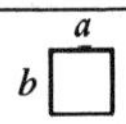 square, $b/a = 1$	$f = \dfrac{14.2}{N_R}$
circle	$f = \dfrac{16}{N_R}$
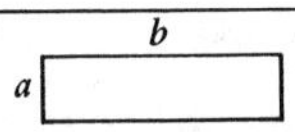 rectangle, $b/a = 4$	$f = \dfrac{18.3}{N_R}$
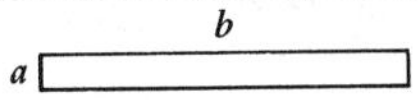 rectangle, $b/a = 8$	$f = \dfrac{20.5}{N_R}$
 parallel plates, $b/a = \infty$	$f = \dfrac{24}{N_R}$

used primarily with air (handwritten annotation next to circle row)

related directly to the Reynolds number (N_R). Table 6.7 shows some typical friction factors for different cross sections that have smooth walls [43]. From Kays & London, Compact Heat Exchangers, published by McGraw-Hill Book Co.

6.25 FIN EFFICIENCY FACTOR

Fins are used extensively with forced convection cooling systems to increase the amount of heat that is removed. The fins extend into the air stream to expose a large amount of surface area, which improves the heat transfer. One of the factors involved in the transfer is the temperature difference between the cooling air and the surface of the fin. More heat can be transferred when there is a large temperature difference.

Sometimes the fin is relatively long, so that there is a temperature gradient along the fin. The base of the fin may be substantially hotter than the tip of the fin, as shown in Figure 6.34.

The tip of the fin should be nearly as hot as the base of the fin, in order to transfer the maximum amount of heat. This requires a large fin thickness (δ), a short length (d), and a high thermal conductivity (K), to reduce the thermal gradient along the fin [53].

To compensate for the difference in the amount of heat that is transferred from a fin with a nonuniform temperature, a fin efficiency factor (η) is used. This fin efficiency is defined as the ratio of heat that is actually transferred from the fin to the heat that would be transferred at the maximum uniform

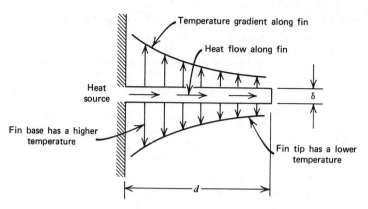

Figure 6.34 Heat flow from a cooling fin.

temperature condition, where the tip of the fin is at the same temperature as the base. The fin efficiency factor is shown in Eq. 5.21 [3, 15, 17].

$$\eta = \frac{\tanh md}{md} \quad \text{(ref. Eq. 5.21)}$$

where   tanh = hyperbolic tangent

$m = \sqrt{2h_c/K\delta}$
h_c = convection coefficient
K = fin thermal conductivity
δ = fin thickness
d = effective height of fin from base

Sometimes fins extend between two surfaces, which are both heat sources. When this condition exists, the height of the fin from the base is half of the total height, as shown in Figure 6.35.

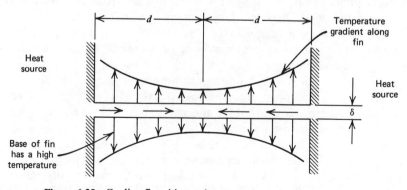

Figure 6.35 Cooling fin with two heat sources on opposite ends.

6.26 Sample Problem—Hollow Core PCB with a Finned Heat Exchanger 225

Figure 6.36 shows an electronic box with multiple fin air cooled heat exchangers in the side panels.

The fin efficiency for a single fin, as shown in Figure 6.35, can be determined from Figure 5.7 when the values of m and d are known.

The use of the heat exchanger convection coefficient equation, pressure drop equations, and fin efficiency equation can be demonstrated with the use of a sample problem.

6.26 SAMPLE PROBLEM—HOLLOW CORE PCB WITH A FINNED HEAT EXCHANGER

An electronic system has many discrete, hybrid, and LSI components that will dissipate a large amount of heat. The size of the electronic box will permit the use of several 4.0 in (10.16 cm) × 6.2 in (15.75 cm) plug-in PCBs. The system must be capable of operating in a class 1X environment, using forced convection cooling at sea level conditions, where the surrounding ambient air temperature is 131°F (55°C). A maximum steady state power

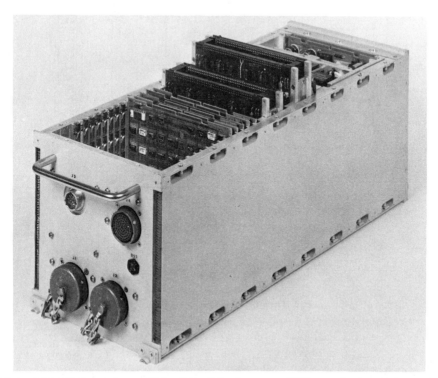

Figure 6.36 Electronic box with plug-in modules that are cooled by conduction of heat to air cooled epoxy bonded heat exchangers in the side walls. (Courtesy Litton Stystems, Inc.)

dissipation of 50 watts is expected for each PCB under certain conditions, with a relatively uniform heat distribution across the component mounting surface. The maximum allowable component mounting surface temperature is 212°F (100°C). Because of the high power dissipation, a hollow core construction is proposed for the PCBs.

Determine the following:

1. The minimum cooling air weight flow required through each PCB.
2. The heat transfer convection coefficient for a multiple fin heat exchanger at the center of the PCB.
3. The maximum component case or surface mounting temperature.
4. The static pressure drop through the PCB heat exchanger.

A sketch of the proposed PCB is shown in Figure 6.37.

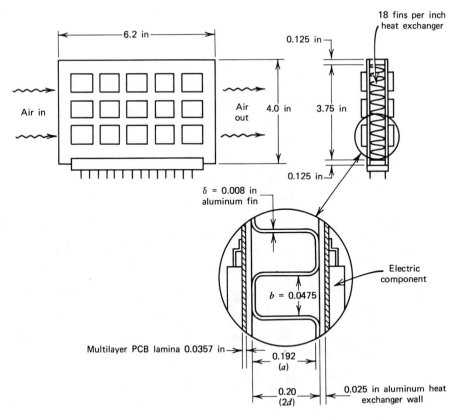

Figure 6.37 Hollow core PCB with a dip brazed aluminum heat exchanger at the center and electronic components mounted on both sides.

6.26 Sample Problem—Hollow Core PCB with a Finned Heat Exchanger

SOLUTION FOR PART 1

An efficient cooling system can be designed by providing an exit air temperature of 160°F (71.1°C). The information required is as follows (see Section 6.17):

$Q = 50$ watts $= 170.6$ Btu/hr $= 11.95$ cal/sec (power dissipation)

$C_p = 0.24$ Btu/lb °F $= 0.24$ cal/g °C (specific heat)

$$\Delta t = 160°F - 131°F = 29°F \quad \text{or} \quad 71.1°C - 55°C = 16.1°C \tag{6.85}$$

Substitute into Eq. 6.7 to determine the minimum required cooling air weight flow through each PCB in English units.

$$W_{min} = \frac{170.6 \text{ Btu/hr}}{(29°F)(0.24 \text{ Btu/lb °F})(60 \text{ min/hr})} = 0.408 \frac{\text{lb}}{\text{min}} \tag{6.86}$$

In metric units:

$$W_{min} = \frac{11.95 \text{ cal/sec}}{(16.1°C)(0.24 \text{ cal/g °C})} = 3.09 \frac{\text{g}}{\text{sec}} \tag{6.86a}$$

An extra 10% should be added to the foregoing cooling air flow rates to cover the possibility of cooling air leakage. A good seal is often difficult to maintain for a long period of time in a severe service environment.

SOLUTION FOR PART 2

The forced convection coefficient for the PCB heat exchanger shown in Figure 6.37 is determined from Eq. 6.9, and the J factor is determined from Eq. 6.81. The Reynolds number (N_R) is also required, and it can be obtained with the use of Eq. 6.11. But first, the cooling air flow through one air duct must be determined.

$$\text{number of ducts} = 18 \text{ fins/in} \times 3.75 \text{ inch} = 67.5 \text{ ducts}$$

Air flow through one duct, in English units:

$$W = \frac{0.408 \text{ lb/min}}{67.5 \text{ ducts}} = 0.00604 \frac{\text{lb}}{\text{min}} \text{ per duct}$$

Air flow through one duct, in metric units:

$$W = \frac{3.09 \text{ g/sec}}{67.5 \text{ ducts}} = 0.0458 \frac{\text{g}}{\text{sec}} \text{ per duct}$$

Hydraulic diameter of one duct ($a = 0.192$ in, $b = 0.0475$ in):

$$D = \frac{4 \times \text{area}}{\text{perimeter}} = \frac{4ab}{2a + 2b} = \frac{2ab}{a + b} \qquad (6.87)$$

$$D = \frac{2(0.192)(0.0475)}{0.192 + 0.0475} = 0.0761 \text{ in} = 0.00635 \text{ ft} = 0.193 \text{ cm}$$

Cross section area of one duct:

$$A = ab = (0.192)(0.0475) = 0.00912 \text{ in}^2 = 0.0000633 \text{ ft}^2 = 0.0588 \text{ cm}^2$$

The weight velocity flow through one duct is determined from Eq. 6.12.

$$G = \frac{0.00604 \text{ lb/min} \times 60 \text{ min/hr}}{0.0000633 \text{ ft}^2} = 5725 \frac{\text{lb}}{\text{hr ft}^2} \text{ (English units)}$$

$$G = \frac{0.0458 \text{ g/sec}}{0.0588 \text{ cm}^2} = 0.779 \frac{\text{g}}{\text{sec cm}^2} \text{ (metric units)}$$

The average cooling air temperature is:

$$t_{av} = \frac{t_{in} + t_{out}}{2} = \frac{131°F + 160°F}{2} = 145.5°F = 63.0°C$$

The physical properties of air (ref. Figure 6.28; conversion between English units and metric units is obtained from Section 1.11) [7, 8, 37, 38, 52].

$K = 0.017$ Btu/hr ft °F $= 0.0000702$ cal/sec cm °C (conductivity)

$C_p = 0.24$ Btu/lb °F $= 0.24$ cal/g °C (specific heat)

$\mu = 0.050$ lb/ft hr $= 0.000206$ g/cm sec (viscosity)

$\rho = 0.0656$ lb/ft^3 $= 0.00105$ g/cm^3 (density)

Substitute into Eq. 6.11 for the Reynolds number. In English units:

$$N_R = \frac{(5725 \text{ lb/hr ft}^2)(0.00635 \text{ ft})}{0.050 \text{ lb/ft hr}} = 727 \text{ (dimensionless)} \qquad (6.88)$$

In metric units:

$$N_R = \frac{(0.779 \text{ g/sec cm}^2)(0.193 \text{ cm})}{0.000206 \text{ g/cm sec}} = 729 \text{ (dimensionless)} \qquad (6.88a)$$

Substitute into Eq. 6.81 for the Colburn J factor.

6.26 Sample Problem—Hollow Core PCB with a Finned Heat Exchanger

$$J = \frac{0.72}{(N_R)^{0.7}} = \frac{0.72}{(727)^{0.7}} = 0.00715 \text{ (dimensionless)}$$

Substitute into Eq. 6.9 for the heat exchanger forced convection coefficient, in English units.

$$h_c = \frac{(0.00715)(0.24)(5725)}{[(0.24)(0.050)/0.017]^{0.666}} = 12.4 \frac{\text{Btu}}{\text{hr ft}^2 \, °F} \tag{6.89}$$

In metric units:

$$h_c = \frac{(0.00715)(0.24)(0.779)}{[(0.24)(0.000206)/0.0000702]^{0.666}} = 0.00168 \frac{\text{cal}}{\text{sec cm}^2 \, °C} \tag{6.89a}$$

The forced convection coefficient will be slightly higher at the cooling air entrance section of the heat exchanger because of the extra turbulence that is developed. This small increase is conservatively ignored here.

The fin efficiency factor shown in Eq. 5.21 must be considered because there will be a temperature gradient along the heat exchanger fins. Since there are electronic components mounted on both sides of the PCB, there will be a heat input to the heat exchanger fins from two different surfaces, as shown in Figure 6.35.

Given $\delta = 0.008$ in $= 0.000666$ ft $= 0.0203$ cm (fin thickness)
$K = 99$ Btu/hr ft °F $= 0.409$ cal/sec cm °C 6061 (aluminum fin)
$d = 0.20$ in/2 $= 0.10$ in $= 0.00833$ ft $= 0.254$ cm (note, heat in from both sides)
$h_c = 12.4$ Btu/hr ft² °F $= 0.00168$ cal/sec cm² °C (convection coeff.)

Substitute into Eq. 5.21 for English units, with dimensions in feet.

$$m = \sqrt{\frac{2(12.4)}{99(0.000666)}} = 19.4$$

$$md = 19.4(0.00833) = 0.1616$$

In metric units:

$$m = \sqrt{\frac{2(0.00168)}{(0.409)(0.0203)}} = 0.636$$

$$md = (0.636)(0.254) = 0.1615$$

Substitute into Eq. 5.21 for fin efficiency.

$$\eta = \frac{\tanh 0.1615}{0.1615} = \frac{0.16011}{0.1615} = 99.1\%$$

The corrected value of the forced convection coefficient in English units is

$$h_c = 12.4(0.991) = 12.3 \frac{\text{Btu}}{\text{hr ft}^2 \, ^\circ\text{F}} \tag{6.90}$$

In metric units:

$$h_c = 0.00168(0.991) = 0.00166 \frac{\text{cal}}{\text{sec cm}^2 \, ^\circ\text{C}} \tag{6.90a}$$

The value of the forced convection coefficient (h_c) is sharply influenced by the hydraulic diameter in the laminar flow regions. With low values of cooling air weightflow velocities, the convection coefficient for a given hydraulic diameter does not show much change. As the hydraulic diameter decreases, the forced convection coefficient increases rapidly, as shown in Figure 6.38. As the fluid flow increases and the turbulent region is approached, the convection coefficient increases very rapidly [70].

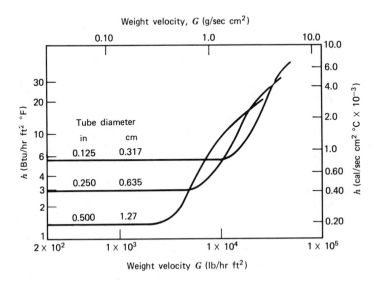

Figure 6.38 Variation of forced convection coefficient with hydraulic diameter and flow rate. (Ref: General Electric Heat Transfer Data Book 1975)

6.26 Sample Problem—Hollow Core PCB with a Finned Heat Exchanger

SOLUTION FOR PART 3

The maximum component case temperatures will occur on the components that are mounted closest to the cooling air exhaust end of the heat exchanger.

A large number of interconnections are required for the circuits, so multilayer circuit laminars will be required. Interconnections between layers are made with plated throughholes, which are filled with solder or epoxy to improve the mechanical strength of the interconnection.

The thickness of the multilayer circuit lamina from the component to the heat exchanger will be about 0.0357 in (0.0907 cm). This thickness must be closely controlled because the component heat will flow through the multilayer circuit lamina to the air cooled heat exchanger at the center of the PCB. A thicker circuit lamina will have a higher thermal resistance, which will increase the temperature rise across the lamina.

Plated throughholes will tend to reduce the thermal resistance through the circuit lamina. An enlarged view through the circuit lamina, showing the heat flow path, is shown in Figure 6.39.

Some of the component heat may be lost by external convection and radiation, but this is conservatively ignored because the PCBs face each other and are close enough to restrict natural convection. It is therefore assumed that all the heat is transferred from the components to the cooling air by conduction through the PCB.

The heat flow path from the component surface (or case) to the cooling

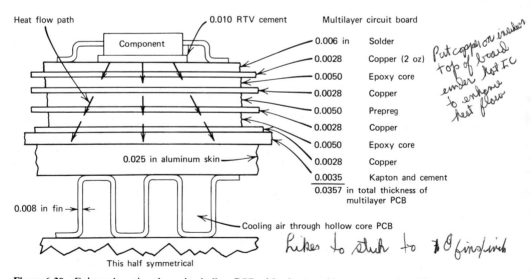

Figure 6.39 Enlarged section through a hollow PCB with a heat exchanger core and multilayer circuit laminas on both sides.

air in the hollow core heat exchanger is evaluated by determining the temperature rise along four segments of the heat flow path as follows:

Δt_1 = temperature rise across component interface, from case to circuit lamina

Δt_2 = temperature rise through the multilayer circuit lamina to the aluminum heat exchanger skin

Δt_3 = temperature rise across the convection air film, from the aluminum heat exchanger surface to the center of cooling air stream

Δt_4 = temperature rise of the cooling air stream from the entrance of the heat exchanger to the exit

Each section along the heat flow path is examined in more detail.

Δt_1: temperature rise across component interface from case to circuit lamina

The electronic types being examined are hybrid and large scale integrated (LSI) components. These usually come in relatively large case sizes, approximately 1.0 in (2.54 cm) × 1.0 in (2.54 cm). These components will dissipate a large amount of heat, since the total power dissipation on the PCB is 50 watts. When the power density on such a component is more than about 4 watts/in², a good thermally conductive cement should be used to prevent an excessive temperature rise across the component interface as shown in Figure 6.40. Room temperature vulcanizing (RTV) cement is often used to permit the component to be removed if it has to be replaced. Epoxy cements are available with higher thermal conductivities, but components bonded with this material are very difficult to replace without damaging the PCB.

Test data show that some heat is conducted from the body of the hybrid component to the electrical lead wires. This heat flows along the lead wires to the circuit lamina solder pads, then through the circuit lamina into the heat exchanger.

The heat loss through the lead wires is usually ignored because the strain relief produces a long lead length with a high resistance, and there is another

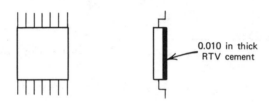

Figure 6.40 Hybrid component fastened to PCB with RTV cement.

6.26 Sample Problem—Hollow Core PCB with a Finned Heat Exchanger

high resistance through the small cross section area through the circuit lamina.

Considering a maximum power dissipation of 4 watts for one hybrid, the temperature rise across an RTV interface can be determined from Eq. 3.1. See Section 1.11 for conversions between English units and metric units.

Given $Q = 4.0$ watts $= 13.65$ Btu/hr $= 0.956$ cal/sec (heat flow)
$L = 0.010$ in $= 0.000833$ ft $= 0.0254$ cm (RTV thickness)
$K = 0.22$ Btu/hr ft °F $= 0.000908$ cal/sec cm °C (RTV conductivity)

$$A = \frac{(1.0)(1.0) \text{ in}^2}{144 \text{ in}^2/\text{ft}^2} = 0.00694 \text{ ft}^2 = 6.45 \text{ cm}^2 \text{ (RTV area)}$$

In English units, with dimensions in feet:

$$\Delta t_1 = \frac{(13.65)(0.000833)}{(0.22)(0.00694)} = 7.4°F \qquad (6.91)$$

In metric units:

$$\Delta t_1 = \frac{(0.956)(0.0254)}{(0.000908)(6.45)} = 4.1°C \qquad (6.91a)$$

Δt_2: temperature rise through the multilayer circuit lamina to the aluminum heat exchanger skin

The multilayer circuit lamina consists of 0.005 in thick epoxy fiberglass sheets, with etched copper circuits on each side. Two of these 0.005 in sheets are bonded together with a prepreg epoxy fiberglass cement using high temperatures and pressures to force the prepreg to flow between the copper runs, eliminating any air bubbles. Holes for interconnecting circuits are drilled through the bonded assembly, which is then plated with a thin copper layer. The plating forms an electrically conducting tube through the drilled holes and also builds up the thickness of the copper runs on both sides of the bonded lamina assembly. The original copper thickness on the lamina is usually 0.0014 in or about 1 ounce. The plating process usually adds about another ounce, which brings the total thickness of the copper runs on the outer surfaces to 0.0028 in.

The copper-plated throughholes (PTH) tend to increase the effective thermal conductivity through the epoxy fiberglass laminas, because copper has a very high thermal conductivity. Typical values are about 20%, depending upon the number of PTHs used.

In addition, the copper circuit layers within the bonded lamina will tend to spread the heat flowing through the lamina if there are many copper circuit runs. This tends to increase the effective cross section area along the heat flow path through the circuit lamina about 15%, depending upon the number of copper circuit runs.

The temperature rise through the circuit lamina is also determined from Eq. 3.1.

Given $Q = 4$ watts $= 13.65$ Btu/hr $= 0.956$ cal/sec (heat)
$L = 0.0357$ in $= 0.00297$ ft $= 0.0907$ cm (thickness of circuit lamina)
$K = 0.20$ Btu/hr ft °F $= 0.000826$ cal/sec cm °C (conductivity with no PTHs)
$K = 0.24$ Btu/hr ft °F $= 0.000991$ cal/sec cm °c (conductivity with PTHs)
$A = (1.0)(1.0) = 1.0$ in² $= 0.00694$ ft² $= 6.452$ cm² (area with no heat spreading)
$A = 1.15$ in² $= 0.00798$ ft² $= 7.419$ cm² (area with heat spreading)

Conservatively ignore the improved conductivity of the PTHs and the spreading effects of the copper runs. In English units:

$$\Delta t_2 = \frac{(13.65)(0.00297)}{(0.20)(0.00694)} = 29.2°F \tag{6.92}$$

In metric units:

$$\Delta t_2 = \frac{(0.956)(0.0907)}{(0.000826)(6.452)} = 16.3°C \tag{6.92a}$$

Δt_3: temperature rise across the convection air film from the aluminum heat exchanger surface to the center of the cooling air stream

Equation 5.7 is used to determine this temperature rise. However, first the surface area of the heat exchanger must be calculated. If the total power dissipation is considered, the total heat exchanger surface area must be used. It is convenient to consider the solid height of the fins to compute the side wall area, as shown in Figure 6.41.

"compress fins, subtract from total duct area.

fin solid height $= (18$ fins/in$)(3.75$ in$)(0.008$ in$) = 0.54$ inch

fin area $= (67.5$ fins$)(0.192$ in$)(6.2$ in$)(2$ surfaces$)$

fin area $= 160.7$ in² $= 1.116$ ft² $= 1036.8$ cm²

sidewall area $= (3.21$ in$)(6.2$ in$)(2$ surfaces$)$

sidewall area $= 39.8$ in² $= 0.276$ ft² $= 256.8$ cm²

total surface area $A = 1.392$ ft² $= 1293.6$ cm²

6.26 Sample Problem—Hollow Core PCB with a Finned Heat Exchanger

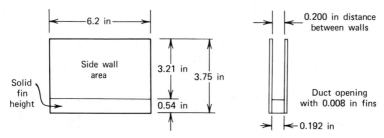

Figure 6.41 Heat exchanger fins shown as a solid height.

Heat:

$$Q = 50 \text{ watts} = 170.65 \frac{\text{Btu}}{\text{hr}} = 11.95 \frac{\text{cal}}{\text{sec}} \text{ (total power)}$$

$$h_c = 12.3 \frac{\text{Btu}}{\text{hr ft}^2 \text{°F}} = 0.00166 \frac{\text{cal}}{\text{sec cm}^2 \text{°C}} \text{ (Ref. Eqs. 6.90 and 6.90a)}$$

For English units, with dimensions in feet:

$$\Delta t_3 = \frac{Q}{h_c A} = \frac{170.65}{(12.3)(1.392)} = 10.0 \text{°F} \qquad (6.93)$$

In metric units:

$$\Delta t_3 = \frac{11.95}{(0.00166)(1293.6)} = 5.6 \text{°C} \qquad (6.93\text{a})$$

Δt_4: temperature rise of the cooling air from the entrance of the heat exchanger to the exit

As the cooling air flows through the PCB heat exchanger, it will pick up the heat dissipated from the electronic components. The temperature rise of the cooling air is determined by considering the total PCB power dissipation with Eq. 6.7.

Given $Q = 50$ watts $= 170.65$ Btu/hr $= 11.95$ cal/sec (total power)
$W = 0.408$ lb/min $= 3.09$ g/sec (ref. Eqs. 6.86 and 6.86a)
$C_p = 0.24$ Btu/lb °F $= 0.24$ cal/g °C (specific heat)

In English units:

$$\Delta t_4 = \frac{Q}{W C_p} = \frac{170.65 \text{ Btu/hr}}{(0.408 \text{ lb/min})(60 \text{ min/hr})(0.24 \text{ Btu/lb °F})} = 29.0 \text{°F} \qquad (6.94)$$

In metric units:

$$\Delta t_4 = \frac{11.95 \text{ cal/sec}}{(3.09 \text{ g/sec})(0.24 \text{ cal/g °C})} = 16.1°C \qquad (6.94a)$$

The maximum component case temperature is obtained by adding all of the temperature rises Δt_1 through Δt_4 to the inlet air temperature of 131°F (55°C).

$$t_{case} = t_{in} + \Delta t_1 + \Delta t_2 + \Delta t_3 + \Delta t_4 \qquad (6.95)$$

For English units, substitute Eqs. 6.91 through 6.94 into Eq. 6.95.

$$t_{case} = 131 + 7.4 + 29.2 + 10.0 + 29.0 = 206.6°F \qquad (6.96)$$

For metric units, substitute Eqs. 6.91a through 6.94a into Eq. 6.95.

$$t_{case} = 55 + 4.1 + 16.3 + 5.6 + 16.1 = 97.1°C \qquad (6.96a)$$

SOLUTION PART 4

The static pressure drop through the PCB heat exchanger only is desired from Eq. 6.82, in terms of the height of a column of water.

The friction factor is obtained from Table 6.7, considering the cross section of one tube of the heat exchanger to be a rectangular duct with an aspect ratio of 4. Using the Reynolds number computed in Eqs. 6.88 and 6.88a, the friction in English units and metric units is shown in Eq. 6.97.

$$f = \frac{18.3}{N_R} = \frac{18.3}{727} = 0.0252 \text{ (dimensionless)} \qquad (6.97)$$

The cooling air velocity (V) through the heat exchanger is shown in Eq. 6.98. It is obtained from the physical properties of the air, which were previously determined for the Reynolds number shown in Eqs. 6.88 and 6.88a. The area (A) of one duct was shown in Eq. 6.87.

$$V = \frac{W}{\rho A} \qquad (6.98)$$

In English units:

$$V = \frac{0.00604 \text{ lb/min}}{(0.0656 \text{ lb/ft}^3)(0.0000633 \text{ ft}^2)(60 \text{ sec/min})}$$

$$= 24.2 \frac{\text{ft}}{\text{sec}} = 290.4 \frac{\text{in}}{\text{sec}} \qquad (6.99)$$

6.26 Sample Problem—Hollow Core PCB with a Finned Heat Exchanger

In metric units:

$$V = \frac{0.0458 \text{ g/sec}}{(0.00105 \text{ g/cm}^3)(0.0588 \text{ cm}^2)} = 741 \frac{\text{cm}}{\text{sec}} \tag{6.99a}$$

Also:

$L = 6.2$ in $= 0.516$ ft $= 15.7$ cm (length)

$D = 0.0761$ in $= 0.00635$ ft

$\quad = 0.193$ cm (ref. Eq. 6.87) (hydraulic diameter)

$g = 32.16$ ft/sec² $= 386$ in/sec² $= 980$ cm/sec² (gravity)

A dimensional analysis shows that Eq. 6.82 will produce the static pressure loss in terms of the height of a column of air. The pressure loss is desired in terms of the height of a column of water. Therefore, the ratio of the air density (ρ_{air}) and the water density (ρ_{H_2O}) is used for the conversion, as shown in Eq. 6.100.

$$H_L = (4f)\left(\frac{L}{D}\right)\left(\frac{V^2}{2g}\right)\left(\frac{\rho_{air}}{\rho_{H_2O}}\right) \tag{6.100}$$

For English units, with dimensions in inches:

$$H_L = (4)(0.0252)\left(\frac{6.2}{0.0761}\right)\frac{(290.4)^2}{(2)(386)}\frac{(0.0656 \text{ lb/ft}^3 \text{ air})}{(62.4 \text{ lb/ft}^3 \text{ H}_2\text{O})}$$

$$= 0.94 \text{ in H}_2\text{O} \tag{6.101}$$

In metric units:

$$H_L = (4)(0.0252)\left(\frac{15.7}{0.193}\right)\frac{(741)^2}{(2)(980)}\frac{(0.00105 \text{ g/cm}^3 \text{ air})}{(1.0 \text{ g/cm}^3 \text{ H}_2\text{O})}$$

$$= 2.39 \text{ cm H}_2\text{O} \tag{6.101a}$$

The foregoing static pressure losses are only for the heat exchanger. If the complete static pressure loss through an electronic box is desired, other flow losses, such as entrance effects, turns, expansions contraction, and exhaust effects, must be added to obtain the final picture, as discussed in Section 6.7.

If the total pressure is required for the heat exchanger section, the velocity head must be added to the static head as shown in Eq. 6.3. The velocity head relation for English units is shown in Eqs. 6.2 and 6.4, where the velocity must be in ft/min. The velocity head in English units becomes:

$$H_v = \left[\frac{(24.2 \text{ ft/sec})(60 \text{ sec/min})}{4005}\right]^2 = 0.13 \text{ in H}_2\text{O} \quad (6.102)$$

The velocity head in metric units is obtained from Eq. 6.5a as follows:

$$H_v = \left(\frac{741}{1277}\right)^2 = 0.33 \text{ cm H}_2\text{O} \quad (6.102a)$$

The total pressure loss through the heat exchanger, in English units, becomes

$$H_t = H_v + H_L = 0.13 + 0.94 = 1.07 \text{ in H}_2\text{O} \quad (6.103)$$

In metric units:

$$H_t = 0.33 + 2.39 = 2.72 \text{ cm H}_2\text{O} \quad (6.103a)$$

Equation 6.84 could also have been used to determine the static pressure loss in terms of $\sigma \Delta P$.

$$\sigma \Delta P = 0.226(4f)\left(\frac{L}{D}\right)\left(\frac{W}{A}\right)^2 \quad (\text{ref. Eq. 6.84})$$

where $W = \dfrac{0.408 \text{ lb/min}}{67.5 \text{ ducts}} = 0.00604$ lb/min (ref. Eq. 6.86)

$L = 6.2$ in $=$ length of duct
$D = 0.0761$ in $=$ hydraulic diameter
$A = (0.192)(0.0475) = 0.00912$ in² (ref. Eq. 6.87)
$f = 0.0252$ (ref. Eq. 6.97)

$$\sigma \Delta P = 0.226(4)(0.0252)\left(\frac{6.2}{0.0761}\right)\left(\frac{0.00604}{0.00912}\right)^2$$

$\sigma \Delta P = 0.813$ in H$_2$O

$$\sigma = \frac{\rho}{\rho_{\text{std}}} = \frac{0.0656}{0.0765} = 0.858 \quad (\text{ref. Eq. 6.65 and Eq. 6.98})$$

$$\Delta P = \frac{\sigma \Delta P}{\sigma} = \frac{0.813}{0.858} = 0.94 \text{ in H}_2\text{O}$$

This checks with Eq. 6.101.

6.27 Undesirable Air Flow Reversals

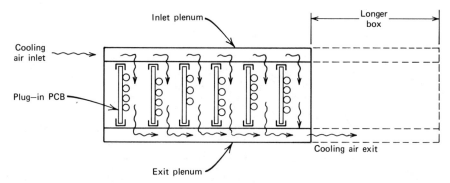

Figure 6.42 Parallel air flow system with an inlet and an exit cooling air plenum.

6.27 UNDESIRABLE AIR FLOW REVERSALS

Under certain sets of conditions it is possible for the air in a forced convection system to reverse its flow direction. If this happens, the cooling efficiency can be sharply reduced and overheating of the electronics may result.

This type of situation can develop in a parallel flow cooling system which incorporates a cooling air distribution plenum as shown in Figure 6.42.

The purpose of an inlet air distribution plenum is to provide a uniform air flow to the PCBs in the system. Ideally, the static pressure will be high and the velocity pressure will be low. A high static pressure will distribute the air equally to all of the PCBs. A low velocity pressure means a low velocity, which results in small entrance and turn losses.

If the box length is increased and the number of PCBs in the system are increased, or if the power dissipation in the system is substantially increased, additional cooling air will be required. If the cross section flow area of the air inlet plenum is not increased, the velocity of the cooling air

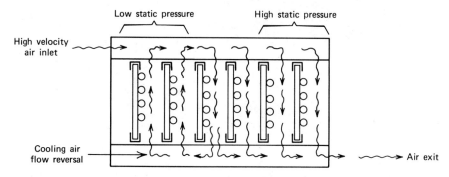

Figure 6.43 Air flow reversals may occur if the cooling air inlet velocity is too high.

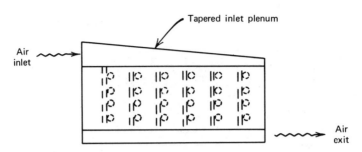

Figure 6.44 Tapered air inlet plenum on an electronic box.

flow must be increased to provide the additional air. As the velocity is increased, the static pressure will decrease. A high air velocity can result in such a low static pressure at the air inlet that it is well below the outside atmospheric pressure. When this happens, a flow reversal may occur. The air will then be drawn in from the exit plenum, as shown in Figure 6.43.

To prevent flow reversal, the inlet air plenum cross section area should

Figure 6.45 Electronic components are cemented to an aluminum core board with RTV cement to improve the conduction heat transfer. (Courtesy Kearfott Division, The Singer Company.)

6.27 Undesirable Air Flow Reversals 241

Figure 6.46 Inertial navigation assembly with a mounting rack that has a cooling air plenum in its base. (Courtesy Kearfott Division, The Singer Company.)

be greater than the sum of all the individual PCB inlet ports used to supply cooling air to the PCBs.

A tapered inlet plenum can also be used for the cooling air, as shown in Figure 6.44. The taper should be gradual so that the cross section area of the plenum at any point is greater than the total cross section area of the downstream air inlet ports to the PCBs.

Figure 6.45 shows a circuit board with many discrete and hybrid components held in position with RTV cement.

Figure 6.46 shows a mounting rack and an inertial navigation system that is packaged in a $\frac{3}{4}$ ATR short chassis.

7

Cooling Minicomputers, Microcomputers, and Microprocessors

7.1 INTRODUCTION

The market for small computers has grown so rapidly that some companies are calling it an explosion. The number of different types of electronic systems that can be assembled for different applications is staggering. Virtually every major industry in the world is turning to some form of small computer to assist its manufacturing, processing, and data handling. Steel mills, automobiles, washing machines, television, sewing machines, and microwave ovens are just a few examples of areas that are turning to the new, small computers.

Changes are occurring very rapidly in the field of small computers. As advances were made in microcomputers, they grew rapidly in capacity until they are more powerful than some of their bigger brothers, the minicomputers.

The heart of the small computers is the microprocessor, which has often been defined as a "computer on a chip." The microprocessor is really an LSI component. It incorporates almost all of the functions of the traditional computer system in one large DIP type of package, as shown in Figure 7.1.

7.2 MINICOMPUTER SYSTEMS

A typical minicomputer is usually a small chassis that will sit on the top of a desk. It is about the size of one filing cabinet drawer. Electronic components are mounted on one side of plug-in types of epoxy fiberglass printed circuit boards (PCBs). Dual inline package (DIP) components are very

7.2 Minicomputer Systems

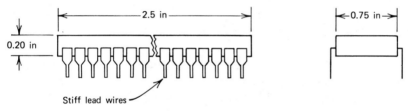

Figure 7.1 Typical microprocessor component.

popular because they can often be plugged right into the PCB without soldering.

Formed sheet metal chassis are usually used to enclose the electronics. In high production systems, die-cast and investment-cast aluminum partitions are very common.

Low cost PCB edge type connectors are often used for hobby and some commercial applications, which will be used in temperature and humidity controlled atmospheres. Pin and socket types of PCB connectors are usually used in more severe environments, where temperatures can vary over a wide range and where high humidity may result in moisture condensation.

Cathode ray tubes (CRTs) are often a part of the display system, together with a typewriter keyboard that is used for data input. The video screen provides an instant display of the input, which can be revised or printed as desired.

Access panels are usually provided at the top or the front of the minicomputers to remove or replace PCBs, which are plugged into a master interconnecting or mother board. Wire wrapped interconnections are quite popular for mother boards. This type of harness can be assembled rapidly with an extremely reliable gastight connection. When changes are required, they can be made quickly and easily. Pins can be wrapped and rewrapped many times without reducing the high reliability of the connection.

Most of the PCBs are oriented vertically in the minicomputer chassis, with access through top or front covers. When floppy disk memories are used, the PCB is often mounted in a horizontal position immediately above the floppy disk and drive mechanism.

Business machine minicomputers and minicomputers used in severe thermal environments will usually have a small, quiet cooling air exhaust fan mounted at the side or rear of the chassis. Small, inexpensive fans with a life of 100,000 hr are readily available from several fan manufacturers to improve the reliability. This type of system is usually completely enclosed, with only air inlet louvers or perforations on the sides of the chassis. Louvers and perforations should not be placed in the top covers of the chassis because dirt and dust can enter and coffee can be spilled into the electronics.

7.3 SAMPLE PROBLEM—PCB MOUNTED ABOVE MINICOMPUTER FLOPPY DISK

A minicomputer with a floppy disk memory requires an 8 in (20.32 cm) × 6 in (15.24 cm) flow soldered circuit board for the electronic components. For good maintenance, it is desirable to mount the PCB in a horizontal position just under the top cover and just above the disk drive mechanism, as shown in Figure 7.2. The maximum ambient temperature expected is about 110°F (43.3°C). The maximum allowable component surface temperature for the commercial grade components is 185°F (85°C). The heat is uniformly distributed across the surface of the PCB and is expected to be about 0.2 watt/in² (0.031 watt/cm²) for a total power dissipation of 9.6 watts. To reduce outside electrical interferences, the PCB will be completely enclosed. The inside and outside surfaces of the cover are painted. Determine if the design is satisfactory without a fan.

SOLUTION

Heat from the PCB must pass through the top cover before it can be transferred to the outside ambient. Since the inside and the outside surfaces of the cover are painted, radiation heat transfer can be considered along with natural convection. A mathematical model of the system is shown in Figure 7.3. The view factor from the PCB to the top cover will be high because the PCB is completely enclosed.

The PCB acts like a heated plate facing upward (shown in Figure 5.2). The inside surface of the cover will pick up heat by convection from the PCB, so that Eq. 5.5 is used to determine the convection coefficient. However, since the temperature rise is not known yet, a convection coefficient is assumed and its value is checked later. Experience with this type of geometry almost always results in a natural convection coefficient less than 1.0. A value of 0.8 Btu/hr ft² °F is assumed to start, and Eq. 5.28 is used to

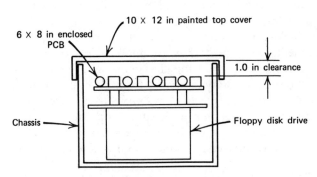

Figure 7.2 Cross section through a minicomputer with a mini floppy disk drive and a PCB.

7.3 Sample Problem—PCB Mounted Above Minicomputer Floppy Disk

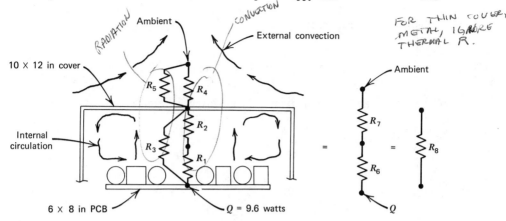

Figure 7.3 Mathematical model of a PCB and its cover.

determine the values of the thermal resistors. Considering the PCB first:

$$h_c = 0.8 \text{ Btu/hr ft}^2 \text{ °F (assumed to start)}$$

$$A = (6)(8) = 48 \text{ in}^2 = 0.333 \text{ ft}^2 \text{ (area)}$$

Substitute into Eq. 5.28.

$$R_1 = \frac{1}{h_c A} = \frac{1}{(0.8)(0.333)} = 3.75 \frac{\text{hr °F}}{\text{Btu}} \qquad (7.1)$$

For the inside surface of the top cover:

$$h_c = 0.8 \frac{\text{Btu}}{\text{hr ft}^2 \text{ °F}} \text{ (assumed to start)}$$

$$A = (10)(12) = 120 \text{ in}^2 = 0.833 \text{ ft}^2$$

$$R_2 = \frac{1}{h_c A} = \frac{1}{(0.8)(0.833)} = 1.50 \frac{\text{hr °F}}{\text{Btu}} \qquad (7.2)$$

The radiation resistance R_3 from the PCB to the cover acts in parallel with the convection resistances R_1 and R_2. Its value can be determined from Figure 5.29. Since the temperature difference from the PCB to the cover is not yet known, a radiation coefficient is assumed and its value is checked later. Experience has shown that the radiation coefficient h_r is usually higher than the convection coefficient h_c. A radiation coefficient of 1.5 is assumed to start.

$$h_r = 1.5 \text{ Btu/hr ft}^2 \text{ °F (assumed to start)}$$

$$A = (6)(8) = 48 \text{ in}^2 = 0.333 \text{ ft}^2 \text{ (area)}$$

Substitute into Eq. 5.28.

$$R_3 = \frac{1}{h_r A} = \frac{1}{(1.5)(0.333)} = 2.0 \frac{\text{hr °F}}{\text{Btu}} \quad (7.3)$$

Combine resistors R_1, R_2, and R_3 as shown in Figure 7.3.

$$\frac{1}{R_6} = \frac{1}{R_1 + R_2} + \frac{1}{R_3}$$

$$\frac{1}{R_6} = \frac{1}{3.75 + 1.50} + \frac{1}{2.0}$$

$$R_6 = 1.45 \frac{\text{hr °F}}{\text{Btu}} \quad (7.4)$$

Convection resistance R_4 and radiation resistance R_5 for the top cover act in parallel, with the same surface area, so that they can be combined directly using a form of Eq. 5.79 as follows:

$$R_7 = \frac{1}{(h_c + h_r)A} \quad (7.5)$$

Again, the values of h_c and h_r are not known, so values are assumed to start, and then checked later.

$$h_c = 0.6 \frac{\text{Btu}}{\text{hr ft}^2 \text{ °F}} \text{ (assumed to start)}$$

$$h_r = 1.3 \frac{\text{Btu}}{\text{hr ft}^2 \text{ °F}} \text{ (assumed to start)}$$

$$A = (10)(12) = 120 \text{ in}^2 = 0.833 \text{ ft}^2$$

$$R_7 = \frac{1}{(0.6 + 1.3)(0.833)} = 0.63 \frac{\text{hr °F}}{\text{Btu}} \quad (7.6)$$

Resistors R_6 and R_7 are in series:

$$R_8 = R_6 + R_7 = 1.45 + 0.63 = 2.08 \frac{\text{hr °F}}{\text{Btu}} \quad (7.7)$$

The overall temperature rise from the outside ambient to the surface of the PCB is determined from Eq. 5.7.

$$\Delta t_1 = Q R_8 \quad (7.8)$$

7.3 Sample Problem—PCB Mounted Above Minicomputer Floppy Disk

where $Q = (9.6 \text{ watts})(3.413 \text{ Btu/hr/watt}) = 32.76 \text{ Btu/hr}$

$R_8 = 2.08 \text{ hr °F/Btu (ref. Eq. 7.7)}$

$$\Delta t_1 = (32.76)(2.08) = 68.1°F \tag{7.9}$$

The assumed convection and radiation coefficient must be checked to make sure they are approximately correct. First find the temperature rise from the ambient to the top cover.

$$\Delta t_{\text{cover}} = QR_7 = (32.76)(0.63) = 20.6°F \tag{7.10}$$

Determine the correct convection coefficient from the top cover to the ambient, using Eq. 5.5, where

$$\Delta t_{\text{cover}} = 20.6°F \text{ (ref. Eq. 7.10)}$$

$$L = \frac{2L \times W}{L + W} \text{ (ref. Table 5.2)}$$

$$L = \frac{2(12)(10)}{12 + 10} = 10.91 \text{ in} = 0.909 \text{ ft}$$

$$h_c = 0.27 \left(\frac{20.6}{0.909}\right)^{0.25} = 0.59 \frac{\text{Btu}}{\text{hr ft}^2 \text{ °F}} \tag{7.11}$$

This is close to the assumed value of 0.6 used in Eq. 7.5, so that the results are valid.

Determine the correct radiation coefficient from the cover to the ambient, using the curve in Figure 5.29, when $\Delta t = 20.6°F$ and the ambient is 110°F. (View factor is about 1, and the emissivity is about .9)

$$h_r = 1.2 \frac{\text{Btu}}{\text{hr ft}^2 \text{ °F}} \tag{7.12}$$

This is close to the assumed value of 1.3 used in Eq. 7.5, so that the results are valid.

Determine the correct radiation coefficient from the PCB to the inside surface of the cover. The cover temperature then becomes

$$t_{\text{cover}} = 110 + \Delta t_{\text{cover}} = 110 + 20.6 = 130.6°F \tag{7.13}$$

The PCB surface temperature becomes

$$t_{\text{PCB}} = 110 + \Delta t_1 = 110 + 68.1 = 178.1°F \tag{7.14}$$

The temperature rise from the PCB to the cover will be

$$\Delta t_{PC} = 178.1 - 130.6 = 47.5°F \qquad (7.15)$$

The radiation coefficient can now be obtained from Figure 5.29, using the value from Eq. 7.15.

$$h_r = 1.45 \frac{\text{Btu}}{\text{hr ft}^2 \, °F} \qquad (7.16)$$

This is close to the value of 1.5 originally assumed for Eq. 7.3, so that the results are valid.

Determine the correct coefficient of convection from the PCB to the inside surface of the cover using the heat flow, which must divide between the parallel flow path developed by resistors R_1 and R_2 (in series), which are in parallel with R_3, as shown in Figure 7.3.

$$Q_{\text{conv}} = \frac{\Delta t_{PC}}{R_1 + R_2} = \frac{47.5}{3.75 + 1.50} = 9.04 \frac{\text{Btu}}{\text{hr}}$$

$$Q_{\text{rad}} = \frac{\Delta t_{PC}}{R_3} = \frac{47.5}{2.0} = 23.75 \frac{\text{Btu}}{\text{hr}}$$

$$\text{Check:} \quad 9.6 \text{ watts} = 32.79 \frac{\text{Btu}}{\text{hr}}$$

Using the convection heat flow (Q_{conv}), the temperature rise from the PCB to the local ambient inside the box becomes

$$\Delta t_2 = Q_{\text{conv}} R_1 = (9.04)(3.75) = 33.9°F \qquad (7.17)$$

The convection coefficient is determined from Eq. 5.5 for the 6 × 8 PCB

$$h_c = 0.27 \left(\frac{33.9}{0.571}\right)^{0.25} = 0.75 \frac{\text{Btu}}{\text{hr ft}^2 \, °F} \qquad (7.18)$$

This is close to the original value of 0.8, which was assumed in Eq. 7.1, so that the results are acceptable.

The PCB surface temperature becomes 178.1°F, as shown in Eq. 7.14. Since this is below the maximum allowable temperature of 185°F, the design is satisfactory without a fan.

7.4 SAMPLE PROBLEM—FAN COOLED MINICOMPUTER

A minicomputer contains 15 plug-in PCBs, with a mother board in a card cage, for a rack mounted system. The system also includes an EMI filter and a small exhaust type of cooling fan, which provides an air flow of 15 cfm through the chassis. Rows of DIPs are mounted on each PCB, which measures 5×9 in and dissipates about 8 watts. The total power dissipation in the box is 120 watts, as shown in Figure 7.4. The maximum allowable component surface temperature is 160°F (71°C) and the ambient temperature is 110°F (43.3°C). Determine if the design is satisfactory.

SOLUTION

Heat transfer by radiation will be small, because each PCB will "see" the adjacent hot PCB, so that radiation heat transfer is ignored. Forced convection heat transfer is therefore considered where the Reynolds number must be computed to obtain the heat transfer coefficient (h_c). The power dissipation of the fan does not add to the total heat load, because an exhaust fan is used to cool the box.

The average free air space between the components and the adjacent PCB is about 0.18 in. For 15 PCBs this results in 16 air spaces, with each air space as shown in Figure 7.5.

The air flow cross section through the PCBs becomes

$$A_{\text{flow}} = (5.0)(0.18)(16 \text{ spaces}) = 14.4 \text{ in}^2 = 0.10 \text{ ft}^2 \tag{7.19}$$

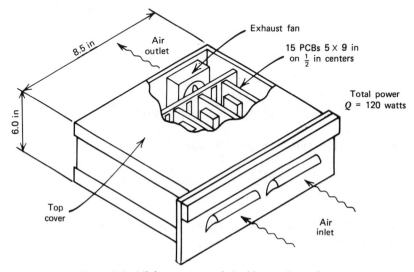

Figure 7.4 Minicomputer cooled with an exhaust fan.

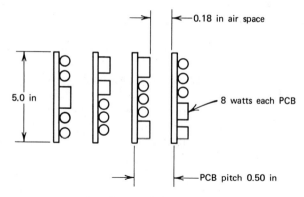

Figure 7.5 Cross section of the air flow path through a PCB card cage.

The average cooling air flow velocity through the card stack is

$$V = \frac{F}{A} = \frac{15 \text{ ft}^3/\text{min}}{0.10 \text{ ft}^2} = 150 \frac{\text{ft}}{\text{min}} \qquad (7.20)$$

The density of the air passing through the minicomputer will change as the air picks up heat. The average density is determined from the average of the inlet and outlet temperatures. The inlet temperature is known to be 110°F. The outlet temperature is obtained from the temperature rise of the cooling air as it passes through the box, using Eq. 6.7.

Given $Q = 120$ watts $= 409.5$ Btu/hr (ref. Table 1.1)
$W = 1.0$ lb/min (approximate flow for 15 ft³/min at 130°F estimated air temperature)

$$C_p = 0.24 \frac{\text{Btu}}{\text{lb °F}} \text{ (specific heat of air)}$$

$$\Delta t_1 = \frac{409.5 \text{ Btu/hr}}{(1.0 \text{ lb/min})(60 \text{ min/hr})(0.24 \text{ Btu/lb °F})} = 28.4°\text{F} \qquad (7.21)$$

The average air temperature can now be obtained.

$$t_{av} = \frac{t_{in} + t_{out}}{2} = \frac{110 + 138.4}{2} = 124.2°\text{F} \qquad (7.22)$$

The air density is obtained from Eq. 6.32 or from Figure 6.28. Using Eq. 6.32, we obtain

$$\rho = \frac{(14.7)(144)}{(53.3)(460 + 124.2)} = 0.068 \frac{\text{lb}}{\text{ft}^3} \qquad (7.23)$$

7.4 Sample Problem—Fan Cooled Minicomputer

The viscosity (μ) and the thermal conductivity (K) of the cooling air are obtained from Figure 6.28.

$$\left. \begin{array}{l} \mu = 0.048 \dfrac{\text{lb}}{\text{ft hr}} \\[6pt] K = 0.0165 \dfrac{\text{Btu}}{\text{hr ft °F}} \end{array} \right\} \quad (7.24)$$

The hydraulic diameter of the air flow passage through the PCB stack, as shown in Figure 7.5, is obtained from Eq. 6.11.

$$D = \frac{2ab}{a+b} = \frac{2(5.0)(0.18)}{5.0 + 0.18} = 0.347 \text{ in} = 0.0289 \text{ ft} \quad (7.25)$$

Substitute into Eq. 6.11 for the Reynolds number.

$$N_R = \frac{VD\rho}{\mu} = \frac{(150 \text{ ft/min})(60 \text{ min/hr})(0.0289 \text{ ft})(0.068 \text{ lb/ft}^3)}{0.048 \text{ lb/ft hr}}$$

$$N_R = 368 \text{ (dimensionless)} \quad (7.26)$$

The Colburn J factor for the forced convection coefficient is determined from Eq. 6.10.

$$J = \frac{6}{(N_R)^{0.98}} = \frac{6}{(368)^{0.98}} = 0.0183 \quad (7.27)$$

The cooling air weight flow (G) through one duct is obtained from Eq. 6.12.

Given $W = 1.0$ lb/min (flow through 16 ducts in box)

$$A = \frac{(5.0)(0.18) \text{in}^2}{144 \text{ in}^2/\text{ft}^2} = 0.00625 \text{ ft}^2 \text{ (area of one duct)}$$

$$G = \frac{W}{A} = \frac{(1.0 \text{ lb/min})(60 \text{ min/hr})}{(16 \text{ ducts})(0.00625 \text{ ft}^2)} = 600 \frac{\text{lb}}{\text{hr ft}^2} \quad (7.28)$$

The forced convection coefficient is now determined from Eq. 6.9.

$$h_c = JC_p G \left(\frac{C_p \mu}{K}\right)^{-2/3} \quad \text{(ref. Eq. 6.9)}$$

$$h_c = \frac{(0.0183)(0.24)(600)}{[(0.24)(0.048)/0.0165]^{0.666}} = 3.3 \frac{\text{Btu}}{\text{hr ft}^2 \text{ °F}} \quad (7.29)$$

The temperature rise across the convection air film from the cooling air to the surface of the component is obtained from Eq. 5.7.

Given $Q = 8$ watts $= 27.3$ Btu/hr (heat on one PCB)
$h_c = 3.3$ Btu/hr ft² °F (ref. Eq. 7.29)
$A = (5.0)(9.0) = 45$ in² $= 0.312$ ft² [PCB surface area (one side)]

$$\Delta t_2 = \frac{Q}{h_c A} = \frac{27.3}{(3.3)(0.312)} = 26.5°F \qquad (7.30)$$

The hot spot surface temperature on the component is determined by the temperature rise of the cooling air and the temperature rise across the PCB convection film.

$$t_{\text{surf}} = 110 + \Delta t_1 + \Delta t_2 \qquad (7.31)$$

$$t_{\text{surf}} = 110 + 28.4 + 26.5 = 164.9°F \qquad (7.32)$$

This is slightly higher than the maximum allowable component surface temperature of 160°F, so that the design may be considered unacceptable. However, when the electronic component lead wires extend through the PCB and they are soldered to copper runs on the back side, a considerable amount of heat may flow through the PCB to the back side. When there are many etched copper circuit runs on the back side, there can be a significant increase in the effective cooling area on the PCB. Test data show that it is possible to increase the effective area as much as 30% with a large number of copper runs on the back side of the PCB.

Assuming a 25% increase in the effective PCB area due to copper runs on the back side:

$$\Delta t_2 = \frac{27.3}{(3.3)(0.312 \times 1.25)} = 21.2°F \qquad (7.33)$$

The hot spot surface temperature on the component now becomes

$$t_{\text{surf}} = 110 + 28.4 + 21.2 = 159.6°F \qquad (7.34)$$

The design is now satisfactory since the component hot spot temperature is below 160°F.

7.5 MICROCOMPUTER SYSTEMS

The microcomputer is a small system, which will typically fit on one small plug-in PCB. It is generally much smaller than the minicomputer, which is considered to be a desk top computer. The microcomputer is normally considered to be a computer that has a microprocessor for its central processing unit (CPU).

One single PCB microcomputer will usually contain the microprocessor and any additional support components, such as the clock and crystal. The PCB will also contain the memory ROM (read-only memory), RAM (random access memory), and PROM (programmable read-only memory). It may also contain the input and output and the asychronous interface control.

Since the microcomputer consists of one PCB, it can be plugged into a minicomputer to perform the functions of the CPU.

Microcomputers are fabricated many different ways, depending upon their application. For military environments, where vibration and shock are as important as the thermal requirements, these PCBs are often fabricated with an aluminum heat sink core. This type of construction is shown in Figure 4.2. When the power densities are high and forced convection cooling air is available, hollow core PCBs similar to those in Figure 6.21 are convenient to use. This type of construction is more expensive, but it brings the cooling air very close to the heat dissipating components, so that it increases the cooling efficiency substantially.

For most commercial applications, the microprocessor PCB usually consists of an epoxy fiberglass board, with plated throughholes and copper runs on both sides.

7.6 SAMPLE PROBLEM—COOLING A MICROCOMPUTER

A microcomputer PCB is plugged into a minicomputer chassis that is cooled by natural convection. Access to the PCB is obtained by removing the top cover, which has louvers that permit the warm air to exit from the chassis. The PCBs in the chassis are spaced on 1 in centers to improve the natural convection cooling, as shown in Figure 7.6. The microcomputer PCB dissipates 5.0 watts and measures 7.5 × 8.5 in, not including the electrical edge connector. The system must operate in a 122°F (50°C) environment. The maximum allowable component surface temperature is 160°F (71°C). Determine if the design is satisfactory.

SOLUTION

Radiation heat transfer is not considered, because the PCBs "see" each other. It is assumed that the PCB is fully populated with electronic components, which represent a uniformly distributed head load.

The natural convection heat transfer coefficient that will be developed by the vertical PCB can be determined from Eq. 5.4. This relation depends upon the temperature difference between the ambient and the surface of the PCB, as well as the vertical height.

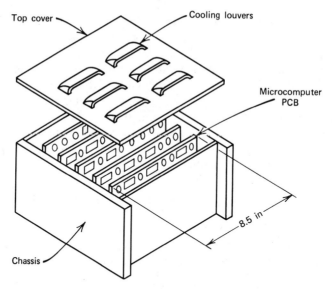

Figure 7.6 Microcomputer PCB within a minicomputer chassis.

Given $\Delta t = 160 - 122 = 38°F$
$L = 7.5$ in/12 $= 0.625$ ft (vertical height of PCB)

$$h_c = 0.29 \left(\frac{\Delta t}{L}\right)^{0.25} = 0.29 \left(\frac{38}{0.625}\right)^{0.25}$$

$$h_c = 0.81 \frac{\text{Btu}}{\text{hr ft}^2 \, °F} \quad (7.35)$$

The heat that can be carried away by natural convection is determined from Eq. 5.7.

Given $h_c = 0.81$ Btu/hr ft² °F
$A = (7.5)(8.5) = 63.75$ in² $= 0.443$ ft²
$\Delta t = 38°F$

$$Q = h_c A \Delta t = (0.81)(0.443)(38) = 13.6 \frac{\text{Btu}}{\text{hr}} = 4.0 \text{ watts} \quad (7.36)$$

The actual power dissipation is really 5.0 watts. Since the PCB is only capable of dissipating 4 watts under the conditions given above, the design is not satisfactory.

It might be possible to add a small fan to the system to improve the cooling, but it will also increase the cost and reduce the reliability. When there are a few hot components that dissipate most of the heat, it may be possible to add cooling fins or large hats to increase the surface area of the hot components for improved heat transfer.

The amount of copper can often be increased on the back side of dip soldered circuit boards. When the electrical lead wires extend through the board for soldering, they also carry heat to the back side of the board. When there are a large number of wide copper runs on the back side, the amount of heat transfer surface is increased. Test data show that it is possible to increase the effective surface area as much as 30% with this method.

If additional copper can be added to the back side of the PCB in this sample problem, to increase the effective surface area by 25%, the amount of heat that can be carried away will also be increased by the same amount. The amount of heat that can now be removed is as follows:

$$Q = 1.25(4.0) = 5.0 \, \text{watts} \tag{7.37}$$

Since this is equal to the PCB power dissipation, the design is now satisfactory.

There will be a slight increase in the temperature of the cooling air as it passes through the box, which must be considered to obtain the hot spot temperature at the top of the PCB. For a small box the increase is small, so that it is ignored. See Chapter 8, Equation 8.11, to determine the temperature rise of the cooling air passing through the box.

7.7 MICROPROCESSOR DEVELOPMENT

Microprocessors are an outgrowth of the integrated circuits that were developed by Texas Instruments and by Fairchild in about 1959. As the space programs pushed toward miniaturization, cost was not considered as important as weight. Many research contracts were awarded by the government to seek methods for reducing the size of electronic systems. Many companies used their own funds to develop new and smaller electronic components.

This intensive dedication led to the integrated circuit, then to small scale integrated (SSI) circuits, medium scale integrated (MSI) circuits, and now to very large scale integrated (VSI) circuits.

The Intel Company produced the first general purpose microprocessor, which was called the 4004, in the early 1970s. This device was used in a desk calculator. The next big event was the introduction of the 8008 microprocessor by Intel, which was later improved to become the 8080 microprocessor. This last device was very popular and sales were brisk. Other manufacturers then climbed on the microprocessor bandwagon and reaped the financial rewards of high sales volume.

Microprocessors are being used in many areas that previously had no electrical applications. They are well suited for automatic control systems required by automobile engines, washing machines, petroleum refineries, chemical plants, food processors, and many others.

A microprocessor system requires very few components, which reduces the size and the weight of the system. In addition, it results in substantially lower power requirements, fewer electrical interconnects, and much lower costs. The microprocessor, truly a "computer on a chip," often incorporates the CPU, the memory, and some input and output functions of the typical computer.

7.8 MOUNTING MICROPROCESSORS TO RESIST VIBRATION FATIGUE

Microprocessors that will be used in severe vibration environments must be mounted very carefully. The large body of the microprocessor, with its large number of stiff lead wires, often produce many lead wire and solder joint failures in improperly mounted components.

In a sinusoidal or random vibration environment, the PCB will tend to bend back and forth during its resonant condition, as shown in Figure 7.7. This action produces relative motion between the mounting surface and the body of the component. This relative motion induces stresses in the lead wires and solder joints, which results in failures when the number of fatigue cycles reaches a critical value.

Fatigue failures in microprocessors can be avoided by mounting the component near the supported edges of a PCB rather than at the center of a plug-in PCB. When the microprocessor must be mounted at the center of a plug-in PCB, a stiffening rib should be placed adjacent to the component to reduce the dynamic displacements during the resonant condition. [1]

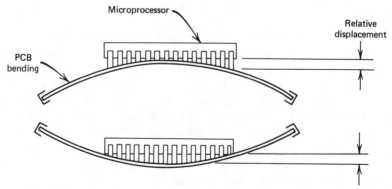

Figure 7.7 Microprocessor mounted on a PCB in a vibration environment.

7.9 MICROPROCESSORS IN SEVERE THERMAL ENVIRONMENTS

Microprocessors that are required to operate in severe thermal environments must be mounted, not only to ensure proper cooling but also to prevent failures in the lead wires and solder joints. Improper mounting of these components can lead to rapid solder joint failures in temperature cycling environments, because of sharp differences in the thermal expansion coefficients of the various materials used in electronic assemblies. A strain relief must be used to prevent strain related failures in the lead wires and solder joints.

The mounting geometry of the microprocessor is very similar to the DIP, except that the microprocessor is much larger. The electrical lead wires in both cases are quite rugged, with a step change in the cross section of the lead wires to stop the body of the component from contacting the PCB when the component is inserted. This normally leaves an air gap under the component body, which is desirable, since it acts like a strain relief. However, when the power dissipation of the microprocessor is quite high or when the thermal environment is quite severe, an air gap under the component body may not be desirable, especially in a conduction cooled system.

Figure 4.15 shows a number of different methods for mounting DIPs. These same mounting methods also apply to microprocessors. The mounting methods must be capable of providing adequate cooling and adequate solder joint strain relief in severe thermal environments.

7.10 SAMPLE PROBLEM—COOLING A MICROPROCESSOR MOUNTED ON A PCB

An exhaust fan is used to cool an electronic box, which is required to operate in an ambient temperature of 122°F (50°C). The maximum allowable surface temperature of a microprocessor (see Figure 7.1) mounted on a PCB within the box is 185°F (85°C). The cooling air velocity through the box is 150 ft/min, with the flow path parallel to the long dimension of the microprocessor. The power dissipation of the component is 0.60 watt. Determine if the cooling will be adequate.

SOLUTION

The convection coefficient for the cooling air passing over the microprocessor must be determined, so that the Reynolds number is required. The physical properties of the air are obtained from Figure 6.28, using a temperature of 150°F as a guess. The actual air temperature will have to be checked to see how accurate the guess was. The arrangement of the exhaust fan in the box is similar to the one shown in Figure 7.4.

Given V = 150 ft/min = 9000 ft/hr (air velocity)
       ρ = 0.065 lb/ft^3 (air density at 150°F)
       μ = 0.048 lb/ft hr (viscosity)
       K = 0.017 Btu/hr ft °F (thermal conductivity)
       L = 2.5 in = 0.208 ft (length)

Substitute into Eq. 6.11 for the Reynolds number.

$$N_R = \frac{(9000)(0.208)(0.065)}{0.048} = 2535 \text{ (dimensionless)} \qquad (7.38)$$

The forced convection coefficient for the microprocessor can be determined from Eq. 6.51.

$$h_c = 0.055 \left(\frac{K}{L}\right)(N_R)^{0.75} \text{ (ref. Eq. 6.51)}$$

$$h_c = 0.055 \left(\frac{0.017}{0.208}\right)(2535)^{0.75} = 1.60 \frac{\text{Btu}}{\text{hr ft}^2 \text{ °F}} \qquad (7.39)$$

The temperature rise across the convection air film is determined from the standard convection relation shown by Eq. 5.7.

Given Q = 0.60 watts = 2.05 Btu/hr (ref. Table 1.1.)
       h_c = 1.60 Btu/hr ft^2 °F (ref. Eq. 7.39)
       A = (2.5)(0.75) + (2)(2.5 + 0.75)(0.20)
       A = 3.175 in^2 = 0.0220 ft^2 (ref. Figure 7.1) surface area does not include the lead wires.

$$\Delta t = \frac{Q}{h_c A} = \frac{2.05}{(1.60)(0.0220)} = 58.2\text{°F} \qquad (7.40)$$

The average air temperature must be chekced to see if the original assumed value of 150°F was valid.

$$t_{av} = \frac{122 + (122 + 58.2)}{2} = 151.1\text{°F}$$

This is close enough to the assumed value, so that the results are acceptable.

The maximum allowable temperature rise from the cooling air to the microprocessor surface is 185 − 122 = 63°F. Since the calculated temperature rise is only 58.2°F, the design is satisfactory.

One note of caution. The foregoing analysis does not include any heat added to the cooling air from the fan or any other electronic components. When the cooling air picks up heat from another source before passing over

7.10 Sample Problem—Cooling a Microprocessor Mounted on a PCB

Figure 7.8 Rack mounted sheet metal electronic box that is cooled with a small air circulating fan. (Courtesy Litton Systems, Inc.)

the microprocessor, the extra temperature rise must be included in the determination of the hot spot temperature of the microprocessor.

Figure 7.8 shows a small electronic chassis that is cooled by a small axial flow fan.

8

Effective Cooling for Large Racks and Cabinets

8.1 INDUCED DRAFT COOLING FOR LARGE CONSOLES

Large electronic enclosures are commonly used by commercial and military groups all over the world. These units are used for ground support, countermeasures, communications, computers, manufacturing processing controls, and many other applications. These systems must be capable of reliable operation for many years, with little or no maintenance. For this reason fans and blowers are often discouraged, and natural convection cooling is often stressed.

Natural convection cooling can be used effectively by making maximum use of the driving force that is available when power dissipating components heat the surrounding air. The heated air expands, so that it becomes less dense. In a gravity field, this causes the hot air to rise because the hot air is lighter. As the air rises, it also carries away the heat. The driving force is the pressure differential that is created by the density change. This driving force depends upon the height of the column of heated air and upon the temperature rise of the heated air. The flow characteristics are similar to the conditions in a chimney. In most naturally cooled systems, the driving force is relatively small, so that the flow resistance through the console must be reduced to a minimum to provide the required cooling.

As the cooling air flow rate through the console increases, the pressure drop through the system will also increase, approximately as the square of the air velocity. As more cooling air flows through the console, more heat can be carried away. These two parameters can be plotted on a curve, as shown in Figure 8.1.

The point of intersection of these two curves represents the actual operating point for the system. This curve is very similar to the one shown in

8.2 Air Flow Losses for Large Cabinets

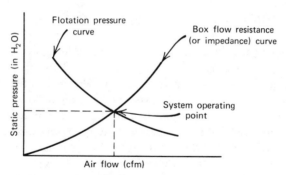

Figure 8.1 Air flow impedance curve for induced draft.

Figure 6.12, where a fan is used to provide the driving pressure through the electronic box.

8.2 AIR FLOW LOSSES FOR LARGE CABINETS

Equipment cabinets, consoles, and card racks will usually have some type of obstruction in the cooling air flow path, which will force the air to change its flow direction. Anytime the air is forced to change its direction, it increases the flow resistance, so that the flow rate decreases. Resistance to the air flow is also developed when the cooling air passes from one compartment in the console to another. When the air first enters a compartment, it will usually be forced to contract slightly or to be squeezed as it passes through an opening or port. After the air enters the compartment, it will usually be free to expand when it enters an enlarged area beyond the bulkhead.

It is convenient to express the static air flow losses in terms of velocity heads, as explained in Section 6.3. Under these conditions, the typical air flow losses experienced with the conditions described above can be roughly approximated as shown in Table 8.1.

Table 8.1 Air Flow Losses through Large Consoles

Condition	Static Pressure Loss (Velocity Heads)
Contraction	$0.50 H_v$
Expansion	$1.0 H_v$
90° turn	$1.5 H_v$

— Not completely accurate, but close approximations generated from reams of ASHRAE data.

— linear ratio for angles other than 90°

8.3 FLOTATION PRESSURE AND PRESSURE LOSS

The flotation pressure (P_f) required to force a column of heated air to rise and flow through a passage is shown by Eq. 8.1. This equation also represents the pressure loss through the system. The main obstructions to the flow will be the inlet and the outlet losses for a straight flow through system. Table 8.1 shows the inlet loss can usually be approximated as about 0.50 velocity head, and the outlet loss can usually be approximated as about 1.0 velocity head. Therefore, the total *static* loss for air flowing through a passage with no other obstructions will be about 1.5 velocity heads.

These approximate relations can be used to determine the approximate air flow velocity through an air duct or an electronic control console when there are no intermediate flow restrictions.

The flow relations are based upon the flotation pressure that is developed by the density change in the cooling air as it rises through the chassis, carrying away the heat. Once a steady state condition has been established, the flotation pressure (P_f) will be the same as the static pressure drop (ΔP) through the chassis, when both are measured in the same units, in or cm. In English units, the flotation pressure, or pressure drop, will be as shown in Eq. 8.1 [18].

$$P_f = \Delta P = 0.192 \rho H \left(1 - \frac{T_{out}}{T_{in}}\right) = \text{in H}_2\text{O} \qquad (8.1)$$

In metric units, the flotation equation is

$$P_f = \Delta P = \rho H \left(1 - \frac{T_{out}}{T_{in}}\right) = \text{cm H}_2\text{O} \qquad (8.1a)$$

The units for each equation are shown in Table 8.2.

The use of the flotation equation is demonstrated with a sample problem.

8.4 SAMPLE PROBLEM—INDUCED DRAFT COOLING FOR A LARGE CABINET

A 6 ft tall (182.9 cm) cabinet must operate in a 100°F (37.8°C) ambient temperature, with a uniformly distributed heat load of 500 watts along its height. The cabinet is 19 in (48.2 cm) wide, with a free flow path of 100 in^2 (645.2 cm^2) area through the center of the cabinet, as shown in Figure 8.2. Determine the temperature rise of the cooling air and the cooling air flow rate through the cabinet.

8.4 Sample Problem—Induced Draft for a Large Cabinet

Table 8.2 Units for Eqs. 8.1 and 8.1a

Item	English Units	Metric Units
P_f or ΔP	in H_2O	cm H_2O
0.192	Conversion $\frac{lb}{ft^2}$ to in H_2O	—
ρ	$\frac{lb}{ft^3}$ Mean density	$\frac{g}{cm^3}$ Mean density
H	Console height in ft if all heat is at bottom. Use ½ of height for uniform heat input	Console height in cm if all heat is at bottom. Use ½ of height for uniform heat input
T_{out}	°R Absolute outlet temperature	°K Absolute outlet temperature
T_{in}	°R Absolute inlet temperature	°K Absolute inlet temperature

SOLUTION

The method of solution is to assume an outlet temperature, which is used to compute the average air density. The flotation pressure is then computed with Eq. 8.1. With a free flow path through the center of the cabinet, the air velocity can be determined by considering only the entry loss of 0.5 velocity head, and the exit loss of 1.0 velocity head, for a total static pressure loss of 1.5 velocity heads, which will also be equal to the flotation pressure.

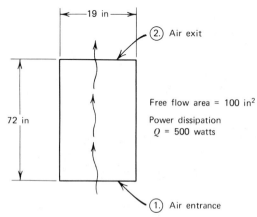

Figure 8.2 Induced draft flow through a large cabinet.

Start by assuming a cooling air temperature rise of 16°F (8.89°C). The outlet air temperature then becomes 100 + 16 = 116°F (46.7°C).

Consider English units first.

Given $T_{out} = 460 + 116 = 576°R$

$T_{in} = 460 + 100 = 560°R$

$H = \dfrac{6}{2} = 3$ ft (equivalent height)

$t_{av} = \dfrac{116 + 100}{2} = 108°F$ (average air temperature)

$\rho = \dfrac{P}{RT}$ (ref. Eq. 6.32)

$$\rho = \dfrac{(14.7 \text{ lb/in}^2)(144 \text{ in}^2/\text{ft}^2)}{(53.3 \text{ ft/°R})(460 + 108)°R} = 0.0699 \dfrac{\text{lb}}{\text{ft}^3} \qquad (8.2)$$

Substitute into Eq. 8.1 for the flotation pressure.

$$P_f = 0.192(0.0699)(3.0)\left(1 - \dfrac{576}{560}\right) = -0.00115 \text{ in } H_2O \qquad (8.3)$$

The negative sign shows that the warm air is moving up, opposite to the direction of gravity. This sign is usually ignored.

Consider metric units next.

Given $T_{out} = 273 + 46.6 = 319.6°K$

$T_{in} = 273 + 37.8 = 310.8°K$

$H = 91.44$ cm (equivalent height)

$t_{av} = \dfrac{46.6 + 37.8}{2} = 42.2°C$ (average air temperature)

$$\rho = \dfrac{1034.4 \text{ g/cm}^2}{(2924 \text{ cm/°K})(273 + 42.2)°K} = 0.00112 \dfrac{\text{g}}{\text{cm}^3} \qquad (8.2a)$$

Substitute into Eq. 8.1a for the flotation pressure.

$$P_f = (0.00112)(91.44)\left(1 - \dfrac{319.6}{310.8}\right) = -0.0029 \text{ cm } H_2O \qquad (8.3a)$$

The velocity of the cooling air can be determined with the use of Eq. 6.2 for English units and Eq. 6.2a for metric units. These equations represent values for one velocity head. The flotation pressure equation represents 1.5

8.4 Sample Problem—Induced Draft for a Large Cabinet

velocity heads, so that a correction must be used. The average density of the air flowing through the console is used to obtain the average velocity, from Eq. 8.3.

$$V = 60 \frac{\text{sec}}{\text{min}} \left[\frac{2(32.14 \text{ ft/sec}^2)(62.4 \text{ lb/ft}^3)(0.00115 \text{ in } H_2O) \left(\frac{1.0 H_v}{1.5 H_v}\right)}{(12 \text{ in/ft})(0.0699 \text{ lb/ft}^3 \text{ air})} \right]^{1/2}$$

← correction

same equation as the one above Eq. 2

$$V = 114.9 \frac{\text{ft}}{\text{min}} \tag{8.4}$$

For metric units, the flotation pressure shown in Eq. 8.3a is used to obtain the average velocity.

$$V = \left[\frac{2(979.6 \text{ cm/sec}^2)(1 \text{ g/cm}^3)(0.0029 \text{ cm}) \left(\frac{1.0 H_v}{1.5 H_v}\right)}{0.00112 \text{ g/cm}^3 \text{ air}} \right]^{1/2}$$

$$V = 58.3 \frac{\text{cm}}{\text{sec}} \tag{8.4a}$$

The flow (F) in cfm through the cabinet is determined from the flow area and the flow velocity.

$$F = AV \tag{8.5}$$

Given $A = 100 \text{ in}^2 = 0.694 \text{ ft}^2 = 645 \text{ cm}^2$ (area)
$V = 114.9 \text{ ft/min} = 58.3 \text{ cm/sec}$ (velocity of air flow)

In English units:

$$F = (0.694 \text{ ft}^2)\left(114.9 \frac{\text{ft}}{\text{min}}\right) = 79.7 \frac{\text{ft}^3}{\text{min}} \tag{8.6}$$

In metric units:

$$F = (645 \text{ cm}^2)\left(58.3 \frac{\text{cm}}{\text{sec}}\right) = 37,600 \frac{\text{cm}^3}{\text{sec}} \tag{8.6a}$$

The cooling air weight flow through the cabinet is determined from Eq. 8.7.

$$W = \rho F \tag{8.7}$$

In English units:

$$W = \left(0.0699 \, \frac{\text{lb}}{\text{ft}^3}\right)\left(79.7 \, \frac{\text{ft}^3}{\text{min}}\right) = 5.57 \, \frac{\text{lb}}{\text{min}} \quad (8.8)$$

In metric units:

$$W = \left(0.00112 \, \frac{\text{g}}{\text{cm}^3}\right)\left(37{,}600 \, \frac{\text{cm}^3}{\text{sec}}\right) = 42.1 \, \frac{\text{g}}{\text{sec}} \quad (8.8a)$$

The temperature rise of the cooling air flowing through the cabinet is determined from Eq. 6.7. (page 175)

Given $Q = 500$ watts $= 1706.5$ Btu/hr $= 119.5$ cal/sec (heat)
$W = 5.57$ lb/min $= 42.1$ g/sec (weight flow)
$C_p = 0.24$ Btu/lb °F $= 0.24$ cal/g °C (specific heat)

In English units:

$$\Delta t = \frac{1706.5 \text{ Btu/hr}}{(5.57 \text{ lb/min})(60 \text{ min/hr})(0.24 \text{ Btu/lb °F})} = 21.2°F \quad (8.9)$$

In metric units:

$$\Delta t = \frac{119.5 \text{ cal/sec}}{(42.1 \text{ g/sec})(0.24 \text{ cal/g °C})} = 11.8°C \quad (8.9a)$$

A temperature rise of 16°F was assumed to start. The resulting temperature rise calculated for the assumed condition turned out to be 21.2°F. The assumed value does not check with the calculated value, so that a second iteration is required to obtain better results.

For the second iteration, a temperature rise of 19°F (10.5°C) is assumed. The equilibrium condition will result in a calculated temperature of 19.5°F. This is close enough for good accuracy, so that the expected temperature rise through the cabinet will be about 19.2°F (10.7°C). The resulting cooling air flow for the second iteration is 125.2 ft/min, with a flow rate of 6.06 lb/min, which results in about 86.9 cfm. Therefore, the temperature rise through the cabinet will be about

$$\Delta t = 19.2°F \, (10.7°C) \quad (8.10)$$

8.5 NATURAL COOLING FOR LARGE CABINETS WITH MANY FLOW RESTRICTIONS

Large cabinets will often have many restrictions along the cooling air flow path. Under these circumstances, it is more convenient to use another approximate relation, which includes the various flow losses directly in terms of the velocity head and the net flow area at each restriction. When Eqs. 8.1, 6.1, and 6.7 are combined, the approximate temperature rise through the cabinet (in English units only) can be made directly as shown in Eq. 8.11.

$$\Delta t = 2.5 \left(\frac{T_a}{100}\right) \left(\frac{Z}{H}\right)^{1/3} \left(\frac{Q}{P'}\right)^{2/3} = °F \quad (8.11)$$

where T_a = °R absolute ambient temperature
 Q = watts, power dissipation
 P' = atmospheric pressure, expressed as the number of atmospheres (dimensionless)
 H = ft = height of cabinet (use full height if all heat is at bottom; use ½ of height if heat is uniformly distributed along height)

$$Z = 0.226 \left(\frac{H_{v1}}{A_1^2} + \frac{H_{v2}}{A_2^2} + \frac{H_{v3}}{A_3^2} + \cdots\right) \quad (8.12)$$

H_{v1} H_{v2} H_{v3} = number of velocity heads lost at flow points 1, 2, and 3 (dimensionless)
A_1 A_2 A_3 = in² = cross section area at flow points 1, 2, and 3

Equations 8.11 and 8.12 can be used to compute the temperature rise of the cooling air flowing through a console, as a result of the induced draft.

The preceding sample problem can be solved using the preceding two equations. The following information is required:

Q = 500 watts (total power dissipation)
T_a = 460 + 100 = 560°R (absolute ambient temperature)
H = 6/2 = 3.0 ft (effective height of console)
P' = 1.0 (1 atmosphere at sea level conditions)
H_{v1} = 0.5 = number of velocity heads lost at point 1 at the console entrance
H_{v2} = 1.0 = number of velocity heads lost at point 2 at the console exhaust
A_1 = 100 in² (air flow cross section, point 1 entrance)
A_2 = 100 in² (air flow cross section, point 2 exit)

Substitute into Eq. 8.12 to determine Z (the static loss factor).

$$Z = 0.226 \left[\frac{0.5}{(100)^2} + \frac{1.0}{(100)^2} \right] = 3.39 \times 10^{-5} \frac{\text{in H}_2\text{O}}{(\text{lb/min})^2}$$

Substitute into Eq. 8.11 for the cooling air temperature rise.

$$\Delta t = 2.5 \left[\left(\frac{560}{100} \right) \left(\frac{3.39 \times 10^{-5}}{3.0} \right)^{1/3} \left(\frac{500}{1.00} \right)^{2/3} \right]$$

$$\Delta t = 19.8°F \tag{8.13}$$

Comparing Eq. 8.10 with Eq. 8.13 shows that there is relatively good agreement with the two different methods.

Restricted openings and perforated panels are often used in bulkheads that separate various compartments in tall consoles. These flow restrictions will increase the pressure drop through the cabinet and reduce the cooling air flow. Other common restrictions are side air entrances and side air exits, which force the cooling air to turn 90° to enter and to exit from a tall cabinet.

Sometimes baffles may be used to change the direction of the cooling air so that it passes over an extremely hot component. Every time the cooling air is forced to change its direction, friction flow losses will occur and the cooling air flow will be reduced. The analysis of these losses can be demonstrated with a sample problem.

8.6 SAMPLE PROBLEM—TEMPERATURE RISE OF COOLING AIR IN A CABINET WITH AN INDUCED DRAFT

A 5.0 ft tall console with a 300 watt uniformly distributed power dissipation must operate in a 122°F ambient. There are three flow restrictions, one at the inlet, one at a center bulkhead, and one at the exit, as shown in Figure 8.3a. The flow restrictions are shown as flow resistors for the mathematical model in Figure 8.3b. Using induced draft cooling, determine the temperature rise of the cooling air as it rises through the cabinet. Also determine the cfm flow and the pressure drop through the cabinet.

SOLUTION

Table 8.1 shows the static pressure losses in terms of velocity heads for induced draft cooling of large racks and cabinets. The pressure loss expected at each flow restriction can be approximated with this information.

8.6 Sample Problem—Temperature Rise of Cooling Air in a Cabinet

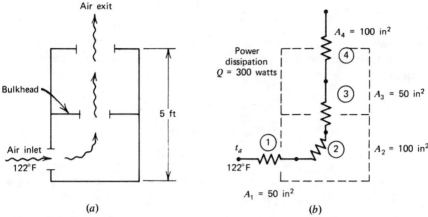

Figure 8.3 Large cabinet with a mathematical model showing the flow restrictions.

Point 1, Entrance

The air must contract as it enters the cabinet. After it enters, assume that the air will expand because the area within the cabinet is greater than the opening.

The resulting static pressure losses are as follows:

$$\text{Contraction} = 0.5H_v$$
$$\underline{\text{Expansion} = 1.0H_v}$$
$$H_1 = 1.5H_{v_1} \quad (8.14)$$

Point 2, 90° Turn

The air is forced to make a 90° turn after it enters the cabinet, in order to continue rising. The resulting static pressure loss is as follows:

$$H_2 = 1.5H_{v_2} \quad (8.15)$$

Point 3, Opening through Bulkhead

The air must contract as it passes through the bulkhead. After passing through, the air expands as it flows into the next compartment. The resulting static pressure loss is

$$\text{Contraction} = 0.5H_v$$
$$\underline{\text{Expansion} = 1.0H_v}$$
$$H_3 = 1.5H_{v_3} \quad (8.16)$$

Point 4, Exit

The air must contract before it passes through the top opening. After passing through, the air expands as it flows to the ambient. The resulting static pressure loss is

$$\text{Contraction} = 0.5 H_v$$
$$\text{Expansion} = 1.0 H_v$$
$$H_4 = 1.5 H_{v_4} \quad (8.17)$$

The free flow cross section area at each section is as follows:

$$A_1 = 50 \text{ in}^2$$
$$A_2 = 100 \text{ in}^2$$
$$A_3 = 50 \text{ in}^2$$
$$A_4 = 100 \text{ in}^2$$

Substitute into Eq. 8.12.

$$Z = 0.226 \left[\frac{1.5}{(50)^2} + \frac{1.5}{(100)^2} + \frac{1.5}{(50)^2} + \frac{1.5}{(100)^2} \right]$$
$$Z = 0.000339 \frac{\text{in H}_2\text{O}}{(\text{lb/min})^2} \quad (8.18)$$

$T_a = 460 + 122 = 582°R$ (absolute ambient temperature)
$Q = 300$ watts (power dissipation)
$P' = 1.0$ (1 atmosphere at sea level)
$H = \dfrac{5.0}{2} = 2.5$ ft (height for uniform heat input)

Substitute into Eq. 8.11 for the cooling air temperature rise through the console.

$$\Delta t = 2.5 \left[\left(\frac{582}{100}\right) \left(\frac{0.000339}{2.5}\right)^{1/3} \left(\frac{300}{1}\right)^{2/3} \right]$$
$$\Delta t = 33.5°F \quad (8.19)$$

The cooling air weight flow through the console is determined from Eq. 6.7. See Table 1.1 for power and heat conversions.

8.6 Sample Problem—Temperature Rise of Cooling Air in a Cabinet

Given $Q = 300$ watts $= 1023.9$ Btu/hr (heat input)

$\Delta t = 33.5°F$ (ref. Eq. 8.19)

$C_p = 0.24$ Btu/lb °F (specific heat of air; ref. Figure 6.28)

$$W = \frac{1023.9 \text{ Btu/hr}}{(33.5)(0.24)(60 \text{ min/hr})} = 2.12 \frac{\text{lb}}{\text{min}} \tag{8.20}$$

The average density of the cooling air flowing through the console is determined from Eq. 6.32, using average temperatures.

Given $t_{av} = \dfrac{t_{in} + t_{out}}{2} = \dfrac{122 + (122 + 33.5)}{2} = 138.75°F$

$$\rho = \frac{(14.7 \text{ lb/in}^2)(144 \text{ in}^2/\text{ft}^2)}{(53.3 \text{ ft/°R})(460 + 138.75)} = 0.0663 \frac{\text{lb}}{\text{ft}^3} \tag{8.21}$$

The cfm flow through the console is determined from Eq. 6.34 as follows: (page 188)

$$F = \frac{W}{\rho} = \frac{2.12 \text{ lb/min}}{0.0663 \text{ lb/ft}^3} = 31.9 \frac{\text{ft}^3}{\text{min}} \tag{8.22}$$

The static pressure drop through the console is determined from Eq. 8.1. (page 262)

$$\Delta P = 0.192(0.0663)(2.5)\left(1 - \frac{460 + 122 + 33.5}{460 + 122}\right)$$

$$\Delta P = 0.00183 \text{ in H}_2\text{O} \tag{8.23}$$

CHECK:
The pressure drop through the cabinet can also be determined from Eq. 6.2. This relation changes with the velocity. Since there are two different cross section areas in the problem, 50 in² and 100 in², two different velocities must be computed.

At points 1 and 3 in Figure 8.3, the area is 50 in². The velocity through this section then becomes

$$V = \frac{F}{A} = \frac{(31.9 \text{ ft}^3/\text{min})(144 \text{ in}^2/\text{ft}^2)}{50 \text{ in}^2} = 91.8 \frac{\text{ft}}{\text{min}} \tag{8.24}$$

Substitute into Eq. 6.2 to obtain the static pressure loss for one velocity head.

$$1.0 H_v = \left(\frac{V}{4005}\right)^2 = \left(\frac{91.8}{4005}\right)^2 = 0.000525 \text{ in H}_2\text{O} \tag{8.25}$$

(points 1 & 3)

At points 2 and 4 in Figure 8.3, the area is 100 in². The velocity through this section then becomes

$$V = \frac{F}{A} = \frac{(31.9)(144)}{100 \text{ in}^2} = 45.9 \frac{\text{ft}}{\text{min}} \qquad (8.26)$$

Substitute into Eq. 6.2 to obtain the static pressure loss for one velocity head.

$$1.0 H_v = \left(\frac{V}{4005}\right)^2 = \left(\frac{45.9}{4005}\right)^2 = 0.000131 \text{ in H}_2\text{O} \qquad (8.27)$$

(points 2 & 4)

Equations 8.14 through 8.17 show there are 1.5 velocity heads lost at each point, resulting in the following static pressure losses:

Point 1 = 1.5(0.000525) = 0.000787 in H$_2$O loss
Point 2 = 1.5(0.000131) = 0.000196 in H$_2$O loss
Point 3 = 1.5(0.000525) = 0.000787 in H$_2$O loss
Point 4 = 1.5(0.000131) = 0.000196 in H$_2$O loss

Total static loss = 0.001966 in H$_2$O (8.28)

Comparing Eq. 8.28 with Eq. 8.23 shows that the two different methods for computing the static pressure loss through the console give approximately the same results.

8.7 WARNING NOTE FOR INDUCED DRAFT SYSTEMS

A console that is cooled with the induced draft technique must be completely enclosed around the perimeter of the unit. If access panels, or side covers, are removed for maintenance or repairs, the induced draft can be sharply reduced. When the draft is reduced, the cooling air flow is reduced and the temperature increases.

A warning note should therefore be placed in some conspicuous place, to alert personnel when a panel is removed. The note might read as follows:

> WARNING: Do not operate equipment more than 30 minutes with access covers removed.

The operating time may be greater or smaller, depending on the amount of heat dissipated, the effective mass of the most critical components, the outside temperature, and the size of the access panel.

ALSO, FILTERS AT INLET WILL REDUCE FLOW.

8.8 TALL CABINETS WITH STACKED CARD BUCKETS

Tall enclosed equipment cabinets often have several circuit card racks (or card buckets) stacked one above the other. Each card bucket may consist of an open wire frame construction, which extends across the width of the cabinet and supports the plug-in PCBs that are oriented in a vertical position. The PCBs should be spaced to provide about 0.75 in (1.905 cm) average air gap between the component surface and the back surface of the adjacent PCB.

The air gap is required to permit the cooling air to flow freely between the circuit boards without choking off the flow. A typical installation is shown in Figure 8.4.

The static pressure loss factor Z resulting from the cooling air flowing through the card buckets can be determined from Eq. 8.12 when the static losses are in terms of the velocity head. When the losses are not known in terms of the velocity head, it is often convenient to determine the loss factor Z by examining the geometry of the air passage. In this case the air passage is the duct formed by the air spaces between the PCBs. When the card buckets are stacked one above the other in an enclosed cabinet, the flow losses through the PCBs can be combined with the entrance and exit flow losses as shown in Eq. 8.29.

$$Z = 0.226 \left(\frac{4f\frac{L}{D}}{A_d^2} + \frac{H_{v1}}{A_1^2} + \frac{H_{v2}}{A_2^2} + \cdots \right) \quad (8.29)$$

where f = Fanning friction factor shown in Table 6.7; For high aspect ratio ducts with laminar flow, $f = 24/N_R$ (dimensionless)
  N_R = Reynolds number (ref. Eq. 6.11)
  L = in (length of air flow path)
  D = in (hydraulic diameter of PCB duct)
  A_d = in² (cross section area of PCB duct)
  H_{v1} H_{v2} = number of velocity heads lost, points 1 and 2
  A_1 A_2 = in² (section area at points 1 and 2)

In Eq. 8.29 the Fanning friction factor (f) is a function of the Reynolds number (N_R) for laminar flow. Since the flow is not really known, the Reynolds number is not known either. Therefore, to start the solution, it is necessary to assume a temperature rise, from which the average temperature of the air and the Reynolds number are computed. The actual temperature rise is then calculated and compared with the assumed value.

If the assumed value is not very close to the calculated value, another temperature rise is assumed and the iteration process is repeated until good

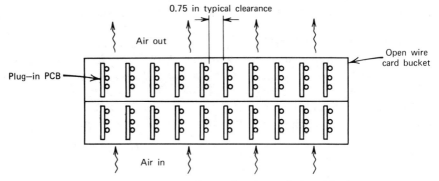

Figure 8.4 Natural convection cooling for stacked card buckets.

agreement is reached. This usually requires two or three iterations for good results. A sample problem is used to demonstrate the procedure.

8.9 SAMPLE PROBLEM—INDUCED DRAFT COOLING OF A CONSOLE WITH SEVEN STACKED CARD BUCKETS

Figure 8.5 shows an enclosed equipment cabinet which contains seven card buckets that are cooled by natural convection. Each bucket holds 19 PCBs 8.0 × 8.0 in, spaced 1 in apart. The average air flow space between the components and the adjacent PCB is 0.75 in. Each PCB dissipates 3 watts,

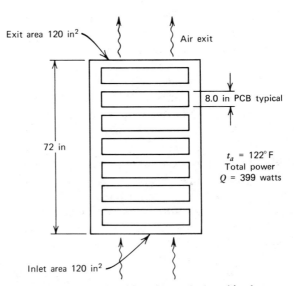

Figure 8.5 Console with seven stacked card buckets.

8.9 Sample Problem—Induced Draft Cooling of a Console with Card Buckets

for a total power dissipation in the entire cabinet of 399 watts. The cabinet must operate in a 122°F ambient at sea level conditions.

Determine the cooling air temperature rise through the cabinet, and the cooling air mass flow, pressure drop, and the component hot spot surface temperature, based upon a uniform heat distribution on a typical PCB.

SOLUTION

When there are 19 PCBs, there are 20 air flow spaces for each card bucket, as shown in Figure 8.6.

Equation 8.29 is used to determine the Z factor. However, since the friction factor (f) is required and the Reynolds number is not known, assume a temperature rise of 29°F through the console to start. Now the average physical properties of the cooling air can be established.

A second iteration may be required to obtain a temperature rise that agrees closely with the assumed value.

Average air temperature:

$$t_{av} = \frac{t_{in} + t_{out}}{2} = \frac{122 + (122 + 29)}{2} = 136.5°F$$

Average air density, ref. Eq. 6.32:

$$\rho = \frac{P}{RT} = \frac{(14.7 \text{ lb/in}^2)(144 \text{ in}^2/\text{ft}^2)}{(53.3 \text{ ft/°R})(460 + 136.5)°R} = 0.0666 \frac{\text{lb}}{\text{ft}^3} \quad (8.30)$$

Total air weight flow through cabinet (ref. Eq. 6.7):

$$W = \frac{Q}{C_p \Delta t} = \frac{(399 \text{ watts})(3.413 \text{ Btu/hr watt})}{(0.24 \text{ Btu/lb°F})(60 \text{ min/hr})(29°F)} = 3.26 \frac{\text{lb}}{\text{min}} \quad (8.31)$$

Cfm flow through the cabinet:

$$F = \frac{W}{\rho} = \frac{3.26}{0.0666} = 48.9 \frac{\text{ft}^3}{\text{min}} \quad (8.31a)$$

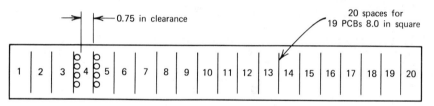

Figure 8.6 Arrangement of each card bucket.

The Reynolds number shown in Eq. 6.11 must be determined for the duct formed by the PCBs.

Given A_d = 20 spaces $(8.0)(0.75) = 120$ in² (total duct area)

$$V = \frac{F}{A_d} = \frac{(48.9 \text{ ft}^3/\text{min})(144 \text{ in}^2/\text{ft}^2)(60 \text{ min/hr})}{120 \text{ in}^2}$$

$V = 3521$ ft/hr (cooling air velocity)

$$D = \frac{2ab}{a+b} = \frac{2(8.0)(0.75)}{8.0 + 0.75}$$

$D = 1.371$ in $= 0.114$ ft (hydraulic diameter of duct)

$\mu = 0.048$ lb/ft hr (viscosity at 136°F; ref. Fig. 6.28) *(page 202)*

$$N_R = \frac{VD\rho}{\mu} = \frac{(3521)(0.114)(0.0666)}{0.0482}$$

$N_R = 554$ (dimensionless) \hfill (8.32)

The Fanning friction factor is obtained from Table 6.7.

$$f = \frac{24}{N_R} = \frac{24}{554} = 0.0433 \text{ (dimensionless)}$$

L = length of air flow path through 7 card buckets, considering only length of bucket

$L = (8)(7 \text{ buckets}) = 56$ in

[margin note: No expansion + contraction losses at inlet + outlet of each card bucket]

$H_1 = 0.5H_{v_1}$ (console entrance loss; ref. Table 8.1)

[margin note: if louvers at inlet/outlet, additional losses added]

$H_2 = 1.0H_{v_2}$ (console exit loss; ref. Table 8.1)

$A_1 = A_2 = 120$ in² (console inlet and exit areas)

Substitute into Eq. 8.29 for the Z factor.

$$Z = 0.226 \left[\frac{4(0.0433)(56 \text{ in}/1.371 \text{ in})}{(120)^2} + \frac{0.5}{(120)^2} + \frac{1.0}{(120)^2} \right]$$

$$Z = 0.000134 \frac{\text{in H}_2\text{O}}{(\text{lb/min})^2} \hfill (8.33)$$

(page 267)

Substitute into Eq. 8.11 for the temperature rise.

$$\Delta t_1 = 2.5 \left(\frac{460 + 122}{100} \right) \left(\frac{0.000134}{6/2} \right)^{1/3} \left(\frac{399}{1} \right)^{2/3}$$

$\Delta t_1 = 28°F$ \hfill (8.34)

8.9 Sample Problem—Induced Draft Cooling of a Console with Card Buckets

A temperature rise of 29°F was assumed to start. Equation 8.11 was then used to calculate the temperature rise for the equilibrium condition, which resulted in a value of 28°F. The calculated value agrees relatively well with the assumed value, so that the results are acceptable.

The natural convection coefficient (h_c) is required to determine the temperature rise across the convection film from the ambient air to the component surface. This is obtained from Eq. 6.9. The Colburn J factor for the ducts formed by the PCBs is determined from Eq. 6.10 for ducts with an aspect ratio greater than about 8 to 1.

$$\text{duct aspect ratio} = \frac{8}{0.75} = 10.7$$

$$J = \frac{6}{(N_R)^{0.98}} \text{ (ref. Eq. 6.10)}$$

$$J = \frac{6}{(554)^{0.98}} = 0.0123 \text{ (dimensionless)} \tag{8.35}$$

The weight velocity flow is determined from Eq. 6.12.

Given $W = 3.26$ lb/min (ref. Eq. 8.31)
$A_d = 120$ in² $= 0.833$ ft² (flow area through PCBs)

$$G = \frac{W}{A_d} = \frac{(3.26 \text{ lb/min})(60 \text{ min/hr})}{0.833 \text{ ft}^2} = 234.8 \, \frac{\text{lb}}{\text{hr ft}^2} \tag{8.36}$$

$\frac{C_p \mu}{K} = 0.70$ (dimensionless Prandtl number; ref. Figure 6.28) for a temperature of 136.5°F

Substitute into Eq. 6.9 for the convection coefficient.

$$h_c = \frac{(0.0123)(0.24)(234.8)}{(0.70)^{2/3}} = 0.88 \, \frac{\text{Btu}}{\text{hr ft}^2 \, °F} \tag{8.37}$$

The temperature rise across the convection air film for one PCB is determined from Eq. 5.7.

Given $Q = 3$ watts $\times$ 3.413 Btu/hr watt $= 10.24$ Btu/hr (heat)

$h_c = 0.88$ Btu/hr ft² °F (convection coefficient)

$A_s =$ surface area of one board face (if there is a lot of copper on the back side of each board and the component lead wires extend through the board, the effective area can be increased as much as 30%)

$$A_s = \frac{(1.3)(8.0)(8.0) \text{ in}^2}{144 \text{ in}^2/\text{ft}^2} = 0.577 \text{ ft}^2 \text{ (each board)}$$

(with annotation: "Area factor" pointing to 1.3)

$$\Delta t_2 = \frac{Q}{h_c A_s} = \frac{10.24}{(0.88)(0.577)} = 20°F \tag{8.38}$$

The hot spot surface temperature (t_{hs}) for the electronic components mounted on the PCB will be the sum of the Δt's plus the ambient air temperature at the console inlet.

$$t_{hs} = 122 + \Delta t_1 + \Delta t_2$$
$$t_{hs} = 122 + 28 + 20 = 170°F \tag{8.39}$$

The pressure drop through the console is determined with the use of Eq. 8.1.

Given $H = \dfrac{6}{2} = 3$ ft (console height for uniform heat)

$\rho = 0.0666$ lb/ft³ (air average density; ref. Eq. 8.30)
$T_{in} = 460 + 122 = 582°R$ (absolute inlet air temperature)
$\Delta t_1 = 28°F$ (temperature rise through console)
$T_{out} = T_{in} + \Delta t_1 = 582 + 28 = 610°R$ (exit temperature)

$$\Delta P = 0.192 \rho H \left(1 - \frac{T_{out}}{T_{in}}\right)$$

$$\Delta P = 0.192(0.0666)(3)\left(1 - \frac{610}{582}\right) = -0.00184 \text{ in H}_2\text{O} \tag{8.40}$$

8.10 ELECTRONICS PACKAGED WITHIN SEALED ENCLOSURES

Electronic equipment is often packaged within sealed enclosures to minimize radio frequency interference (RFI) or electromagnetic interference (EMI) or to protect equipment that must operate in very dusty or dirty areas.

In order to determine how the internal ambient temperatures are affected by natural convection and radiation, James K. Tierney and Eugene Koczkur ran a series of tests on a completely enclosed aluminum cabinet [54] which measured 30 in high, 20 in wide and 15 in deep. Several heat sources were used, which could be moved around within the cabinet. The interior walls were painted with a black paint that had an emissivity of 0.98. The outside walls were left in their natural finish, with an emissivity of 0.2 for the first

8.10 Electronics Packaged within Sealed Enclosures

tests. The outer surfaces were then finished with the high emissivity paint, and the tests were repeated.

Wiring and heaters occupied about 46% of the free cabinet volume, leaving 54% of the cabinet volume free. Temperatures were obtained from thermocouples placed throughout the system. A summary of the test results is shown in Figures 8.7 and 8.8 [54].

When filters and louvers on conventionally designed, naturally cooled electronic systems become plugged with dust and dirt, the results shown in the figures give the approximate temperature rise that may be expected under these conditions.

Figures 8.7 and 8.8 show how the internal ambient air temperature will rise as the power dissipation within the cabinet increases. This is for a uniform heat distribution within the cabinet, with and without a high emissivity paint on the outside surface [54].

Figure 8.9 shows what happens when too much heat is concentrated near the top of the enclosed cabinet. The hot component heats the air locally, and the reduced density forces the air to rise (in a gravity field). Convection is suppressed and the air is trapped, because it cannot move down to replace the cooler air below as it would in normal convection.

Figure 8.10 shows how natural convection can cool a hot component when it is placed near the bottom of a totally enclosed cabinet. Gravity will

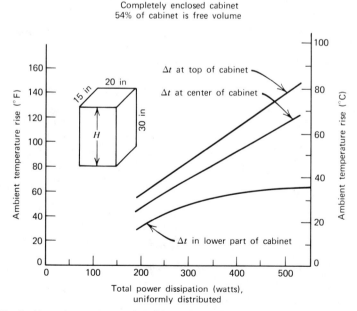

Figure 8.7 Ambient air temperature in cabinet when outside skin emissivity $e = 0.2$, with no paint on the outside.

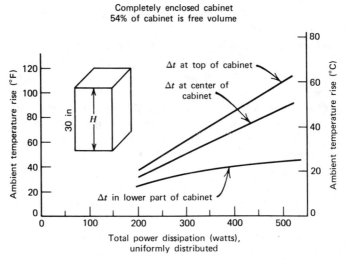

Figure 8.8 Ambient air temperature in cabinet when outside skin emissivity $e = 0.98$, with black paint on the outside.

permit the warm air to rise and pass across the internal wall surfaces, for more effective heat removal [54].

When the power dissipation is approximately uniform along the height of the console, Figure 8.11 shows how the internal ambient air temperature will vary along the height of the console [54].

These tests show that increasing the emissivity of the outer skin from 0.2 to 0.98 will reduce the internal ambient air temperature rise as much as 25°F

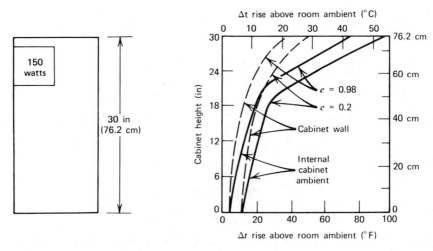

Figure 8.9 Temperature rise of internal ambient air when the heat source is located near the top of the cabinet.

8.11 Small Enclosed Modules within Large Consoles

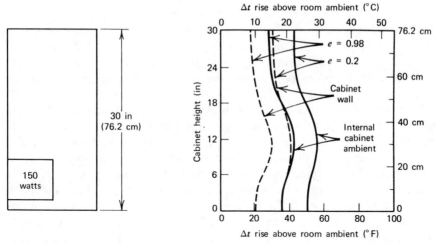

Figure 8.10 Temperature rise of internal ambient air when the heat source is located near the bottom of the cabinet.

and will reduce the outer skin temperature by the same amount. Radiation is therefore an important factor in controlling the interior temperatures of sealed electronic consoles.

8.11 SMALL ENCLOSED MODULES WITHIN LARGE CONSOLES

Printed circuit boards are often enclosed within small sheet metal boxes which are then installed in large consoles. The sheet metal enclosures

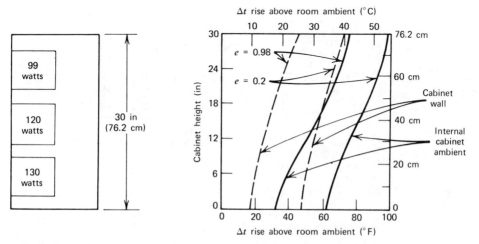

Figure 8.11 Temperature rise of internal ambient air when the heat source is approximately uniform along the height.

minimize RFI and EMI problems, but they also prevent any form of direct ventilation for cooling. Heat generated by the components must be removed by a combination of internal conduction, convection, and radiation to the inside walls of the enclosure. The heat must then pass through the walls, where it is removed by convection and some radiation to the outside ambient, as shown in Figure 8.12.

When all of the components on the PCBs are approximately the same height, it is possible to place the components very close to the sides of the enclosed PCB module. Heat from the components can then be transferred by conduction and radiation across the small air gap for more effective cooling. In most systems, however, there will be a large variation in the size of the components. When tall components are spaced close to the side walls, the air gap between the walls and the small components may become quite large. A large air gap will increase the conduction resistance from the small components to the side walls, resulting in a large temperature rise. Also, tall components may block the internal convection path, so that this mode of heat transfer may be sharply reduced. Therefore, when there are large variations in the heights of various components on PCBs, an air gap of about 0.75 in should be provided between the tallest components and side walls. This will ensure the transfer of heat by natural convection in addition to radiation within the PCB enclosure.

If a smaller air gap than 0.75 in is used, pinching may occur and the internal natural convection will be reduced. Test data show that an air gap

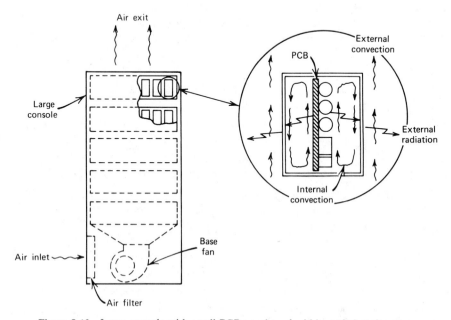

Figure 8.12 Large console with small PCBs packaged within sealed enclosures.

8.12 Sample Problem—Small PCB Sealed within an RFI Enclosure

of about 0.50 in between the PCB components and the inside surface of the sealed enclosure can reduce the internal natural convection coefficient as much as 50%.

A mathematical model of the internal and external convection and radiation thermal resistors can be established for the sealed enclosure, as shown in Figure 8.13. The radiation thermal resistors are shown in parallel with the convection thermal resistors, since they both transfer heat between the same nodes. Very little heat will be conducted across an air gap of 0.75 in, so that conduction heat transfer is ignored.

Heat transferred by radiation from the PCB to the walls of the enclosure can be increased by painting the inside surfaces of the enclosure or by anodizing the surfaces if they are aluminum. Most organic paints and finishes of any color, including clear lacquer, will increase the emittance of the surface. However, aluminum paint, will not provide a very large increase in the emittance, so that it should be avoided.

External heat transfer by radiation from the enclosed module will be negligible when the modules are close to one another. Under these conditions the modules can "see" each other, so that no heat will be effectively lost by radiation.

The module surfaces that face the end walls of the large cabinet will transfer more heat by radiation, because the cabinet walls will normally be much cooler than the walls of the small sealed modules.

8.12 SAMPLE PROBLEM—SMALL PCB SEALED WITHIN AN RFI ENCLOSURE

A PCB is mounted within a sealed enclosure to provide protection against radio frequency interference (RFI). The small sealed enclosure is mounted within a large console that is cooled by forced convection. The average velocity of the cooling air through the console is about 150 ft/min. The

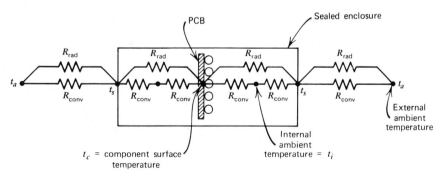

Figure 8.13 Mathematical model using a thermal resistor network to represent a small PCB within a sealed enclosure.

modules are spaced 0.10 in (0.254 cm) apart. Each module measures 5.0 in (12.7 cm) high × 5.5 in (13.97 cm) deep × 1.5 in (3.81 cm) thick and dissipates about 4.0 watts, as shown in Figure 8.14. The maximum cooling air temperature expected at sea level conditions is 122°F (50°C). The inside surface of the PCB enclosure is painted black to improve its radiation heat transfer. Determine the approximate hot spot temperature of the PCB surface and the enclosure surface when the cooling air temperature is 122°F.

SOLUTION

External radiation cooling is negligible, since the external surface of one sealed enclosure is adjacent to the other. A mathematical model of the system is shown in Figure 8.15. The model has been simplified by conservatively assuming that all the heat is removed from only one surface of the sealed enclosure. This condition occurs when the PCB within the enclosure has very little copper on the back side, or when the component lead wires do not extend through the PCB to the back side. Under these circumstances, the back side of the PCB will dissipate relatively little heat.

The external convection resistance R_3 is determined first. Its value can be obtained from the forced convection coefficient (h_c) and the external surface area (A_s), using Eq. 5.28. The Reynolds number must be calculated first, using Eq. 6.11. (The average air temperature is estimated to be about 130°F)

Given $V = 150$ ft/min $= 9000$ ft/hr $= 76.2$ cm/sec (velocity)

$$D = \frac{2ab}{a+b} = \frac{2(5.5)(0.10)}{5.5 + 0.10}$$

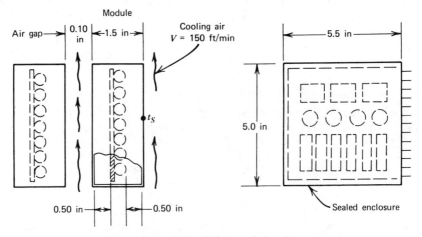

Figure 8.14 PCB within a sealed enclosure.

8.12 Sample Problem—Small PCB Sealed within an RFI Enclosure

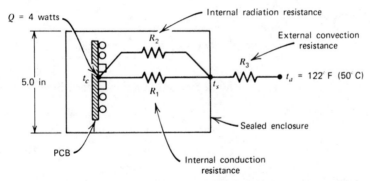

Figure 8.15 Simplified mathematical model for a PCB in a sealed enclosure.

$= 0.196$ in (hydraulic diameter of air gap, 0.10 in)
$D = 0.0163$ ft $= 0.498$ cm
$\rho = 0.067$ lb/ft³ $= 0.00107$ g/cm³ (air density; ref. Figure 6.28) at 130°F
$\mu = 0.048$ lb/ft hr $= 0.000198$ g/cm sec (viscosity of air at 130°F)

Substitute into Eq. 6.11 for the Reynolds number in English units

$$N_R = \frac{(9000 \text{ ft/hr})(0.0163 \text{ ft})(0.067 \text{ lb/ft}^3)}{0.048 \text{ lb/ft hr}}$$

$$N_R = 205 \text{ (dimensionless)} \quad (8.41)$$

In metric units:

$$N_R = \frac{(76.2)(0.498)(0.00107)}{0.000198} = 205 \quad (8.41a)$$

For duct aspect ratios greater than about 8, Eq. 6.10 can be used to determine the Colburn J factor.

$$J = \frac{6}{(N_R)^{0.98}} = \frac{6}{(205)^{0.98}} = 0.0325 \text{ (dimensionless)}$$

The CFM flow between the sealed modules can be determined for English units. The duct formed by the space between the sealed modules is 0.10×5.5 in for a vertical flow of air through the system, with a velocity of 150 ft/min.

$$F = AV = \frac{(0.10)(5.5)\text{in}^2}{144 \text{ in}^2/\text{ft}^2} \times 150 \frac{\text{ft}}{\text{min}} = 0.57 \frac{\text{ft}^3}{\text{min}}$$

In metric units:

$$F = (0.254 \text{ cm})(13.97 \text{ cm})\left(76.2 \frac{\text{cm}}{\text{sec}}\right) = 270 \frac{\text{cm}^3}{\text{sec}}$$

The weight flow between the sealed modules is determined for English units.

$$W = \rho F = \left(0.067 \frac{\text{lb}}{\text{ft}^3}\right)\left(0.57 \frac{\text{ft}^3}{\text{min}}\right) = 0.038 \frac{\text{lb}}{\text{min}} \quad (8.42)$$

In metric units:

$$W = \left(0.00107 \frac{\text{g}}{\text{cm}^3}\right)\left(270 \frac{\text{cm}^3}{\text{sec}}\right) = 0.289 \frac{\text{g}}{\text{sec}} \quad (8.42\text{a})$$

Substitute into Eq. 6.12 for the weight velocity, using English units.

$$G = \frac{W}{A} = \frac{(0.038 \text{ lb/min})(144 \text{ in}^2/\text{ft}^2)(60 \text{ min/hr})}{(0.10)(5.5) \text{ in}^2} = 597 \frac{\text{lb}}{\text{hr ft}^2} \quad (8.43)$$

In metric units:

$$G = \frac{0.289 \text{ g/sec}}{(0.254)(13.97) \text{ cm}^2} = 0.0814 \frac{\text{g}}{\text{sec cm}^2} \quad (8.43\text{a})$$

Substitute into Eq. 6.9 for the convection coefficient, using the Prandtl number of 0.70 for air at 130°F, as shown in Figure 6.28, in English units.

$$h_c = JC_p G \left(\frac{C_p \mu}{K}\right)^{-2/3} = \frac{(0.0325)(0.24)(597)}{(0.70)^{2/3}}$$

$$h_c = 5.9 \frac{\text{Btu}}{\text{hr ft}^2 \text{ °F}} \quad (8.44)$$

In metric units:

$$h_c = \frac{(0.0325)(0.24)(0.0814)}{(0.70)^{0.666}} = 0.000805 \frac{\text{cal}}{\text{sec cm}^2 \text{ °C}} \quad (8.44\text{a})$$

The thermal resistance R_3 is determined from Eq. 5.28 for English units. The surface area (A_s) is 5.0×5.5 in².

$$R_3 = \frac{1}{h_c A_s} = \frac{144 \text{ in}^2/\text{ft}^2}{(5.9 \text{ Btu/hr ft}^2 \text{ °F})(5.0)(5.5) \text{in}^2}$$

$$R_3 = 0.887 \frac{\text{hr °F}}{\text{Btu}} \quad (8.45)$$

8.12 Sample Problem—Small PCB Sealed within an RFI Enclosure

In metric units:

$$R_3 = \frac{1}{(0.000805 \text{ cal/sec cm}^2 \text{ °C})(12.7)(13.97)\text{cm}^2} = 7.0 \frac{\text{sec °C}}{\text{cal}} \quad (8.45a)$$

The temperature rise across the forced convection film from the ambient to the surface of the enclosure is the same as the temperature rise across resistor R_3, and it is obtained from Eq. 3.15 for English units.

$$\Delta t_3 = QR_3 = (4 \text{ watts})\left(3.413 \frac{\text{Btu}}{\text{hr watt}}\right)\left(0.887 \frac{\text{hr °F}}{\text{Btu}}\right)$$

$$\Delta t_3 = 12.1°F \quad (8.46)$$

In metric units:

$$\Delta t_3 = (4 \text{ watts})\left(0.239 \frac{\text{cal}}{\text{sec watt}}\right)\left(7.0 \frac{\text{sec °C}}{\text{cal}}\right) = 6.7°C \quad (8.46a)$$

The surface temperature of the sealed enclosure is now obtained for English units, where the ambient temperature is 122°F.

$$t_s = 122 + \Delta t_3 = 122 + 12.1 = 134.1°F \quad (8.47)$$

In metric units, where the ambient temperature is 50°C:

$$t_s = 50 + 6.7 = 56.7°C \quad (8.47a)$$

To determine the values of resistors R_1 and R_2, it is necessary to assume a value for the temperature rise from the module surface (t_s) to the PCB component surface (t_c). The assumed value is compared with the calculated value, and a correction is made with a second iteration if there is a poor match. A good approximation of the temperature rise can be obtained with two or three iterations. Start by assuming a temperature rise of 40°F (22.2°C).

The air gap between the PCB electronic components and the inside wall of the enclosure is only 0.50 in (1.27 cm), so that there is not enough space for fully developed convection currents. Therefore, an equivalent convection coefficient can be obtained based on the gaseous conduction of heat across the 0.50 in gap, as shown by Eq. 5.32, considering an average air gap temperature of about 150°F (65.5°C).

$$h_{AG} = \frac{K}{L} \quad \text{(ref. Eq. 5.32)}$$

Given $K = 0.017$ Btu/hr ft °F (ref. Figure 6.28)
$L = 0.50$ in $= 0.0417$ ft (length of air gap)

$$h_{AG} = \frac{0.017}{0.0417} = 0.41 \frac{\text{Btu}}{\text{hr ft}^2 \text{ °F}} = 0.000055 \frac{\text{cal}}{\text{sec cm}^2 \text{ °C}} \quad (8.48)$$

The internal convection thermal resistance is obtained from Eq. 5.28 for English units.

$$R_1 = \frac{1}{h_{AG} A} \text{ (ref. Eq. 5.28)}$$

Given $A = \frac{(5.0)(5.5)}{144} = 0.191 \text{ ft}^2 = 177.4 \text{ cm}^2$

$$R_1 = \frac{1}{(0.41)(0.191)} = 12.7 \frac{\text{hr °F}}{\text{Btu}} \quad (8.49)$$

In metric units:

$$R_1 = \frac{1}{(0.000055)(177.4)} = 102.5 \frac{\text{sec °C}}{\text{cal}} \quad (8.49a)$$

The radiation coefficient (h_r) is obtained from Figure 5.29 for a view factor of 1.0; an emissivity of 0.9, based upon the assumed temperature rise of 40°F; and a temperature of the heat receiver wall of 134°F, as shown by Eq. 8.47.

$$h_r = 1.4 \frac{\text{Btu}}{\text{hr ft}^2 \text{ °F}} = 0.000189 \frac{\text{cal}}{\text{sec cm}^2 \text{ °C}} \quad (8.50)$$

Radiation resistor R_2 is computed from Eq. 5.28 for English units.

$$R_2 = \frac{1}{h_r A_s} = \frac{144 \text{ in}^2/\text{ft}^2}{(1.4 \text{ Btu/hr ft}^2 \text{ °F})(5.0)(5.5) \text{ in}^2}$$

$$R_2 = 3.74 \frac{\text{hr °F}}{\text{Btu}} \quad (8.51)$$

In metric units:

$$R_2 = \frac{1}{(0.000189 \text{ cal/sec cm}^2 \text{ °C})(12.7)(13.97) \text{ cm}^2}$$

$$R_2 = 29.8 \frac{\text{sec °C}}{\text{cal}} \quad (8.51a)$$

8.12 Sample Problem—Small PCB Sealed within an RFI Enclosure

Resistor R_1 is in parallel with resistor R_2. They can be combined using Eq. 3.17 for English units.

$$\frac{1}{R_c} = \frac{1}{R_1} + \frac{1}{R_2} = \frac{1}{12.7} + \frac{1}{3.74}$$

$$R_c = 2.89 \frac{\text{hr °F}}{\text{Btu}} \tag{8.52}$$

In metric units:

$$\frac{1}{R_c} = \frac{1}{102.5} + \frac{1}{29.8}$$

$$R_c = 23.1 \frac{\text{sec °C}}{\text{cal}} \tag{8.52a}$$

The calculated temperature rise is determined from Eq. 3.15 for English units.

$$\Delta t = QR_c = (4 \text{ watts})\left(3.413 \frac{\text{Btu}}{\text{hr watt}}\right)\left(2.89 \frac{\text{hr °F}}{\text{Btu}}\right) = 39.5 \text{°F} \tag{8.53}$$

In metric units:

$$\Delta t = (4 \text{ watts})\left(0.239 \frac{\text{cal}}{\text{sec watt}}\right)\left(23.1 \frac{\text{sec °C}}{\text{cal}}\right) = 22.1 \text{°C} \tag{8.53a}$$

A temperature rise of 40°F was assumed to start. The calculated value was shown by Eq. 8.53 to be 39.5°F, which agrees quite well, so that the assumed value is acceptable.

The component surface temperature can now be calculated using Eqs. 8.47 and 8.53, for English units.

$$t_{\text{comp}} = t_s + \Delta t = 134.1 + 39.5 = 173.6 \text{°F} \tag{8.54}$$

In metric units:

$$t_{\text{comp}} = 56.7 + 22.1 = 78.8 \text{°C} \tag{8.54a}$$

It should be pointed out that the power dissipation of 4 watts was assumed to be uniformly distributed over the face of one side on the circuit board. If the same circuit board is only half populated with components but has the same power dissipation of 4 watts, only half of the circuit board area may be effective. Under these circumstances, the temperature rise of 39.5°F, shown by Eq. 8.53 will be increased substantially. The value of h_c

will increase, as will the value of h_r. The effective PCB heat transfer area will, however, decrease about 50%.

The component surface temperature shown by Eq. 8.54 represents an average temperature at the center of the PCB. Components at the top of the PCB will be a few degrees hotter, and components at the bottom of the PCB a few degrees cooler than the average temperature.

8.13 TEST DATA FOR SMALL ENCLOSED MODULES

Thermal tests were run on a series of PCBs that were individually sealed in small aluminum enclosures, as shown in Figure 8.16. Resistors were mounted on one face of each PCB to generate a measured amount of heat. Thermocouples were placed throughout the sealed enclosure and on the resistors, to determine the temperature profile for different power dissipations. The outer surfaces of the enclosures were iridited, to prevent corrosion. The inside surfaces of the enclosures were painted black on half of the units and iridited on half of the units. All the tests were run with 2 different sets of painted and nonpainted covers to make sure that the data

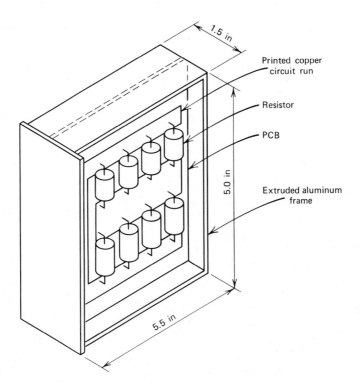

Figure 8.16 Sealed test unit with its cover removed.

8.13 Test Data for Small Enclosed Modules

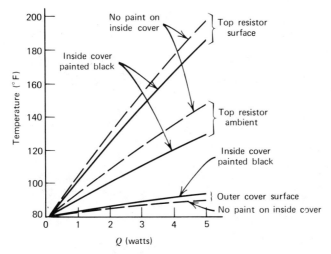

Figure 8.17 Temperature profile of the top resistors when the enclosed module is sitting in free air.

were consistent and repeatable. The results of the tests are shown in Figures 8.17 through 8.21.

The following conclusions can be drawn from these tests:

1. The black paint (or any paint, except aluminum paint) on the inside surface of the covers increases the radiation heat transfer and reduces the average internal ambient temperatures about 10%.

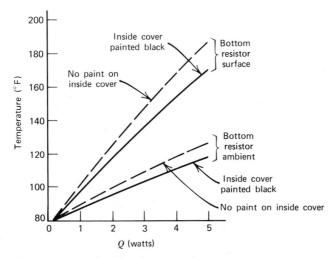

Figure 8.18 Temperature profile of the bottom resistors when the enclosed module is sitting in free air.

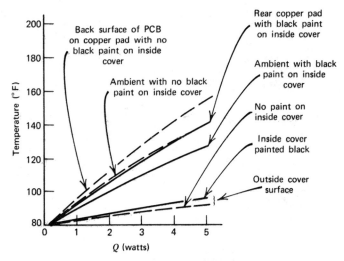

Figure 8.19 Temperature profile of the PCB back surface when the enclosed module is sitting in free air.

2. A substantial amount of heat is conducted through the circuit board to the back side of the board, which increases the overall area available for heat removal. It appears about 33% of the total heat is conducted through the PCB. Most of this is through the electrical lead wires, which were soldered to extensive copper runs on both sides of the PCB.

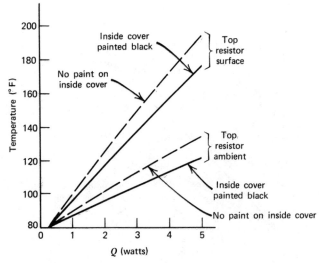

Figure 8.20 Temperature profile of the top resistors when the ambient air over the enclosed module has a velocity of 150 ft/min.

8.14 Pressure Losses in Series and Parallel Air Flow Ducts

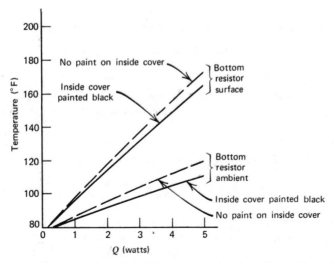

Figure 8.21 Temperature profile of the bottom resistors when the ambient air over the enclosed module has a velocity of 150 ft/min.

3 The top row of resistor were about 10% hotter than the bottom row of resistors. Natural convection appeared to be working in the 0.50 in space between the components and the inside surface of the enclosure. However, it was suppressed about 50%.

4 Using forced convection cooling on the external surfaces of the module, with an air velocity of about 150 linear ft/min, reduces the average temperature within the module about 6%, compared to the values obtained with the same module sitting in free air.

8.14 PRESSURE LOSSES IN SERIES AND PARALLEL AIR FLOW DUCTS

Cooling air flow ducts can usually be considered as systems that consist of series and parallel air flow paths. As the air is forced through a typical electronic chassis or console, the air will be forced to turn, contract, and expand [55]. These changes require energy, which results in flow losses that increase the pressure drop through the system. A very complex air flow path can have branches with high losses, which will sharply reduce the amount of cooling air passing through those branches. If these branches are not identified in advance, and if design changes are not made, overheating of the electronic equipment can occur and critical parts may burn out.

It is convenient to express the flow resistance in terms of the static pressure loss factor (Z) as shown in Eq. 8.12, since it is related to the pressure loss in terms of velocity heads (H_v).

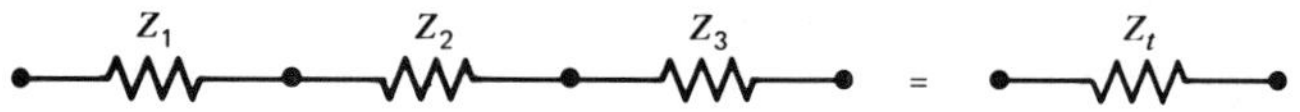

Figure 8.22 Series flow resistor network.

Two basic resistance patterns, series and parallel, are used to generate analog resistor networks of cooling air flow paths. A simple series network is shown in Figure 8.22.

The total effective resistance (Z_t) for the series flow network is determined by adding all the individual resistances as shown in Eq. 8.55.

(same as electrical resistors) →

$$Z_t = Z_1 + Z_2 + Z_3 + \cdots \qquad (8.55)$$

A simple parallel flow resistor network is shown in Figure 8.23.

The total effective resistance (Z_t) for the parallel flow network is determined from Eq. 8.56.

(not same as electrical resistors) →

$$\frac{1}{\sqrt{Z_t}} = \frac{1}{\sqrt{Z_1}} + \frac{1}{\sqrt{Z_2}} + \frac{1}{\sqrt{Z_3}} + \cdots \qquad (8.56)$$

The use of these resistance network flow equations is demonstrated with a sample problem.

8.15 SAMPLE PROBLEM—SERIES AND PARALLEL AIR FLOW NETWORK

A cooling air flow network with six restrictions has a combination of series and parallel air flow paths as shown in Figure 8.24. The cross section data at each restriction is 10 in². The loss at each restriction is 3.0 velocity heads (H_v). The average air density (ρ) is 0.076 lb/ft³. Determine the air weight flow (W) and the pressure drop (ΔP) through each branch, when the total cfm flow (F) through the system is 100 cfm.

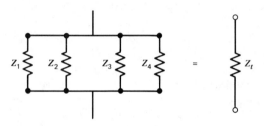

Figure 8.23 Parallel flow resistor network.

8.15 Sample Problem—Series and Parallel Air Flow Network

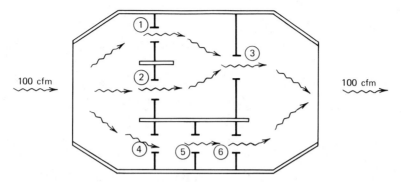

Figure 8.24 Cooling air flow network with six flow restrictions.

SOLUTION

A mathematical model of the flow network is shown in Figure 8.25.

The cooling air flow loss factor Z for each restriction is determined from Eq. 8.12.

$$Z_1 = Z_2 = Z_3 = Z_4 = Z_5 = Z_6 = \frac{0.226 H_v}{A^2}$$

$$Z_1 \text{ through } Z_6 = \frac{(0.226)(3.0)}{(10)^2} = 0.00678 \frac{\text{in H}_2\text{O}}{(\text{lb/min})^2} \tag{8.57}$$

Z_1 and Z_2 are in parallel, as shown in Figure 8.26. They can be combined with the use of Eq. 8.56.

$$\frac{1}{\sqrt{Z_7}} = \frac{1}{\sqrt{Z_1}} + \frac{1}{\sqrt{Z_2}} = \frac{1}{\sqrt{0.00678}} + \frac{1}{\sqrt{0.00678}}$$

$$Z_7 = 0.00169 \frac{\text{in H}_2\text{O}}{(\text{lb/min})^2} \tag{8.58}$$

The resistor network will now appear as shown in Figure 8.27. Since the

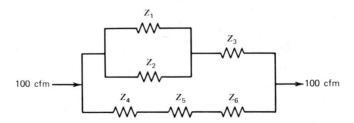

Figure 8.25 Analog resistor network for six flow restrictions.

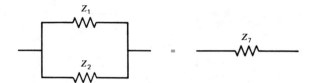

Figure 8.26 Parallel flow circuit for Z_1 and Z_2.

resistors are in series, Eq. 8.55 is used to determine the combined value.

$$Z_8 = 0.00169 + 0.00678 = 0.00847$$
$$Z_9 = (0.00678)(3) = 0.02034$$

Z_8 and Z_9 are combined as resistors in parallel.

$$\frac{1}{\sqrt{Z_t}} = \frac{1}{\sqrt{0.00847}} + \frac{1}{\sqrt{0.02034}}$$

$$Z_t = 0.00313 \frac{\text{in } H_2O}{(\text{lb/min})^2} \qquad (8.59)$$

The total weight flow through the system is determined from Eq. 6.34.

$$W = \rho F = \left(0.076 \frac{\text{lb}}{\text{ft}^3}\right)\left(100 \frac{\text{ft}^3}{\text{min}}\right) = 7.6 \frac{\text{lb}}{\text{min}} \qquad (8.60)$$

The pressure drop through the system can be determined from Eq. 8.61.

$$\Delta P = ZW^2 = \text{in } H_2O \qquad (8.61)$$

Given Z = pressure loss factor = $\dfrac{\text{in } H_2O}{(\text{lb/min})^2}$

W = flow in branch = $\dfrac{\text{lb}}{\text{min}}$

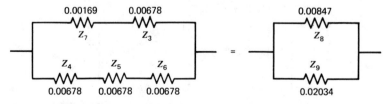

Figure 8.27 Combined resistor network for flow pattern.

8.15 Sample Problem—Series and Parallel Air Flow Network

The weight flow through each branch in the system is determined from the pressure drop and the flow loss (Z).

The pressure drop through the complete system is obtained by substituting Eqs. 8.60 and 8.59 into Eq. 8.61.

$$\Delta P = (0.00313)(7.6)^2 = 0.181 \text{ in } H_2O \tag{8.62}$$

The weight flow through branches Z_8 and Z_9 in Figure 8.27 is obtained by considering the characteristics of a parallel flow system. In this system, the pressure drop across each element will be the same, or 0.181 in H_2O. Substitute this value into Eq. 8.61.

$$W_8 = \sqrt{\frac{\Delta P}{Z_8}} = \sqrt{\frac{0.181}{0.00847}} = 4.62 \frac{\text{lb}}{\text{min}} \tag{8.63}$$

$$W_9 = \sqrt{\frac{\Delta P}{Z_9}} = \sqrt{\frac{0.181}{0.02034}} = 2.98 \frac{\text{lb}}{\text{min}} \tag{8.64}$$

The sum of W_8 and W_9 must add up to 7.60, which checks with the required total flow.

The flow through branches Z_7 and Z_3 must be the same as Z_8, since they are in series.

$$W_7 = W_3 = W_8 = 4.62 \frac{\text{lb}}{\text{min}} \tag{8.65}$$

The flow through branches Z_4, Z_5, and Z_6 must all be the same as Z_9, since they are in series.

$$W_4 = W_5 = W_6 = W_9 = 2.98 \frac{\text{lb}}{\text{min}} \tag{8.66}$$

Branch Z_7 must be investigated to find the flow in branches Z_1 and Z_2, as shown in Figure 8.26. The pressure drop across branch Z_7 is determined from Eqs. 8.58, 8.61, and 8.65.

$$\Delta P_7 = Z_7 W_7^2 = (0.00169)(4.62)^2 = 0.0361 \text{ in } H_2O \tag{8.67}$$

The pressure drop is the same for branches Z_1 and Z_2, since they are in parallel. The weight flow in each branch is determined from Eq. 8.57.

$$W_1 = \sqrt{\frac{\Delta P_7}{Z_1}} = \sqrt{\frac{0.0361}{0.00678}} = 2.31 \frac{\text{lb}}{\text{min}} \tag{8.68}$$

Figure 8.28 Tall Modcomp 7860 series computer with internal fans for cooling the electronics. (Courtesy Modular Computer Systems, Inc.)

8.15 Sample Problem—Series and Parallel Air Flow Network

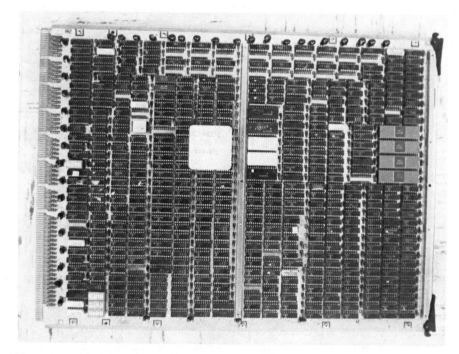

Figure 8.29 Large plug-in PCB for modcomp computer uses wire wrapped interconnections for approximately 400 DIPs. (Courtesy Modular Computer Systems, Inc.)

Since $Z_1 = Z_2$, the flow will be equal, so $W_1 = W_2$.

$$W_2 = W_1 = 2.31 \frac{\text{lb}}{\text{min}} \tag{8.69}$$

As a final check, Figures 8.25 and 8.27 show the following:

$$W_1 + W_2 = W_3 = W_8 = 4.62$$
$$2.31 + 2.31 = 4.62 \frac{\text{lb}}{\text{min}} \text{ (checks)} \tag{8.70}$$

Figure 8.28 shows a tall cabinet with large circuit boards that are cooled with small axial flow fans. One of the large circuit boards, with approximately 400 DIPs, is shown in Figure 8.29.

9

Transient Cooling for Electronic Systems

9.1 SIMPLE INSULATED SYSTEMS

Electronic systems will experience transient heating conditions when the power is first turned on, when there is a change in the power, or when the cooling system is shut off while the electronics is still operating. Spacecraft will experience transient heating conditions due to changing attitudes with respect to the sun, which changes the solar heat load.

When a body is heated, its temperature will rise if the heat is not removed. The body temperature will continue to rise as long as the rate at which the heat is applied is greater than the rate at which the heat is removed. If the body is perfectly insulated, all of the applied heat will be available to raise the temperature. The temperature increase will then be linear, as shown in Figure 9.1.

The temperature rise with respect to time can then be determined from Eq. 6.7, which is repeated here for convenience.

$$\Delta t = \frac{Q}{WC_p} = \frac{°F}{hr} \text{ or } \frac{°C}{sec} \quad \text{(ref. Eq. 6.7)}$$

where Q = heat input = Btu/hr or cal/sec
 W = weight = lb or g
 C_p = specific heat = Btu/lb °F or cal/g °C

FOR ENTIRE BOX
total POWER
TOTAL WEIGHT
AVERAGE C_p

Equation 6.7 is often convenient for determining the worst-case temperature rise in a simple system. Since all real systems have some heat transferred as they are heated internally, the worst-case condition is obtained by assuming no heat is lost by conduction, convection, or radiation. All the

9.2 Sample Problem—Transient Temperature Rise of a Transformer

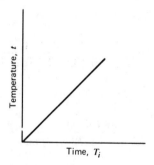

Figure 9.1 Uniform temperature rise for insulated heated body.

heat is therefore available for increasing the temperature, so that the maximum theoretical temperature is obtained.

9.2 SAMPLE PROBLEM—TRANSIENT TEMPERATURE RISE OF A TRANSFORMER

A transformer is rigidly bolted to an aluminum bracket which weighs 0.50 lb (227 g). The transformer dissipates 10 watts and weighs 2.5 lb (1135 g). The transformer is mounted near the corner of a box that has poor ventilation. Also, the surrounding walls are aluminum, which has a shiny surface, so that they will not pick up much heat by radiation. The heat transfer from the transformer by conduction and convection is also poor. Determine approximately how long the transformer can operate before its temperature reaches the maximum allowable of 239°F (115°C), starting from room temperature of 80°F (26.6°C).

SOLUTION

Since there are two different materials in intimate contact, Eq. 6.7 is written in a slightly different form, as shown in Eq. 9.1.

$$\Delta t = \frac{Q}{W_1 C_{p1} + W_2 C_{p2}} \quad (9.1)$$

Given $Q = 10$ watts = 34.13 Btu/hr = 2.39 cal/sec (heat)
$W_1 = 2.5$ lb = 1135 g (transformer weight)
$C_{p1} = 0.15$ Btu/lb °F = cal/g °C (transformer specific heat)
$W_2 = 0.50$ lb = 227 g (aluminum bracket weight)
$C_{p2} = 0.22$ Btu/lb °F = cal/g °C (bracket specific heat)

Substitute into Eq. 9.1 using English units.

$$\Delta t = \frac{34.13 \text{ Btu/hr}}{(2.5)(0.15) + (0.50)(0.22) \text{ Btu/°F}} = 70.4 \frac{°F}{hr} \quad (9.2)$$

Substitute into Eq. 9.1 using metric units.

$$\Delta t = \frac{2.39 \text{ cal/sec}}{(1135)(0.15) + (227)(0.22) \text{ cal/°C}} = 0.0108 \frac{°C}{sec} \quad (9.2a)$$

In English units, the time (T_i) it will take for the transformer to reach a temperature of 239°F is

$$T_i = \frac{(239 - 80) \text{ °F}}{70.4 \text{ °F/hr}} = 2.26 \text{ hrs} \quad (9.3)$$

In metric units, the time it will take for the transformer to reach 115°C is

$$T_i = \frac{(115 - 26.6) \text{ °C}}{0.0108 \text{ °C/sec}} = 8185 \text{ sec} = 2.27 \text{ hr} \quad (9.3a)$$

9.3 THERMAL CAPACITANCE

All real systems have the ability to absorb some of the heat as it is applied. A system with a high thermal capacity, or capacitance can absorb more heat for the same temperature rise than can a system with a low thermal capacity. The thermal capacitance (C) is defined in Eq. 9.4. The units are the same as those shown in Eq. 9.1 [3, 15, 17].

$$C = WC_p = \frac{\text{Btu}}{°F} \text{ or } \frac{\text{cal}}{°C} \quad (9.4)$$

When the power is first turned on in an electronic system, nearly all of the heat is available to raise the temperature, because very little of the heat is lost from the system. As the temperature rises, the heat will find a number of different flow paths to other areas or to the outside ambient. This will tend to reduce the rate of the temperature increase. In a well-designed system, the temperature rise will slowly decrease until a steady state temperature is reached. In a poorly designed system, the temperature rise will continue until some component overheats and burns out. In either case the temperature rise will be more rapid when the power is first turned on; then the slope will gradually decrease until the steady state condition is reached, as shown in Figure 9.2.

9.4 Time Constant

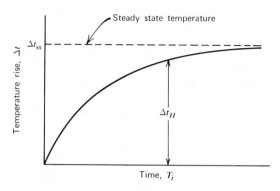

Figure 9.2 Transient temperature rise with respect to time during heating cycle.

9.4 TIME CONSTANT

The time constant (τ) determines how fast the temperature rise will occur with respect to time. A large time constant shows a large mass or a large resistance in the heat flow path, so that the rise is gradual. A small time constant shows a small mass or a small resistance in the heat flow path, so that the rise is more rapid. The temperature rise characteristics are shown in Figure 9.3.

The time constant (τ) is simply the product of the thermal resistance and the capacitance, as shown in Eq. 9.5.

$$\tau = RC = \text{hr, min, or sec} \qquad (9.5)$$

where R = thermal resistance as defined by Eqs. 3.14 and 5.28
C = thermal capacitance as defined by Eq. 9.4.

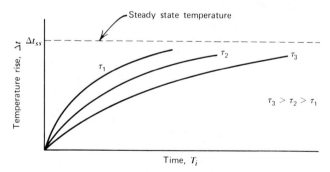

Figure 9.3 Variations in time constant τ and temperature rise Δt.

at time $= 1\,\tau$, system has reached 63% of final Δt

9.5 HEATING CYCLE TRANSIENT TEMPERATURE RISE

The temperature rise that occurs during the heating cycle (Δt_H) in a transient condition can be determined when the steady state temperature rise (Δt_{ss}) is known, as shown by Eq. 9.6.

$$\Delta t_H = \Delta t_{ss}(1 - e^{-T_i/\tau}) = \text{°F or °C} \tag{9.6}$$

where T_i = heating time = hr, min, or sec
  τ = time constant = hr, min, or sec
  Δt_{ss} = temperature rise required to reach a steady state condition

9.6 SAMPLE PROBLEM—TRANSISTOR ON A HEAT SINK

The power transistor shown in Figure 9.4 is cooled by natural convection and dissipates 10 watts. It is mounted on an aluminum heat sink with fins that have a surface area of 0.50 ft². The heat sink has an iridite (chromate) finish, which has a low emittance, so that heat transferred by radiation is small.

The transistor must be capable of handling a power dissipation of 30 watts for 15 min in an ambient temperature of 131°F (55°C). The maximum allowable component surface temperature is 239°F (115°C). Is the design satisfactory?

SOLUTION

The steady state transistor surface temperature is first determined for the 10 watt power dissipation with natural convection cooling. The temperature rise across the convection air film for the steady state condition is determined by combining Eqs. 5.4 and 5.7.

$$h_c = 0.29 \left(\frac{\Delta t}{L}\right)^{0.25} \quad \text{(ref. Eq. 5.4)}$$

$$Q = h_c A \Delta t \quad \text{(ref. Eq. 5.7)}$$

$$Q = 0.29 \left(\frac{\Delta t}{L}\right)^{0.25} A \Delta t = 0.29 A \left(\frac{\Delta t^{1.25}}{L^{0.25}}\right)$$

$$\Delta t_{ss} = \left[\frac{QL^{0.25}}{0.29 A}\right]^{0.8} \quad (\Delta t \text{ across air film}) \tag{9.7}$$

Given Q = 10 watts = 34.13 Btu/hr = 2.39 cal/sec
  L = 3.0 in = 0.25 ft = 7.62 cm
  A = 0.50 ft² = 464.5 cm²

9.6 Sample Problem—Transistor on a Heat Sink

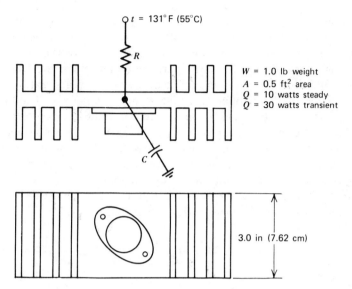

Figure 9.4 Power transistor mounted on a heat sink. Fins are oriented in the vertical position.

The steady state temperature rise for condition 1 where the power dissipation is 10 watts is as follows:

$$\Delta t_{ss_1} = \left[\frac{(34.13)(0.25 \text{ ft})^{0.25}}{0.29(0.50 \text{ ft}^2)}\right]^{0.8} = 59.8°F = 33.2°C \quad (9.8)$$

The steady state surface temperature of the transistor for condition 1 when it dissipates 10 watts in a 131°F (55°C) ambient is as follows:

$$t_{ss_1} = 131°F + 59.8°F = 190.8°F \ (88.2°C) \quad (9.9)$$

When the transistor dissipates 30 watts (102.4 Btu/hr), its steady state temperature rise for condition 2 is again determined from Eq. 9.7.

$$\Delta t_{ss_2} = \left[\frac{(102.4)(0.25 \text{ ft})^{0.25}}{(0.29)(0.50 \text{ ft}^2)}\right]^{0.8} = 144.1°F \ (62.3°C) \quad (9.10)$$

The steady state surface temperature of the transistor for condition 2 when it dissipates 30 watts in a 131°F ambient is as follows:

$$t_{ss_2} = 131°F + 144.1°F = 275.1°F \ (135°C) \quad (9.11)$$

Since the maximum allowable transistor surface temperature is only 239°F (115°C), it is obvious that the transistor must not be allowed to reach a steady state condition when it dissipates 30 watts in a 131°F ambient.

The temperature rise the transistor will experience when the power dissipation is increased from 10 watts to 30 watts for 15 min can be determined from Eq. 9.6. This requires the determination of the average natural convection coefficient during the short period of time when the transistor is dissipating 30 watts. This value cannot be determined accurately because the exact value of the temperature rise is not yet known. Therefore, an estimate is made, the calculations are performed, and the convection coefficient is obtained from the estimate. A temperature rise is computed from this information and is compared with the estimate. If the match is good, the problem is solved. If the match is not good, another estimate is made and the process is repeated until the match is good.

Start by assuming an average temperature rise of 100°F from the ambient to the transistor heat sink surface. Substitute into Eq. 5.4.

$$h_c = 0.29 \left(\frac{\Delta t}{L}\right)^{0.25} = 0.29 \left(\frac{100}{0.25}\right)^{0.25} = 1.29 \frac{\text{Btu}}{\text{hr ft}^2 \, °\text{F}} \quad (9.12)$$

The thermal resistance (R) is determined from Eq. (5.28).

$$R = \frac{1}{h_c A} = \frac{1}{(1.29 \text{ Btu/hr ft}^2 \, °\text{F})(0.50 \text{ ft}^2)} = 1.55 \frac{\text{hr } °\text{F}}{\text{Btu}} \quad (9.13)$$

The thermal capacitance is determined from Eq. 9.4.

Given $W = 1.0$ lb total weight
            $C_p = 0.22$ Btu/lb °F (specific heat)

Substitute into Eq. 9.4.

$$C = WC_p = (1.0)(0.22) = 0.22 \frac{\text{Btu}}{°\text{F}} \quad (9.14)$$

The time constant is determined by substituting Eqs. 9.13 and 9.14 into Eq. 9.5.

$$\tau = \left(1.55 \frac{\text{hr } °\text{F}}{\text{Btu}}\right) \left(0.22 \frac{\text{Btu}}{°\text{F}}\right) = 0.341 \text{ hr} \quad (9.15)$$

The steady state temperature rise (Δt_{ss}) is required for the 30 watt power dissipation condition in Eq. 9.6. The difference in the transient conditions between the 10 watt and the 30 watt power dissipation must be used, as shown in Figure 9.5. The Δt_{ss} from 10 watts to 30 watts is determined from Eqs. 9.8 and 9.10.

$$\Delta t_{ss} = 144.1 - 59.8 = 84.3°\text{F} \quad (9.16)$$

9.6 Sample Problem—Transistor on a Heat Sink

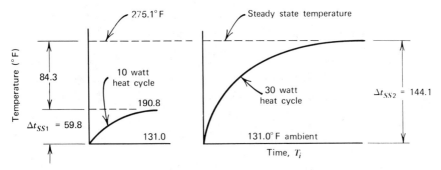

Figure 9.5 Transient temperature change of transistor and heat sink.

Substitute Eqs. 9.15 and 9.16 into Eq. 9.6 using a time $T_i = 0.25$ hr

$$\Delta t_H = 84.3(1 - e^{-0.25/0.341}) = 84.3\left(1 - \frac{1}{2.081}\right)$$

$$\Delta t_H = 43.8°F \text{ after } 0.25 \text{ hr} \quad (9.17)$$

The surface temperature of the transistor and heat sink after 15 min with a power dissipation of 30 watts is shown in Eq. 9.18.

$$t_s = 190.8 + 43.8 = 234.6°F \ (112.5°C) \quad (9.18)$$

The transient temperature curve for the 10 watt and the 30 watt power dissipation is shown in Figure 9.6. Since the transistor surface temperature after 15 min is less than 239°F (115°C), the design should be satisfactory.

An average temperature rise of 100°F was assumed across the convection film to determine the average convection coefficient in Eq. 9.12. The actual

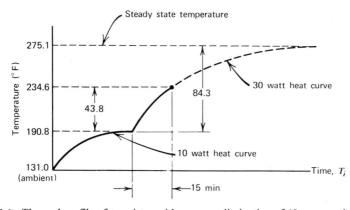

Figure 9.6 Thermal profile of transistor with a power dissipation of 10 watts and 30 watts.

average temperature rise for that film can be determined by taking the ⅔ point along the temperature rise for the approximate average, for the curve generated, based upon the value shown in Eq. 9.17. This is added to the stabilized temperature of 190.8°F for the 10 watt condition shown in Figure 9.6. Since the average temperature rise is required, the value must be compared to the original ambient temperature of 131°F (55°C), as follows:

$$\Delta t_{av} = \left[190.8 + \left(\frac{2}{3}\right)(43.8) \right] - 131 = 89°F \qquad (9.19)$$

This Δt is 11°F off of the original 100°F estimate. However, Eq. 9.12 shows that the natural convection coefficient changes very slowly with Δt. If the 89°F value is used in Eq. 9.12:

$$h_c = 0.29 \left(\frac{89}{0.25}\right)^{0.25} = 1.26 \frac{\text{Btu}}{\text{hr ft}^2 \text{ °F}} \qquad (9.20)$$

The percent error involved is

$$\% \text{ error} = \frac{1.29 - 1.26}{1.26} = 2.38\% \text{ (small)} \qquad (9.21)$$

9.7 TEMPERATURE RISE FOR DIFFERENT TIME CONSTANTS

Sometimes it is desirable to evaluate a thermal design in terms of the time constant τ. When the time constant is known, it is possible to estimate the thermal response of the system. A convenient reference point is one (1) time constant. This condition is obtained from Eq. 9.6 when the time T_i is equal to the time constant τ, which gives a ratio of unity. Then

$$\Delta t_H = \Delta t_{ss}\left(1 - \frac{1}{e}\right) = 0.632 \Delta t_{ss} \qquad (9.22)$$

This shows that one time constant represents a temperature increase that is 63.2% of the steady state temperature rise. When the time constant (τ) is 0.50 hr and the steady state temperature rise is 100°F, then after 0.50 hr the temperature rise will be 63.2°F. An approximate temperature rise curve can then be drawn, since three points are known.

point 1: initial starting point, assume 0°F
point 2: at time of 0.50 hr, t = 63.2°F
point 3: after three time constants, or 1.5 hr, about 95% of the steady state temperature is reached, so at time 1.5 hr, t = 95°F (this is shown in the following section)

9.7 Temperature Rise for Different Time Constants

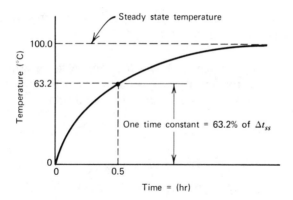

Figure 9.7 Temperature change with respect to time.

The three points shown above will determine a transient curve as shown in Figure 9.7.

Another convenient reference point is the time (T_i) required to reach 95% of the stabilized steady state temperature. Using Eq. 9.6 again:

$$\frac{\Delta t_H}{\Delta t_{ss}} = 0.95 = (1 - e^{-T_i/\tau})$$

$$e^{T_i/\tau} = \frac{1}{0.05} = 20$$

$$\frac{T_i}{\tau} \log_e e = \log_e 20$$

and since $\log_e e = 1.0$:

$$\frac{T_i}{\tau} = 2.996 \cong 3 \quad \text{or} \quad T_i = 3\tau \text{ at the 95\% point}$$

Equation 3.15 shows that $\Delta t = QR$, so that $R = \Delta t_{ss}/Q$. Since

$$\tau = RC, \quad \tau = \frac{\Delta t_{ss} C}{Q}$$

then

$$T_{95} = 3 \left(\frac{\Delta t_{ss}}{Q} \right) C \quad (9.23)$$

9.8 SAMPLE PROBLEM—TIME FOR TRANSISTOR TO REACH 95% OF ITS STABILIZED TEMPERATURE

In the previous sample problem, shown in Figure 9.4, determine the time it takes for the transistor and its heat sink to reach 95% of its stabilized temperature for the 10 watt power dissipation condition.

SOLUTION

$$\Delta t_{ss} = 59.8°F \text{ (ref. Eq. 9.8)}$$

$$Q = 10 \text{ watts} = 34.13 \text{ Btu/hr}$$

$$C = 0.22 \text{ Btu/°F (ref. Eq. 9.14)}$$

Substitute into Eq. 9.23.

$$T_{95} = 3 \left[\frac{(59.8°F)(0.22 \text{ Btu/°F})}{34.13 \text{ Btu/hr}} \right] = 1.156 \text{ hr} \qquad (9.24)$$

9.9 COOLING CYCLE TRANSIENT TEMPERATURE CHANGE

When power is shut off or reduced or when the cooling air is turned on after the electronic system has been operating for a while, the system will cool down. The cooling is rapid at first, then it becomes more gradual, as shown in Figure 9.8.

The temperature change for the cooling cycle (Δt_c) is shown in Eq. 9.25. Notice that the equation is based upon the temperature rise that occurs during the heating cycle (Δt_H), as shown in Eq. 9.6. Therefore, to use Eq. 9.25 the value of Δt_H for the heating cycle must be determined first.

$$\Delta t_c = \Delta t_H (e^{-T_d/\tau}) \qquad (9.25)$$

9.10 SAMPLE PROBLEM—TRANSISTOR AND HEAT SINK COOLING

A transistor is mounted on a heat sink as shown in Figure 9.4, in an environment of 131°F (55°C). The transistor, which dissipates 30 watts, is turned on for 30 min, then shut off. Determine the surface temperature of the transistor 10 min and 20 min after the power has been shut off.

SOLUTION

Equation 9.6 is used to determine the temperature rise of the transistor during its heating cycle. The thermal resistance across the convection film

9.10 Sample Problem—Transistor and Heat Sink Cooling

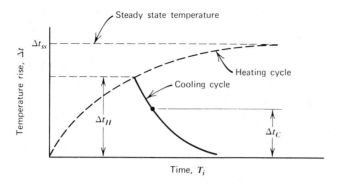

Figure 9.8 Transient temperature reduction during the cooling cycle.

must be known to obtain the RC constant. Since the convection coefficient is not yet known, assume a temperature rise of 70°F from the ambient to the transistor surface and substitute into Eq. 5.4 for h_c.

$$h_c = 0.29 \left(\frac{\Delta t}{L}\right)^{0.25} = 0.29 \left(\frac{70}{0.25}\right)^{0.25} = 1.18 \frac{\text{Btu}}{\text{hr ft}^2 \, °\text{F}}$$

The thermal resistance is determined from Eq. 5.28.

$$R = \frac{1}{h_c A} = \frac{1}{(1.18)(0.50)} = 1.69 \frac{\text{hr °F}}{\text{Btu}}$$

Then $\tau = RC = (1.69)(0.22) = 0.372$ hr
$T_i = 0.50$ hr heating time
$\Delta t_{ss} = 144.1°\text{F}$ (temperature rise required to reach the steady state condition with 30 watts; ref. Eq. 9.10)

Substitute into Eq. 9.6 to determine the temperature rise of the transistor after 30 min of operation.

$$\Delta t_H = 144.1(1 - e^{-0.50/0.372}) = 144.1(0.739)$$
$$\Delta t_H = 106.4°\text{F} \tag{9.26}$$

The average temperature rise can be obtained by taking $\frac{2}{3}$ of the computed rise or $\frac{2}{3}$ of 106.4°F, which gives 71°F. This is close to the assumed value of 70°F, so that the approximate calculations are valid.

The surface temperature of the transistor in a 131°F environment is

$$t_s = 131 + 106.4 = 237.4°\text{F} \, (114.1°\text{C}) \tag{9.27}$$

Equation 9.25 is used to determine the temperature of the transistor surface during the cooling cycle after 10 min (0.1666 hr) and 20 min (0.333 hr). The following information is required:

$$\tau = RC \text{ of system} = 0.372$$

$$\Delta t_H = 106.4°F \text{ (ref. Eq. 9.26)}$$

When $T_i = 0.1666$ hr after power is cut off:

$$\Delta t_c = (106.4)(e^{-0.1666/0.372}) = 67.9°F \quad (9.28)$$

The surface temperature of the transistor in a 131°F environment after 10 min of cooling is

$$t_s = 131 + 67.9 = 198.9°F \ (92.7°C) \quad (9.29)$$

When $T_i = 0.333$ hr after power is cut off:

$$\Delta t_c = (106.4)(e^{-0.333/0.372}) = 43.4°F \quad (9.30)$$

The surface temperature of the transistor in a 131°F environment after 20 min of cooling is

$$t_s = 131 + 43.4 = 174.4°F \ (79.1°C) \quad (9.31)$$

For a more accurate evaluation of the transient temperature during the cooling cycle, the natural convection coefficient should be computed for the average temperature over the time increment.

9.11 TRANSIENT ANALYSIS FOR TEMPERATURE CYCLING TESTS [56]

More electronic equipment is being used now than ever before to provide entertainment, safety, comfort, and convenience. The trend in modern electronics is to miniaturize the equipment while expanding the functional capability. This leads to higher concentrations of heat in smaller packages, which leads to higher operating temperatures. This requires more sophisticated techniques for removing heat, more sophisticated techniques for determining the temperatures of critical components, and means for testing these components to determine their reliability.

One method of testing that has received wide acceptance is the temperature cycling test, which is often called the "Shake and Bake" test. The accepted title for this type of test is the AGREE test, which is an acronym

9.11 Transient Analysis for Temperature Cycling Tests

for "Advisory Group on the Reliability of Electronic Equipment." This is a reliability test that subjects the electronic equipment to a combination of environmental conditions which simulate the actual environments the equipment will see in service.

The "shake" part of the "Shake and Bake" test usually consists of a 2 G peak sinusoidal vibration input to the electronic system for 10 min every hour, or for 10 min at the high temperature phase of the test. The vibration is usually applied at some nonresonant frequency between 20 and 60 Hz. If there is a major structural resonance anywhere near the frequency selected, it is possible to damage a sensitive electronic system. It is therefore very important to know where the major structural resonances are located before starting such a test.

Some of the later AGREE tests are using low level random vibration over a bandwidth of 2000 Hz, to try to improve the vibration simulation found in the actual environment. For more information on the subject of vibration for electronic equipment, see reference [1].

In a typical temperature cycling test, the ambient temperature within the test chamber is usually cycled at a constant rate from $-65°F$ $(-54°C)$ to $+131°F$ $(+55°C)$, and sometimes to $160°F$ $(71°C)$. As the chamber temperature is being cycled over its temperature extremes, the power to the electronics may also be cycled through its full range from zero power to its maximum power. The maximum power condition is usually made to coincide with the highest chamber temperature, to produce the maximum component temperature for the maximum thermal stress. A typical temperature cycling test is shown in Figure 9.9.

This is the age of the high speed digital computer, which is the best and most accurate method for determining the transient thermal response characteristics of a complex electronic system. However, not everyone has a

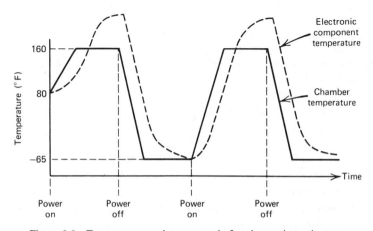

Figure 9.9 Temperature and power cycle for electronic equipment.

large digital computer handy. Also, there are times when the computer is not available. This may be due to a heavy work load by another department, such as payroll, or perhaps to a computer system malfunction. There are also times when it is necessary to double check the results of the computer printout, because of possible programming errors or data entry errors. An alternative method of analysis is always advisable, to guarantee the accuracy of a critical calculation. Small hand calculators are very convenient and ideally suited for solving the equations presented here for that alternative method.

Consider the general case of an electronic component that has heat generated internally (Q_{gen}), the ability to store heat in its mass (Q_{stor}), and a path where heat can flow (Q_{flow}) through a thermal resistance (R) to the outside environment, which has a temperature t_e as shown in Figure 9.10. The value of R is assumed to be constant. Since it may include convection and radiation, it may really change. However, the changes are normally small. When large changes are involved, additional iterations must be used to obtain good accuracy.

A heat flow balance can be made for the system by considering the heat gained and the heat lost. Since the heat can flow in either direction, from the outside environment to the component or from the component to the outside environment, start with the heat flowing to the component.

The heat gained is due to the internally generated heat, plus heat flowing to the component from the outside environment. The heat lost is due to the heat going into storage as a result of the thermal capacitance of the component mass. A component with a large thermal capacity can store more heat, so that it will take longer to heat up. The heat balance is shown below.

Heat Gained	Heat Lost
$Q_{gen} = Q$	$Q_{stor} = WC_p \dfrac{dt}{dT_i}$
$Q_{flow} = (t_e - t)k'$	

Since the heat gained must be equal to the heat lost, an equation with this balance can be written.

$$Q_{gen} + Q_{flow} = Q_{stor}$$

$$Q + (t_e - t)k' = WC_p \frac{dt}{dT_i} \qquad (9.32)$$

The temperature of the component is shown as t, the time is shown as T_i, and the conductance is shown as k', which is defined by Eq. 3.26.

The temperature change inside the AGREE chamber is usually represented as a straight line for each phase of the temperature cycle, with respect to time, as shown in Figure 9.11.

9.11 Transient Analysis for Temperature Cycling Tests

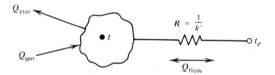

Figure 9.10 Thermal model of an electronic component.

The temperature change can then be represented by a straight line using the slope intercept relation shown in Eq. 9.33.

$$t_e = b + ST_i \tag{9.33}$$

The intercept on the vertical temperature axis for each phase is represented by b_1 and b_2 and b_3. The slope is the change of temperature with respect to time and is shown as S. The component thermal capacitance is shown as C and is defined by Eq. 9.4. Substituting these values into Eq. 9.32, with a little rearranging, produces the following differential equation:

$$\frac{dt}{dT_i} + \frac{k'}{C} t = S \frac{k'}{C} T_i + \frac{Q + bk'}{C} \tag{9.34}$$

This equation can be solved in several ways. One convenient method is to multiply both sides with an integrating factor.

$$\text{Integrating factor: } e^{k'T_i/C} \tag{9.35}$$

To further reduce the complexity of the equation, let

$$D = \frac{Q + k'b}{C} \tag{9.36}$$

Incorporating these factors will lead to Eq. 9.37.

$$\frac{dt}{dT_i} e^{k'T_i/C} + \frac{k't}{C} e^{k'T_i/C} = \frac{Sk'T_i}{C} e^{k'T_i/C} + D e^{k'T_i/C} \tag{9.37}$$

Integrate both sides of the equation and simplify.

$$t = S\left(T_i - \frac{C}{k'}\right) + \frac{DC}{k'} + I e^{-k'T_i/C} \tag{9.38}$$

The integration constant is shown as I, and it is determined from the initial conditions when the time (T_i) is zero.

$$\text{When } T_i = 0, \quad t = \text{initial temperature } t_0 \tag{9.39}$$

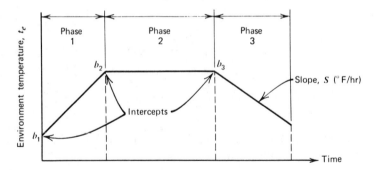

Figure 9.11 Temperature change with respect to time.

Solving for the integration constant, we obtain

$$I = t_0 + \frac{C}{k'}(S - D) \qquad (9.40)$$

Substitute into Eq. 9.38 and simplify.

$$t = S\left(T_i - \frac{C}{k'}\right) + \left(b + \frac{Q}{k'}\right) + \left(t_0 + \frac{CS}{k'} - b - \frac{Q}{k'}\right) e^{-k'T_i/C} \qquad (9.41)$$

This equation represents the transient temperature (t) of the electronic component mass at any time (T_i).

The maximum transient temperature reached by the component can be determined by taking the first derivative of the temperature, with respect to the time and setting the value equal to zero. This will produce the point at which the component temperature slope is zero, which is the maximum temperature point the component reaches during the cycling test [57].

$$\frac{dt}{dT_i} = S + \left(t_0 + \frac{CS}{K'} - b - \frac{Q}{k'}\right)\left(-\frac{k'}{C}\right) e^{-k'T_i/C} \qquad (9.42)$$

Solve the equation for the time at which the zero point occurs.

$$T_i = \frac{C}{k'} \log_e \left(\frac{t_0 k' + SC - bk' - Q}{SC}\right) \qquad (9.43)$$

This equation represents the time point for a zero slope in the component temperature curve. The component temperature is maximum at this point. This condition occurs in phase 3 of Figure 9.11, where the slope (S) is negative.

Although Eq. 9.41 represents a transient condition, it can also be used to evaluate a steady state thermal condition by letting the slope become zero

9.12 Sample Problem—Electronic Chassis in a Temperature Cycling Test

and the time become infinite. The b intercept then becomes the environment temperature (t_e).

$$t = b + \frac{Q}{k'}$$

This is simply the temperature of the ambient (or environment) plus the temperature rise across the resistor, which is the correct relation for the steady state condition.

9.12 SAMPLE PROBLEM—ELECTRONIC CHASSIS IN A TEMPERATURE CYCLING TEST

An electronic chassis is to be subjected to a temperature cycling test in an AGREE chamber from $-65°F$ ($-54°C$) to $+160°F$ ($+71°C$), as shown in Figure 9.12. The test will start at a room temperature of 80°F (26.6°C). The maximum allowable PCB hot spot surface temperature under any electronic component mounted in the chassis is to be 212°F (100°C).

The chassis contains three plug-in PCBs which have solid aluminum cores 0.050 in (0.127 cm) thick. The power dissipation for each board is 4 watts. The chassis is shown in Figure 9.13. The components are cooled by conduction from the boards to the chassis side walls and part of the end walls. The chassis is cooled by natural convection to the outside environment, which is the test chamber during the AGREE test. The surface of the chassis has an iridited finish (chromate), which has a low emissivity. Bolted covers are used on the top and bottom surfaces of the chassis.

Determine the maximum expected circuit board surface temperature that

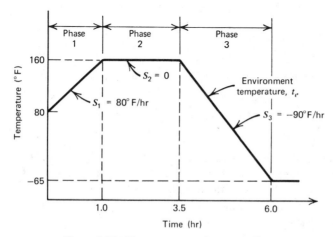

Figure 9.12 Temperature cycle test profile.

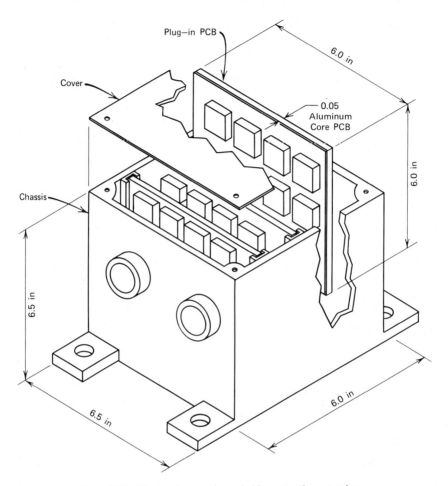

Figure 9.13 Electronic chassis cooled by natural convection.

will be reached during the AGREE test. (The temperatures of the components will be slightly higher than the circuit board.)

SOLUTION

The symmetry of the chassis can be used to simplify the model of the chassis by considering only half of the structure as shown in Figure 9.14. This model shows the heat flow path from the center of the PCB to the outside chamber environment, which is broken up into three individual resistors in series.

9.12 Sample Problem—Electronic Chassis in a Temperature Cycling Test

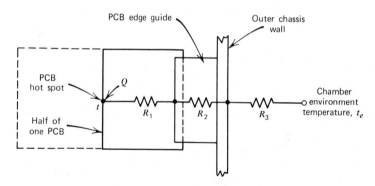

Figure 9.14 Thermal resistance network from the circuit board to the chamber environment.

The individual resistances are determined as follows:

R_1 = conduction resistance along half of the PCB to the edge

$$R_1 = \frac{L}{KA} \tag{9.44}$$

Given $L = 3.0$ in $= 0.25$ ft (length of half a board)

$K = 83.3 \dfrac{\text{Btu}}{\text{hr ft °F}}$ (conductivity of aluminum)

$A = \dfrac{(0.050)(6.0)\text{in}^2}{144 \text{ in}^2/\text{ft}^2} = 0.00208 \text{ ft}^2$ (PCB cross section area)

$$R_1 = \frac{0.25}{(83.3)(0.00208)} = 1.44 \frac{\text{hr °F}}{\text{Btu}} \tag{9.45}$$

R_2 = interface resistance across PCB edge guide to chassis, for beryllium copper spring clips that have a low interface pressure

$$R_2 = \frac{1}{h_i A} \tag{9.46}$$

Given $h_i = 30 \dfrac{\text{Btu}}{\text{hr ft}^2 \text{°F}}$ (interface conductance on guides)

$A = \dfrac{(0.10)(6.0)(2 \text{ surfaces})}{144 \text{ in}^2/\text{ft}^2} = 0.00833 \text{ ft}^2$ (surface area)

$$R_2 = \frac{1}{(30)(0.00833)} = 4.0 \frac{\text{hr °F}}{\text{Btu}} \tag{9.47}$$

R_3 = natural convection resistance from outside surface of the chassis to the environment, neglecting radiation

$$R_3 = \frac{1}{hA} \tag{9.48}$$

Given $h = 1.0 \dfrac{\text{Btu}}{\text{hr ft}^2 \, °F}$ (typical and average natural convection coefficient for a small chassis)

L_1 = 6.5 in (height of chassis side wall)
L_2 = 3.4 in (effective width of chassis per PCB half, which includes the side wall and part of the end walls)
$A = (6.5)(3.4) = 22.1 \text{ in}^2 = 0.153 \text{ ft}^2$ (surface area)

$$R_3 = \frac{1}{(1.0)(0.153)} = 6.5 \, \frac{\text{hr °F}}{\text{Btu}} \tag{9.49}$$

The total resistance from the center of the PCB to the chamber environment is the sum of the individual resistances.

$$R = R_1 + R_2 + R_3 = 1.44 + 4.0 + 6.5 = 11.94 \, \frac{\text{hr °F}}{\text{Btu}} \tag{9.50}$$

Using the conductance, which is the reciprocal of the resistance, we obtain

$$k' = \frac{1}{R} = \frac{1}{11.94} = 0.084 \, \frac{\text{Btu}}{\text{hr °F}} \tag{9.51}$$

Each of the three temperature cycling phases shown in Figure 9.12 must be considered separately to obtain the required answers. The PCB temperature obtained at the end of phase 1 becomes the initial PCB temperature for phase 2. Also, the PCB temperature obtained at the end of phase 2 becomes the initial PCB temperature for phase 3. Starting with phase 1 as shown in Figure 9.12, the required data for the problem are listed below.

DATA FOR PHASE 1

$S = 80°F/hr$ (slope of phase 1 heating cycle)
$T_i = 1.0$ hr (length of time for phase 1)
$W = 0.50$ lb (weight of half a PCB plus part of the chassis)
$C_p = 0.25$ Btu/lb °F (combined specific heat of mass)   (9.52)
$C = WC_p = (0.50)(0.25) = 0.125$ Btu/°F (thermal capacitance)
$b = 80°F$ (intercept at start of phase 1)
$t_0 = 80°F$ (initial PCB temperature for phase 1)

9.12 Sample Problem—Electronic Chassis in a Temperature Cycling Test

$Q = 2$ watts $= 6.82$ Btu/hr (heat on half a board)
$k' = 0.084$ Btu/hr °F (conductance, board to environment)

Substitute these values into Eq. 9.41 to obtain the PCB temperature at the end of phase 1.

$$t = 80(1.0 - 1.488) + 80 + 81.2 + (80 + 119 - 80 - 81.2)e^{-0.672}$$
$$t = 141.5°F \tag{9.53}$$

Next going to phase 2, as shown in Figure 9.12, note the slope of the environmental line goes to zero. The required new data for this phase are as follows:

DATA FOR PHASE 2

$$\left.\begin{array}{l} S = 0°F/hr \text{ (slope for phase 2 cycle)} \\ T_i = 2.5 \text{ hr (length of time for phase 2)} \\ b = 160°F \text{ (intercept value of slope line)} \\ t_0 = 141.5°F \text{ (initial temperature from Eq. 9.53)} \end{array}\right\} \tag{9.54}$$

Substitute into Eq. 9.41 to obtain the PCB temperature at the end of phase 2.

$$t = 0 + 160 + 81.2 + (141.5 + 0 - 160 - 81.2)e^{-1.68}$$
$$t = 222.5°F \tag{9.55}$$

The maximum temperature reached in phase 3 is determined from the point where the slope of the temperature line is zero, using the following data.

DATA FOR PHASE 3

$$\left.\begin{array}{l} S = -90°F/hr \text{ (slope of phase 3 temperature cycle)} \\ b = 160°F \text{ (intercept value of slope line)} \\ t_0 = 222.5°F \text{ (initial PCB temperature from Eq. 9.55)} \end{array}\right\} \tag{9.56}$$

Substitute into Eq. 9.43 to obtain the time where the slope is zero and the temperature is maximum.

$$T_i = 1.488 \log_e \left(\frac{18.69 - 11.25 - 13.44 - 6.82}{-11.25} \right)$$
$$T_i = 0.194 \text{ hr} \tag{9.57}$$

Substitute Eqs. 9.56 and 9.57 into Eq. 9.41 to obtain the maximum PCB hot spot temperature reached in the temperature cycling test.

$$t = -90(0.194 - 1.488) + 160 + 81.2$$
$$+ (222.5 - 134 - 160 - 81.2)e^{-0.1303}$$
$$t = 223.7°F \tag{9.58}$$

Since the maximum PCB temperature exceeds the allowable value of 212°F, the conditions are not acceptable, so that some changes must be made. These changes can be made in the structural design, or changes can be made in the temperature cycle.

When there is a choice between changing the temperature cycle or changing the structural design, the simpler solution is to change the temperature cycle. Design changes are expensive and time consuming. However, if the thermal design of the equipment is marginal, it is better to spend the time and money to make the improvements rather than to risk a failure.

One other area must be evaluated very carefully and that is the power dissipation. Since the source of the heat is the power, it is often possible to reduce the power by simply changing the duty cycle or by going to a component that performs the same function but at a lower power dissipation. Many components can be obtained in the form of a flat pack or in the form of a DIP, with exactly the same functions.

When the thermal analysis is performed, the power dissipation is usually obtained from the electrical engineer. If the engineer's estimate of the heat generated is too high, the mechanical design may end up being larger, heavier, and more expensive than necessary.

9.13 SAMPLE PROBLEM—METHODS FOR DECREASING HOT SPOT TEMPERATURES

What changes can be made to reduce the PCB hot spot temperature to a value below 212°F (100°C) for the electronic system shown in Figure 9.13?

SOLUTION

A number of changes can be made to improve the thermal characteristics of the mechanical design. It may even be possible to reduce the power dissipation by changing the type of components used. However, before any of these changes are made, it may be simpler to alter the temperature cycling test. A revised temperature cycling test may be proposed which

9.13 Sample Problem—Methods for Decreasing Hot Spot Temperatures

reduces the temperature cycling time at the maximum level of 160°F as shown in Figure 9.15.

Using the same methods outlined in the previous sample problem, the PCB temperature at the end of each phase will be as follows:

$$
\left.\begin{array}{ll}
\text{At the end of phase 1:} & t = 115°F \\
\text{At the end of phase 2:} & t = 176°F \\
\text{Point of zero temperature slope:} & T_i = 0.381 \text{ hr} \\
\text{Maximum PCB temperature reached:} & t = 184°F
\end{array}\right\} \quad (9.59)
$$

Since the PCB hot spot temperature is less than 212°F, the conditions are now acceptable.

When the temperature cycling test profile cannot be changed and the power dissipation cannot be changed, it will be necessary to change the thermal design of the system.

An investigation of the thermal resistances previously evaluated shows they can all be reduced, but there will be some penalties. The thermal resistance R_1 for the circuit board shown in Figures 9.13 and 9.14 can be reduced 20% by increasing the aluminum core thickness from 0.050 in to 0.060 in. This will result in a 20% increase in the weight of the aluminum heat sink core. The interface resistance R_2 across the circuit board edge guide can be reduced more than 50% by increasing the contact area and the interface pressure. This will require a new design, with new tooling costs for the change. The natural convection resistance R_3 for the chassis can be reduced as much as 35% by adding fins to increase the surface area. This will require a chassis design modification, with added size, weight, and tooling costs.

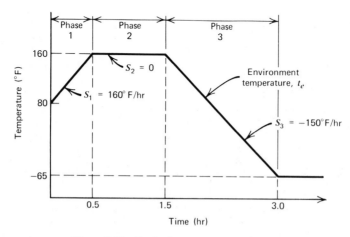

Figure 9.15 Revised temperature cycle test.

These suggested changes will increase the thermal conductance and the thermal mass capacitance to the new values shown in Eq. 9.60.

$$\left. \begin{array}{l} k' = 0.150 \dfrac{\text{Btu}}{\text{hr °F}} \\ \\ C = 0.136 \dfrac{\text{Btu}}{\text{°F}} \end{array} \right\} \qquad (9.60)$$

Using the methods outlined previously in Section 9.12, the PCB temperature can be determined at the end of each phase, using the values given above with the temperature cycling test shown in Figure 9.12 for the modified chassis.

$$\left. \begin{array}{ll} \text{At the end of phase 1:} & t = 141.9°F \\ \text{At the end of phase 2:} & t = 201.5°F \\ \text{Point of zero temperature slope:} & T_i = 0.0434 \text{ hr} \\ \text{Maximum PCB temperature reached:} & t = 201.6°F \end{array} \right\} \qquad (9.61)$$

Since the PCB hot spot temperature is less than 212°F, the conditions are now acceptable.

In the previous sample problems, the power dissipation Q was kept constant during all three phases of the temperature cycling test. It is a simple matter to change the power dissipation at any time, just by altering the value of Q in Eqs. 9.41 and 9.43. When the power is shut off, the value of Q goes to zero.

When the outside environment or the heat sink temperature (t_e) is constant but the transient condition still exists, the slope (S) goes to zero. Let $S = 0$ and $b = t_e$ in Eq. 9.41. This results in the transient hot spot temperature for a system in a constant environment as shown in Eq. 9.62.

$$t = t_e + \frac{Q}{k'} + \left(t_0 - t_e - \frac{Q}{k'} \right) e^{-k'T_i/C} \qquad (9.62)$$

The use of this equation is demonstrated with a sample problem.

9.14 SAMPLE PROBLEM—TRANSIENT ANALYSIS OF AN AMPLIFIER ON A PCB

Several potted amplifier modules are mounted on an 0.052 in (0.132 cm) thick aluminum core PCB that is mounted within an aluminum chassis with rigid side walls. The chassis is cold soaked with power off in a chamber at −65°F (−54°C). The chassis is removed from the cold chamber and quickly clamped to a large fixture in the hot chamber, which is maintained at a temperature of 138°F (58.9°C). The electrical power is turned on at the same time. Determine the approximate amplifier hot spot temperature rise time

9.14 Sample Problem—Transient Analysis of an Amplifier on a PCB

curve when each amplifier dissipates 2 watts. A cross section of the chassis is shown in Figure 9.16. Most of the heat is removed by conduction to the side walls, since radiation and convection heat transfer are small.

SOLUTION

The system is symmetrical, so that only one amplifier is considered for the analysis. A model of the system to be analyzed is shown in Figure 9.17.

The conductance (k') of the heat flow path from the amplifier to the chassis cold plate is determined by breaking up the heat flow in several increments as described below.

k'_1 = conductance across the PCB edge guide from the chassis side walls to the PCB, using the edge guide shown in Figure 3.23b; first the resistance is determined:

$$R_1 = \frac{(8°C \text{ in/watt})(1.8°F/°C)}{4.31 \text{ in}} = 3.34 \frac{°F}{\text{watt}} \times \frac{\text{watt}}{3.413 \text{ Btu/hr}} = .978 \frac{\text{hr °F}}{\text{Btu}}$$

$$k'_1 = \frac{1}{R_1} = 1.02 \frac{\text{Btu}}{\text{hr °F}} = 0.128 \frac{\text{cal}}{\text{sec °C}} \qquad (9.63)$$

k'_2 = conductance along the aluminum core PCB from the edge guide to the amplifier (see Figures 9.16 and 9.17)

$$k'_2 = \frac{KA}{L} = \frac{(100 \text{ Btu/hr ft °F})[4.4 \times 0.052 \text{ in}^2/(144 \text{ in}^2/\text{ft}^2)]}{2.4 \text{ in}/(12 \text{ in/ft})}$$

$$k'_2 = \frac{0.80 \text{ Btu}}{\text{hr °F}} = 0.101 \frac{\text{cal}}{\text{sec °C}} \qquad (9.64)$$

k'_3 = conductance across the resilient interface pad (0.015 in thick) under the amplifier, between the PCB and the amplifier (see Figure 9.18)

$$k'_3 = \frac{KA}{L} = \frac{(1.25 \text{ Btu/hr ft °F})[1.0 \text{ in}^2/(144 \text{ in}^2/\text{ft}^2)]}{0.015 \text{ in}/(12 \text{ in/ft})}$$

$$k'_3 = 6.95 \frac{\text{Btu}}{\text{hr °F}} = 0.876 \frac{\text{cal}}{\text{sec °C}} \qquad (9.65)$$

k'_4 = conductance from bottom of module to center of internal component hot spot (see Figure 9.18)

$$k'_4 = \frac{KA}{L} = \frac{(0.666 \text{ Btu/hr ft °F})[1.0 \text{ in}^2/(144 \text{ in}^2/\text{ft}^2)]}{0.375 \text{ in}/(12 \text{ in/ft})}$$

$$k'_4 = 0.148 \frac{\text{Btu}}{\text{hr °F}} = 0.0139 \frac{\text{cal}}{\text{sec °C}}$$

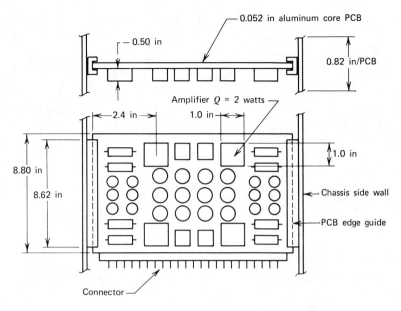

Figure 9.16 Cross section through the chassis with an amplifier on a PCB.

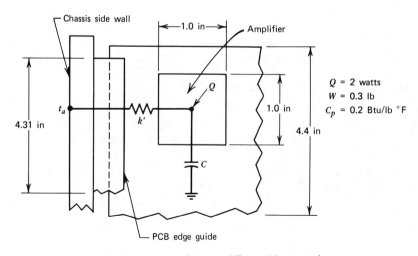

Figure 9.17 Model of the amplifier and its mounting.

9.14 Sample Problem—Transient Analysis of an Amplifier on a PCB

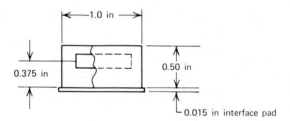

Figure 9.18 Potted amplifier module.

All of the conductances are in series, so that

$$\frac{1}{k'} = \frac{1}{k'_1} + \frac{1}{k'_2} + \frac{1}{k'_3} + \frac{1}{k'_4}$$

$$\frac{1}{k'} = \frac{1}{1.02} + \frac{1}{0.80} + \frac{1}{6.95} + \frac{1}{0.148}$$

$$k' = \frac{1}{9.13} = 0.110 \frac{\text{Btu}}{\text{hr °F}} = 0.0139 \frac{\text{cal}}{\text{sec °C}} \qquad (9.67)$$

The thermal capacitance of the PCB with the amplifier is determined from Eq. 9.4.

$$C = WC_p = (0.30 \text{ lb}) \left(0.20 \frac{\text{Btu}}{\text{lb °F}}\right)$$

$$C = 0.060 \frac{\text{Btu}}{\text{°F}} = 27.24 \frac{\text{cal}}{\text{°C}} \qquad (9.68)$$

Also:

$$t_0 = -65\text{°F} = -54\text{°C (initial temperature)}$$

$$Q = 2 \text{ watts} = 6.826 \frac{\text{Btu}}{\text{hr}}$$

$$t_e = 138\text{°F} = 58.9\text{°C (heat sink temperature)}$$

The amplifier hot spot transient temperature curve is determined by using different time (T_i) increments. Equation 9.62 is used to compute the temperature after 0.50 hr, 1.0 hr, and 2.0 hr. using English units.

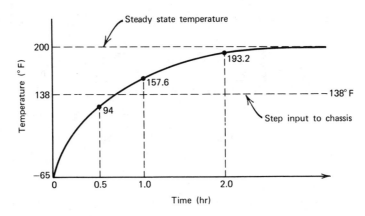

Figure 9.19 Transient hot spot temperature curve for the amplifier.

Figure 9.20 Power supply components mounted on a heavy aluminum heat sink plate. (Courtesy Litton Systems, Inc.)

Table 9.1 Specific Heat (C_p) of Solids–English Units (Btu/lb °F) and Metric Units (cal/g °C)

Solid	32°F
Metals	
Lead	0.031
Solder	0.040
Zinc	0.092
Aluminum	0.221
Silver	0.056
Gold	0.030
Copper	0.092
Nickel	0.102
Iron	0.105
Cobalt	0.102
Steel	0.114
Titanium	0.130
Inconel	0.130
Magnesium	0.230
Nonmetals	
Glass	
Normal	0.190
Crown	0.160
Quartz	0.17–0.27
Carbon, graphite	0.12–0.16
Alumina	0.183
Magnesia	0.222
Silica	0.191
Parafin wax	0.69
Wood, pine	0.67
Epoxy	0.28–0.32
Epoxy fiberglass	0.28–0.32
Nylon	0.40
Rubber	0.45
Bakelite	0.38

When $T_i = 0.50$ hr:

$$t = 138 + \frac{6.826}{0.110} + \left(-65 - 138 - \frac{6.826}{0.110}\right) e^{-(0.110)(0.50)/0.060}$$

$$t = 94°F\ (34.4°C) \qquad (9.69)$$

When $T_i = 1.0$ hr:

$$t = 200 - (265)e^{-(0.110)(1.0)/0.060}$$

$$t = 157.6°F\ (69.8°C) \qquad (9.70)$$

When $T_i = 2.0$ hr:

$$t = 200 - (265)e^{-(0.110)(2.0)/0.060}$$

$$t = 193.2°F\ (89.5°C) \qquad (9.71)$$

The temperature will reach a steady state condition when the time is infinite. Substitute this value into Eq. 9.62:

$$t = 138 + \frac{68.26}{0.110} = 138 + 62 = 200°F\ (93.3°C) \qquad (9.72)$$

The transient hot spot temperature curve for the amplifier is shown in Figure 9.19.

The specific heat of various solids, metals and nonmetals, is shown in Table 9.1.

Figure 9.20 shows a power supply module with high power components mounted on thick aluminum heat sinks.

10

Special Applications for Tough Cooling Jobs

10.1 NEW TECHNOLOGY—APPROACH WITH CAUTION

Design engineers are always seeking better methods for packaging high-power-dissipating electronic equipment. As the power dissipations continue to increase, standard conduction and forced air convection techniques no longer provide adequate cooling for sophisticated electronic systems. The reliability of the electronic system will suffer if high temperatures are permitted to develop. This is extremely important in mass transportation systems such as airplanes, ships, and trains. The failure of a critical electronic control element in these systems can lead to extensive property damage and the loss of many lives.

When standard cooling methods are no longer adequate, exotic new methods are often utilized. New cooling methods should be approached with some caution, because they often bring in new technology, which may contain hidden dangers and new failure mechanisms that are not well known or easily recognized. New materials, together with new manufacturing processes and new technology, will always lead to new problems.

10.2 HEAT PIPES

A heat pipe is a hollow tube type of enclosed structure, containing a fluid that transfers large quantities of heat when it evaporates and a wick that brings the fluid back to its starting point when it condenses. This entire process is accomplished with no outside power, no mechanical moving parts, and no noise. The design is extremely simple and very efficient, since it can transfer heat hundreds of times better than any solid metal conductor [58–61].

The birth of the heat pipe dates to about 1942. Very little attention was paid to this device until about 1963, when it was suggested for spacecraft applications. Since then, many different heat pipe applications have found their way into cooking food, heating homes, cooling motorcycle engines, and recovering heat in industrial exhaust systems.

Heat pipes can be made in many different shapes and sizes. The most common shape is the hollow cylinder or tube. They are often made in flat shapes, with S turns, and in spirals.

A typical heat pipe consists of a sealed tube that has been partially evacuated, so that its internal pressure is below the standard atmosphere of 14.7 psia. The inside walls of the tube are usually lined with a capillary wick structure and a small amount of a fluid, which will vaporize. When heat is applied at one end of the tube, the fluid within the pipe vaporizes or boils. This generates a force that drives the vapor to the opposite end of the tube, where the heat is removed. Removing the heat forces the vapor to condense, and the wick draws the fluid back to the starting point, where the process is repeated [62, 63].

There is a small pressure drop between the heating (or evaporating) end and the cooling (or condensing) end of the pipe. Therefore, the boiling and condensing cycle takes place over a very narrow temperature band. As a result, there is only a small temperature difference between the heat source and the heat sink. A 24 in (60.9 cm) long pipe, 0.50 in (1.27 cm) in diameter can pump 220 watts of heat at 212°F (100°C), with about a 2.4°F (1.3°C) temperature difference along the length of the pipe. A solid copper bar for the same length and power, but with a 100°F (55.5°C) temperature difference along its length, would require a cross section area of about 10.8 in^2 (69.7 cm^2), and would weigh about 75 lb (34,050 g), compared to the heat pipe weight of about 0.75 lb (340 g).

The wick is probably the most critical part of the heat pipe design. It determines the capillary sucking action available for drawing the condensed liquid back to the evaporator end. The porosity and the continuity of the internal passages determine the fluid resistance along the wick. The wick must have sufficient capacity to supply liquid to the heat input end. An inadequate fluid return will result in the drying up of the wick at the heat input end, which results in a breakdown of the wick operation [64, 65].

Capillary action of the wick in the heat pipe permits it to operate in any orientation in a gravity or acceleration field. A typical wick on earth, which has an acceleration of $1.0G$, must be capable of raising the fluid against gravity and with a small internal resistance to flow at the same time. A small pore size is required to draw the liquid up to a high level in a capillary tube. However, a small pore size increases the internal flow resistance. A balance must therefore be made to provide good fluid flow over long distances in the heat pipe wick design, depending upon the orientation in a gravity or acceleration field. A typical section through the length of a heat pipe is shown in Figure 10.1.

10.3 Degraded Performance in Heat Pipes

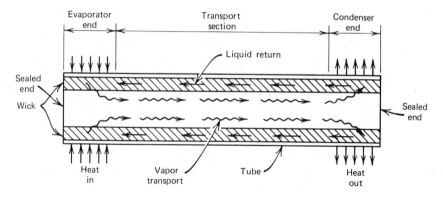

Figure 10.1 Section through the length of a typical heat pipe.

Heat pipe wicks are often made of porous ceramic or woven stainless steel wire mesh. Sometimes the wick is an integral part of the tube housing, formed by extruding small grooves along the inner surface of the wall structure, as shown in Figure 10.2.

10.3 DEGRADED PERFORMANCE IN HEAT PIPES

Heat pipes appear to have the ability to solve many thermal problems involving high heat densities or the transport of heat over long distances. However, it should be pointed out that many of the old heat pipe manufacturers have gone out of the business, and only a handfull are left. This points out how difficult it is to make good heat pipes, with good quality control, which will give good performance for a long period of time, and which can be sold at a good price.

One of the biggest problems with heat pipes is that many of them often exhibit degraded performance after they have been in operation for about 3 to 6 months. This degraded performance occurs slowly, so it may not be noticed in the operating hardware. The operating temperatures of the equipment will increase over a period of time, until malfunctions and failures occur. Since thermocouples are not usually mounted in operating hardware,

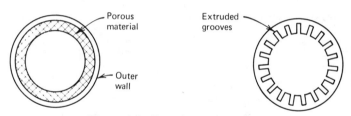

Figure 10.2 Heat pipe wick designs.

the gradual temperature increase over a long time span usually goes unnoticed.

The single greatest cause of this degraded performance appears to be contamination that affects the vapor pressure within the heat pipe. Contamination often results from the type of operation that is used to seal the ends of the tube after it is assembled. Many sealing methods can be used effectively if the processes are properly controlled, the parts are adequately cleaned, and the assembly takes place in a clean room. Electron beam welding is reported to be one of the best known methods for sealing heat pipes today. This techniques appears to provide a good seal with very little contamination. Many other sealing methods are still used, because they are less expensive. If the processes are properly controlled, a good seal can be obtained.

Other sources of outgassing are often found in the wick. Since the wick must utilize many small orifices to pump the fluid from the condenser to the evaporator, this section can easily contain trapped gasses that may affect the vapor pressure. Wicks must therefore be manufactured and cleaned very carefully before they are installed in the heat pipe.

A clean room should be used to assemble heat pipes, to minimize possible contamination. The maximum airborne particle contamination should be about 100,000 particles per cubic foot, when the particle size is around 0.5 μm. When the particle size is around 5.0 μm, the maximum airborne particle contamination should be limited to about 700 particles per cubic foot.

Burn-in tests should be performed on all heat pipes, for at least 100 hr. The burn-in should be performed with 100% of the required design load but in an ambient temperature at least 20% higher than the highest specified operating temperature, to ensure reliable operation.

The cleaning process is extremely important. One heat pipe manufacturer reported an attempt to reduce manufacturing costs by using the extruded grooves in place of the porous material for the wick, as shown in Figure 10.2. The fluid that was used in the extrusion process was very difficult to remove completely, and the resulting contamination affected the internal vapor pressure. This reduced the performance of the heat pipe after several months of operation. Since the particular program required very precise temperature controls for very long time periods, it was decided that the lower cost extruded grooves should not be used for this application.

10.4 TYPICAL HEAT PIPE PERFORMANCE

A heat pipe can work in any orientation, but its performance may be degraded when it is forced to work against gravity. This condition occurs when the heat input end (evaporator end) is higher than the heat output end (condenser end). In this position, the fluid in the wick is forced to move up the pipe, against the direction of gravity. Figure 10.3 shows approximately

10.4 Typical Heat Pipe Performance

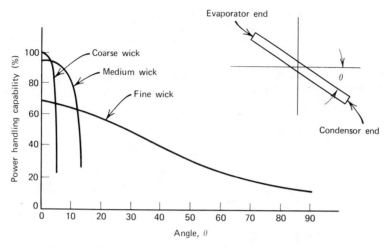

Figure 10.3 Changes of the heat pipe capability when the condensor end is below the evaporator end.

how the performance of a 48 in (122 cm) long water heat pipe with a coarse, medium, and fine wick changes with the angle of orientation (θ).

The coarse wick is capable of handling much more power while it is operating in a horizontal position. However, when the condenser end is angled down slightly, its pumping capacity is sharply reduced. The fine wick cannot pump as much heat in the horizontal position. However, its pumping capacity does not drop off as fast when the condenser end is angled down. The typical performance that can be expected from various heat pipes is shown in Table 10.1.

A wide variety of fluids can be used with heat pipes. A few of these fluids and their typical operating ranges are given in Table 10.2.

Table 10.1 Typical Heat Pipe Performance

Outside Diameter (in)	Length (in)	Power (watts)
1/4	6	300
	12	175
	18	150
3/8	6	500
	12	375
	18	350
1/2	6	700
	12	575
	18	550

Handwritten annotations:
PIPES FIT INTO TIGHT TOLERANCE HOLES

→ LITTON : DESIGNS 15/W FOR A PIPE THIS LENGTH.

EXPECT PIPES TO DEGRADE.

Table 10.2 Operating Temperatures for Several Heat Pipe Fluids

Fluid	Operating Temperature Range (°F)
Ammonia	−50–125
Sulfur dioxide	15–110
Water	40–450
FC 43	250–430
Mercury	400–820
Cesium	750–1830
Sodium	920–2200
Lithium	1500–3000

(handwritten annotations: "these 3 are most common" pointing to Ammonia, Water; "alcohol")

10.5 HEAT PIPE APPLICATIONS

Heat pipes are generally used to transport heat from the source directly to the sink without any external power. Heat pipes can eliminate hot spots and can accept heat that has a high power density. Most heat pipes are external to the device they are cooling. Some new applications are available where the heat pipe is an integral part of the case on an electronic component.

A typical application may involve a component or a subassembly, such as a PCB, which has a high power dissipation and a relatively long heat flow path to the heat sink. The resulting temperature rise from the electronic components to the heat sink is excessive, so forced air cooling and then liquid cooling are examined. When there is no room for fans and ducts, and when the size, weight, and price of a liquid system is evaluated, the price of one or two small heat pipes may look more attractive.

Many compact airborne electronic systems utilize air-cooled cold plates for the side walls of the chassis. Plug-in PCBs are then used to support the electronic components. These are cooled by conducting the heat, along metal strips laminated to the PCB, to the side walls of the chassis, as shown in Figure 10.4.

Considering a typical $\frac{3}{4}$ ATR size chassis, the cross section dimensions will be approximately as shown in Figure 10.4. A plug-in PCB with an aluminum core might be capable of dissipating about 15 watts, based upon a maximum component surface temperature of 212°F (100°C) and a cooling air temperature of 131°F (55°C) flowing through the side wall cold plates.

What happens when the same size PCB has a power dissipation of 70 watts? A metal heat sink core on the PCB would have to be so large and heavy that it is simply unacceptable. A hollow core air cooled PCB, similar to those shown in Figure 6.21, might be used with a multiple fin heat exchanger at the center. However, even this type of construction is only capable of dissipating about 50 to 55 watts in the same environment. A

10.5 Heat Pipe Applications

liquid cooling system could be used, and it would do an excellent job. However, there would be a large increase in the size and weight of the system.

Heat pipes are ideal for the conditions described above. They can be added to the back surface of the PCB to sharply increase the heat transferred from the center of the PCB to the edges. A high temperature rise may still occur at the interface of the PCB with the chassis cold plate, unless a high interface pressure device such as a wedge clamp, shown in Figure 3.23d, is used. Aluminum or copper heat sinks should be used under the heat pipes on the PCB to improve the heat flow to the heat pipes, as shown in Figure 10.5.

Heat pipes have been made with 90° bends to improve the heat transfer from circuit boards that do not plug in. When cold plate side walls are used in a chassis, the 90° bend permits the heat pipe to carry the heat directly from the PCB to the side wall of the chassis, as shown in Figure 10.6. This sharply reduces the temperature rise across the interface from the PCB to the chassis.

Sometimes cooling fins must be extended to improve the cooling by increasing the effective surface area. The heat transfer efficiency of a long fin can be improved substantially by using heat pipes along the length of the fin, as shown in Figure 10.7. The temperature gradient along the fin will be quite small with the heat pipe, producing an isothermal fin. The temperature difference between the fin and the ambient air will be increased, so that the heat transfer from the fin is also increased.

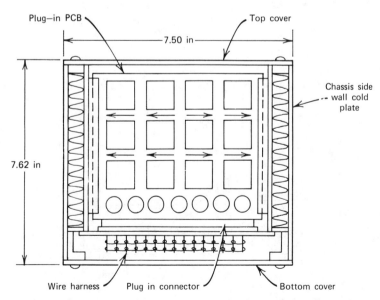

Figure 10.4 Cross section through a chassis with cold plate side walls.

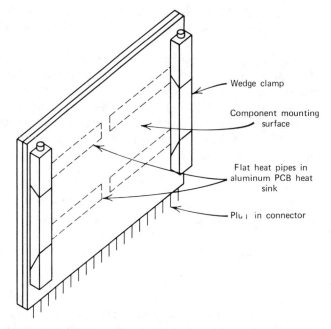

Figure 10.5 Flat heat pipes in an aluminum composite plug-in PCB.

Cooling fins can be added to the condenser end of the heat pipe to improve natural and forced convection cooling. The heat pipe can be bent, if necessary, to reach hot components located in remote areas of a chassis or console. Figure 10.8 shows a tall console with finned heat pipes that extend through the rear panel to provide the required cooling.

Experimental models have been made with small heat pipes that are fabricated into the structure of the electronic component itself. Power transistors have been built with a porous dielectric wick that is placed in contact with the active chip substrate and the inside surfaces of a standard TO type

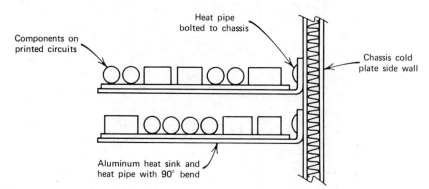

Figure 10.6 Heat pipe with a 90° bend for cooling circuit boards.

10.5 Heat Pipe Applications

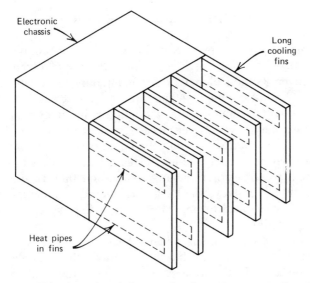

Figure 10.7 Using heat pipes in long cooling fins to improve cooling efficiency.

of container. The unit is evacuated, the wick is saturated with a suitable fluid, and the assembly is hermetically sealed. The evaporation process of the heat pipe takes place on the transistor substrate, and the condensation takes place on the cooler surfaces of the transistor walls. Capillary action in the wick brings the fluid back to the substrate, where the evaporation process is repeated.

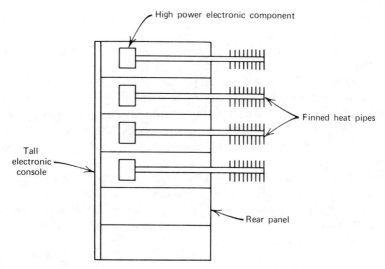

Figure 10.8 Section through a tall console showing heat pipes extending through the rear panel to improve cooling.

10.6 DIRECT AND INDIRECT LIQUID COOLING

Liquid cooling, in general, is much more effective for removing heat than air cooling. Therefore, when high power densities are involved, liquid cooling may be the only practical method for maintaining reasonable component temperatures. Liquid cooling systems are basically classified as direct or indirect. In a direct cooling system, the liquid is in direct contact with the electronic components, which permits the coolant to pick up the heat and carry it away. In an indirect system, the liquid coolant does not come into direct contact with the components. Instead, heat is transferred from the hot component to some intermediate system and then to the liquid. The intermediate system conveys the heat to the liquid by conduction, convection, or radiation. Typical examples of intermediate systems are heat exchangers and fans [6].

Direct liquid cooling systems usually have the electronic components completely immersed in a fluid that has no effect upon the electrical operation of the system. Heat is transferred directly to the fluid by conduction and convection, with very little radiation. Sometimes a low pressure pump is added to circulate the fluid through the electronics to increase the effective cooling. Sometimes the pump is used to spray the liquid coolant directly on the electronic components to carry away the heat.

Indirect liquid cooling systems usually shift the heat transfer problem from the electronic components to another, more remote location. Since the heat must be dumped somewhere, it is often convenient to pump the coolant fluid to a remote air-to-liquid heat exchanger. Large fans can then be used to drive cooling air through the heat exchanger, which will cool the fluid and heat the air. The cooler fluid is pumped back to the electronics to pick up more heat, and the hot air is exhausted to the surrounding ambient.

Indirect cooling systems are often used in missiles and satellites that must operate in the hard vacuum of outer space. The cooling liquid is usually circulated through cold plates, to pick up heat from components mounted on the cold plates. The liquid is then pumped to remote space radiators that are located on the surface of the spacecraft, facing away from the sun toward deep space, where the sink temperature is absolute zero ($-460°R$ or $-273°K$). Heat from the liquid is picked up by the space radiator and dumped into space. The cooler liquid is pumped back to the electronics to pick up more heat.

The U.S. Navy makes extensive use of indirect cooling to remove heat from electronic systems. Electronic components are often mounted on cold plates that use fresh distilled deionized water as the coolant. Forced convection cooling with fresh water is much more effective than is forced convection cooling with air. Typical heat transfer coefficients for fresh water can easily be as much as 100 times greater than typical heat transfer coefficients for air. In addition, water has a specific heat that is more than four times greater than air. Water can therefore absorb four times as much heat as air for the same temperature rise and weight flow.

The fresh water is cooled by circulating it through a remote heat exchanger, which has fresh water on one side and salt water (seawater) on the opposite side. After the fresh water has been cooled, it is returned to the electronic cold plate. Salt water is used to cool the fresh water. The salt water, which has picked up the heat from the fresh water, is dumped back into the sea. Special care must be used to prevent the fresh water from freezing during the winter, or when operating in cold climates.

10.7 FORCED LIQUID COOLING SYSTEMS

High power dissipating electronic systems often make use of forced liquid cooling techniques to control hot spot temperatures. One very common type of cooling device is the cold plate, which provides cooling by conduction and by forced liquid convection. Electronic components are mounted on a metal plate, through which a cooling liquid is circulated, to carry away the heat. The electronic components are fastened directly to the cold plate, which is made of aluminum or copper for high heat conduction. This provides a good heat conduction path from the components to the cooling liquid in the cold plate [6, 66, 68].

The methods for analyzing the thermal characteristics of a liquid cooling system are very similar to the methods shown in Chapter 6 for analyzing forced air cooling for electronics. Many of the equations are the same, with only the values for such parameters as density, viscosity, thermal conductivity, and specific heat changing.

A pump replaces the fan in a liquid cooled system. To select a suitable pump, it is necessary to be able to calculate the total pressure loss through any given system. Three major pressure drops are usually evaluated: (1) friction, which is determined by the liquid velocity and surface roughness within the pipe; (2) difference in elevation; and (3) fitting losses due to elbows, tees, and transitions. These fittings are usually the major source of the pressure drop through an electronic liquid cooling system.

As in the air cooled systems, an exact calculation of the pressure drop is very difficult to obtain. Therefore, it is convenient to use approximate methods that are well documented. One of the most common methods makes use of an equivalent pipe length for various fittings such as elbows and tees. Since the pressure drop through straight pipes with various degrees of roughness is well documented, the calculations are simplified. To obtain a common basis for calculating the equivalent pipe length for various fittings, the effective length is expressed as the number of diameters of a round pipe. In this way, the equivalent length of many fittings can be determined. Table 10.3 shows the effective length of pipe for several fittings

Table 10.3 Effective Length for Various Pipe Fittings [3, 6, 24]

Type of Fitting	Effective Length (Number of Diameters)
45° elbow	15
90° elbow miter, zero radius	60
90° elbow, 1.0 diameter radius	32
90° elbow, 1.5 diameter radius	26
90° elbow, 2.0 diameter radius	20
180° bend, 1.5 diameter radius	50
180° bend, 4-8 diameter radius	10
Globe valve, fully open	300
Gate valve, fully open	7
Gate valve, ¼ closed	40
Gate valve, ½ closed	200
Gate valve, ¾ closed	800
Coupling union	0

expressed in terms of their equivalent diameters. The values are for turbulent flow conditions, but they can also be used to approximate the pressure drop for laminar flow systems.

Pressure losses will also occur at the pipe inlet, depending upon the geometry of the inlet. Some typical losses are shown in Table 10.4. The entry losses are shown as an effective pipe length, which is expressed in terms of the equivalent diameter of a round pipe [3, 6, 24].

10.8 PUMPS FOR LIQUID COOLED SYSTEMS

The pump must be capable of circulating the proper amount of coolant through the system against the total pressure drop through all the fittings, the heat exchanger, and the cold plate. The electric motor driving the pump will usually be matched to the pump so that the motor cannot be overloaded.

Many different types of pumps are available, such as gear, reciprocating, and centrifugal. Gear type pumps do not become airbound very easily, as do centrifugal pumps. However, some centrifugal pumps have been developed to operate without rotating shaft seals, so that a long term leakproof system can be obtained [67].

A number of different safety devices are available to turn off the pump if the flow of the liquid is blocked or if the fluid leaks out. Protective devices are also available for preventing damage due to excessive pressure in a liquid cooling system.

Table 10.4 Effective Length for Various Pipe Entrances

Type of Opening	Description	Effective Length (Number of Diameters)
	Square edge entry	20
	Well–rounded entry	2
	Protruding entry (Borda's mouthpiece)	40
	Tee: elbow entering run at point A	60
	Tee: elbow entering branch at point B	90

10.9 STORAGE AND EXPANSION TANK

Totally enclosed cooling systems are often used to cool electronic equipment. Therefore, some provisions must be made to accommodate the thermal expansion of the fluid as the temperature increases. Also, air must be removed from the coolant and some type of cushion should be provided to reduce pressure surges in the system. This is all accomplished with an expansion tank, which also acts as the storage tank for the liquid coolant.

10.10 LIQUID COOLANTS

Water is probably the best cooling liquid available in terms of density, viscosity, thermal conductivity, and specific heat. For good long term operation, distilled water that has been deionized should be used. When the temperature is expected to drop below freezing or when the surface temperature is expected to exceed 212°F (100°C), ethylene glycol should be added to the water. Ethylene glycol has a lower thermal conductivity than water, so its addition will degrade the thermal performance of the water. However, when both aluminum and copper are being used in the cooling circuit, the addition of ethylene glycol will prevent corrosion of the passages.

Many other liquids are available for cooling electronic systems, such as silicones, but they have a much lower thermal conductivity than water. Coolanol 45, a silicone ester made by Monsanto Chemical Company, is a very effective coolant that has a useful cooling range from −58°F (−50°C) to 392°F (200°C). Its viscosity is substantially higher than that of water. This may increase the pressure drop, which may require the use of a larger pump motor in a system where the water is replaced by Coolanol [68].

A liquid that has approximately the same viscosity as water is FC 75, a fluorochemical made by the 3M Company. This liquid has a specific heat about one fourth that of water. Tables 10.5 and 10.5A show the thermal properties of several different coolant fluids.

10.11 SIMPLE LIQUID COOLING SYSTEM

A simple liquid cooling system can be designed by combining the various items described in the previous sections. A simplified schematic of an elementary liquid cooling system is shown in Figure 10.9.

10.12 MOUNTING COMPONENTS FOR INDIRECT LIQUID COOLING

Electronic components must always be mounted so that there is a low thermal resistance heat flow path to the heat sink. This is true for any

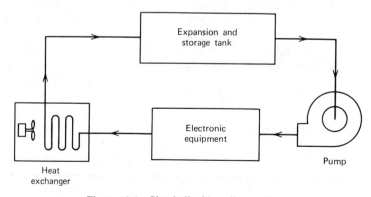

Figure 10.9 Simple liquid cooling system.

Table 10.5 Properties of Various Cooling Fluids: English Units

	Water		FC 75		Coolanol 25		Coolanol 45	
	77°F (25°C)	140°F (60°C)	77°F (25°C)	140°F (60°C)	77°F (25°C)	140°F (60°C)	77°F (25°C)	140°F (60°C)
Viscosity (lb/ft hr)	2.17	1.14	3.49	2.05	10.89	5.93	43.32	16.84
Density (lb/ft³)	62.21	61.40	109.8	103.6	56.19	54.63	55.87	54.31
Thermal conductivity (Btu/hr·ft·°F)	0.353	0.378	0.037	0.035	0.076	0.074	0.078	0.075
Specific heat (Btu/lb·°F)	0.998	0.998	0.248	0.263	0.45	0.48	0.45	0.49
Boiling	212°F (100°C)		214°F (101°C)		355°F (179°C)		355°F (179°C)	
Freezing	32°F (0°C)		−171°F (−113°C)		−125°F (−87°C)*		−88°F (−67°C)*	

Pour Point.

Table 10.5A Properties of Various Cooling Fluids: Metric Units

	Water		FC 75		Coolanol 25		Coolanol 45	
	25°C (77°F)	60°C (140°F)	25°C (77°F)	60°C (140°F)	25°C (77°F)	60°C (140°F)	25°C (77°F)	60°C (140°F)
Viscosity (poise) (g/cm·sec)	0.00896	0.00470	0.01441	0.00847	0.045	0.0245	0.179	0.0696
Density (g/cm³)	0.995	0.982	1.757	1.658	0.900	0.875	0.895	0.870
Thermal conductivity (cal/sec·cm·°C)	0.00146	0.00156	0.000153	0.000144	0.000314	0.000306	0.000322	0.000311
Specific heat (cal/g·°C)	0.998	0.998	0.248	0.263	0.45	0.48	0.45	0.49
Boiling	100°C (212°F)		101°C (214°F)		179°C (355°F)		179°C (355°F)	
Freezing	0°C (32°F)		−113°C (−171°F)		−87°C (−125°F)*		−67°C (−88°F)*	

Pour Point

electronic system that must provide reliable operation for steady state conditions. This may not be necessary for transient conditions, where the power is not on long enough for the temperature to stabilize.

Components mounted on liquid cooled cold plates should be provided with flat smooth surfaces, to improve the transfer of heat across the mounting interface. Power devices such as transistors, resistors, diodes, and transformers can be fastened to the cold plate with bolts that are capable of applying high forces. Bolted components can be removed easily if they have to be replaced.

Silicone grease can be applied at the interface of the component to the cold plate to further reduce the thermal resistance on high power dissipating devices.

Silicone grease should not be used anywhere on a cold plate if there are cemented joints in the area. Silicone grease tends to migrate, so that it will contaminate the cemented surfaces and reduce the bond strength. In a vibration or shock environment, the cemented interface can fracture at a very low stress level, producing a catastrophic failure.

For very high power dissipating components, it is often necessary to reduce the interface temperature rise to an absolute minimum. Under these circumstances, the component may be soldered directly to the cold plate mounting surface. This is done by first pretinning both surfaces with a low temperature solder. The solder is then reflowed using a high temperature for a short period of time to prevent heat from soaking into the component and damaging it. A heat sink may be used on the body of the component to pull away some of the heat during the reflow process, to reduce the possibility of damage.

When a large number of similar high power components must be cooled, it is convenient to mount them on a liquid cooled cold plate, as shown in Figure 10.10. Additional methods for mounting high power components are discussed in Sections 4.4 through 4.10.

Plug-in PCBs are often cooled with the use of liquid cooled cold plates, which are part of the side walls on an electronic chassis. Aluminum or copper heat sinks are used on the PCB to conduct the heat from the

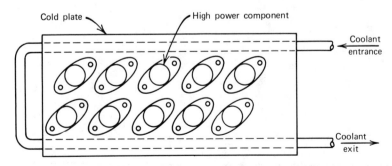

Figure 10.10 Forced liquid cooled cold plate heat sink.

10.13 Basic Forced Liquid Flow Relations

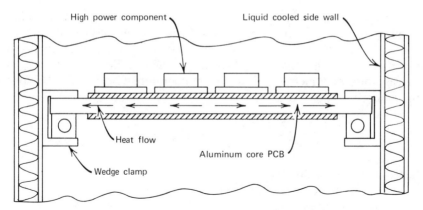

Figure 10.11 Liquid cooled chassis with conduction cooled PCB.

components to the side wall cold plates, as shown in Figure 10.11. This technique is very similar to the air cooled cold plates described in Sections 6.23 through 6.26, except that the cooling fluid is a liquid instead of air.

10.13 BASIC FORCED LIQUID FLOW RELATIONS

The standard fluid flow equations can be used to solve heat transfer problems for liquids as well as for air. Most of the air flow equations can also be used for liquid flow, with some minor modifications. The symbols used for air flow are the same as the symbols used for liquid flow, except that the values are different for liquids and gases.

The two parameters most often evaluated in forced air cooling, as well as in forced liquid cooling, are the temperature rise and the pressure drop. The temperature rise is measured in °F for English units and in °C for metric units. The pressure drop in air cooled systems is measured in inches or centimeters of water. However, with liquid cooled units the pressure drops are much higher, so that they are normally expressed in lb/in² or in g/cm².

The temperature rise in a liquid cooled system can be determined from Eq. 6.7. The symbols remain the same as those defined by that equation, except that everything now relates to the flow of the liquid instead of the air flow. Consistent sets of units must be used for English and metric systems.

$$\Delta t = \frac{Q}{W C_p} \qquad \text{(ref. Eq. 6.7)}$$

where Q = power dissipation
  W = liquid coolant flow rate
  C_p = specific heat of the liquid coolant

A convection film will develop in liquid cooled systems as well as in air cooled systems. This film clings to the heat transfer surface and restricts the flow of heat. The forced convection coefficient for liquid cooling is also shown as h_c and its characteristics are defined in Eqs. 3.53 and 5.7 for English units and metric units. The temperature rise across the forced convection film then becomes:

$$\Delta t = \frac{Q}{h_c A} \quad \text{(ref. Eqs. 3.53 and 5.7)}$$

where Q = power dissipation
h_c = forced convection coefficient
A = surface area

The value of the liquid forced convection coefficient can be determined from Eq. 6.9. The symbols are the same, except that they now must reflect the values of the liquid instead of the air.

$$h_c = JC_p G \left(\frac{C_p \mu}{K}\right)^{-2/3} \quad \text{(ref. Eq. 6.9)}$$

where J = Colburn factor
C_p = specific heat
G = weight flow velocity
K = thermal conductivity
μ = viscosity

The Reynolds number, which is shown in Eq. 6.11 must be obtained before the Colburn J factor can be determined. The symbols are the same as for air; only the values change.

$$N_R = \frac{VD\rho}{\mu} = \frac{GD}{\mu} \quad \text{(ref. Eq. 6.11)}$$

For laminar flow conditions through smooth tubes with Reynolds numbers less than about 2000, the Colburn J factor can be determined from Eq. 10.1.

$$J = \frac{1.6}{\left(\frac{L}{D}\right)^{0.333} (N_R)^{0.666}} \quad (10.1)$$

The forced convection coefficient h_c has relatively little change in the laminar flow region, with low weight flow velocity (G) values. In this range the forced convection coefficient is really determined by the hydraulic diameter, as shown in Figure 10.12. The forced convection coefficient in-

10.13 Basic Forced Liquid Flow Relations

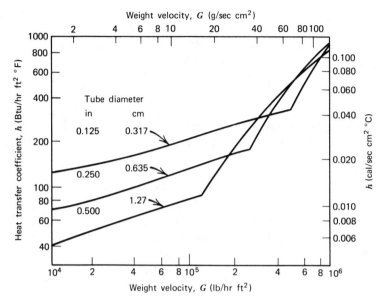

Figure 10.12 Forced convection coefficient for water in smooth pipes. (Ref: General Electric Heat Transfer Data Book, 1975)

creases rapidly as the hydraulic diameter decreases. When the weight flow velocity increases in the turbulent flow range, there is a corresponding increase in the convection coefficient [70].

For turbulent flow conditions through smooth pipes, where the Reynolds number is greater than about 7000, the J factor can be determined from Eq. 10.2 [3, 15].

$$J = \frac{0.025}{(N_R)^{0.2}} \quad (10.2)$$

The relation above applies for low viscosity liquids where the Prandtl number $(C_p \mu / K)$ is between values of about 1.5 to 20.

The weight flow velocity G through pipes and ducts is determined from Eq. 6.12. The symbols for liquids are the same as those for air.

$$G = \frac{W}{A} \quad \text{(ref. Eq. 6.12)}$$

where W = weight flow of liquid
 A = cross section flow area

Pressure drop relations for liquid flow through smooth pipes and ducts are determined from the same Darcy flow equations for air. When the pressure loss is to be expressed in the height of a column of water, Eq. 6.82 is used. The symbols for liquids are the same as those for air. The equation can be used for laminar or turbulent flow conditions.

$$H_L = 4f\left(\frac{L}{D}\right)\left(\frac{V^2}{2g}\right) \quad \text{(ref. Eq. 6.82)}$$

For laminar flow conditions in round tubes, the Fanning friction factor is used in Eq. 6.82, as shown in Table 6.7.

$$f = \frac{16}{N_R} \quad \text{(ref. Table 6.7)}$$

Sometimes it is more convenient to write the head loss using the Hagen-Poiseulle friction factor. The head loss relation is then shown by Eq. 10.3.

$$H_L = f\left(\frac{L}{D}\right)\left(\frac{V^2}{2g}\right) \tag{10.3}$$

For laminar flow in round tubes, the Hagen-Poiseulle friction factor is then used, as shown in Eq. 10.4.*

$$f = \frac{64}{N_R} \quad \text{used primarily with liquids} \tag{10.4}$$

Note that the friction constant in front of *both* Eq. 6.82 and Eq. 10.3 will have a value of 64.

For turbulent flow conditions with Reynolds numbers up to 100,000, the friction factor to be used with Eq. 10.3 is shown in Eq. 10.5 [15].

$$f = \frac{0.316}{(N_R)^{0.25}} \tag{10.5}$$

For turbulent flow conditions with Reynolds numbers up to 300,000, the friction factor to be used with Eq. 10.3 is shown in Eq. 10.6.

$$f = \frac{0.184}{(N_R)^{0.2}} \tag{10.6}$$

Ducts with circular cross sections are not always used in liquid cooling systems. Therefore, it is convenient to use a hydraulic diameter for evalu-

---

* Do not become confused with the Darcy equation for laminar flow in round pipes, shown in Eq. 6.82. In the Darcy equation, the Fanning friction factor is used where $f = 16/N_R$.

ating liquid systems. The hydraulic diameter is defined as shown in Eq. 10.7.

$$D_H = \frac{4 \times \text{cross section area}}{\text{wetted perimeter}} \tag{10.7}$$

For a circular cross section, this is simply the pipe diameter.

$$D_H = \frac{4(\pi D^2/4)}{\pi D} = D = \text{pipe diameter}$$

For a rectangular cross section with inside dimensions $a \times b$, the hydraulic diameter is

$$D_H = \frac{4ab}{2a + 2b} = \frac{2ab}{a + b}$$

Sometimes the hydraulic radius is required for some flow problems. The hydraulic radius is defined in Eq. 10.8.

$$R_H = \frac{\text{cross section area}}{\text{wetted perimeter}} \tag{10.8}$$

For a circular cross section, the hydraulic radius is one fourth of the diameter of the pipe.

$$R_H = \frac{\pi D^2/4}{\pi D} = \frac{D}{4} \tag{10.9}$$

10.14 SAMPLE PROBLEM—TRANSISTORS ON A WATER COOLED COLD PLATE

A water cooled cold plate supports 16 stud mounted transistors, which dissipate 37.5 watts each, for a total power dissipation of 600 watts. The coolant flow rate is 1.0 gal/min (62.8 g/sec) with an inlet temperature of 95°F (35°C), flowing through a tube that has a 0.312 in (0.792 cm) inside diameter, as shown in Figure 10.13. The maximum allowable component mounting surface temperature is 160°F (71°C). Determine the component surface temperature when silicone grease is used at the transistor mounting interface. Also determine the pressure drop through the system resulting from a flow rate of 1 gal/min.

SOLUTION—COMPONENT SURFACE TEMPERATURE

The component mounting surface temperature is determined first. This is obtained by calculating the temperature rise along individual segments along

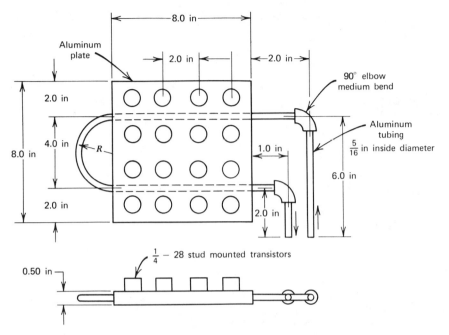

Figure 10.13 Water cooled cold plate with 16 transistors.

the heat flow path from the component to the coolant as follows:

Δt_1 = temperature rise across the transistor mounting interface from the transistor case to the heat sink surface, using silicone grease at the interface

Δt_2 = temperature rise through the aluminum cold plate from the transistor to the coolant tubing

Δt_3 = temperature rise across the liquid coolant convection film from the walls of the tube to the coolant

Δt_4 = temperature rise of the coolant as it flows through the cold plate picking up heat from the transistors

The physical properties of water are shown in Table 10.6 for English units and in Table 10.6A for metric units.

Starting with Δt_1 the temperature rise across the transistor interface is obtained from Table 3.3, using thermal grease at the mounting interface of the $\frac{1}{4}$-28 stud mounted transistor.

$$\Delta t_1 = 0.30 \frac{°C}{watt} (37.5 \text{ watts}) = 11.2°C\ (20.2°F) \qquad (10.10)$$

Symmetry is used to determine the temperature rise through the aluminum cold plate from one transistor to the coolant tube, as shown in Figure 10.14.

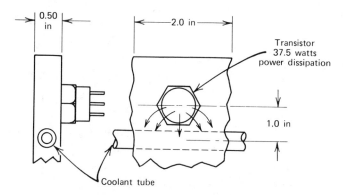

Figure 10.14 Heat transfer path from transistor to coolant tube.

Table 10.6 Properties of Water: English Units

t (°F)	C_p $\left(\dfrac{\text{Btu}}{\text{lb °F}}\right)$	μ $\left(\dfrac{\text{lb}}{\text{ft hr}}\right)$	K $\left(\dfrac{\text{Btu}}{\text{hr ft °F}}\right)$	$\dfrac{C_p \mu}{K}$ (dimensionless)
32	1.009	4.33	0.327	13.4
40	1.005	3.75	0.332	11.3
50	1.002	3.17	0.338	9.4
60	1.000	2.71	0.344	7.9
70	0.998	2.37	0.349	6.8
80	0.998	2.08	0.355	5.8
90	0.997	1.85	0.360	5.1
100	0.997	1.65	0.364	4.5
110	0.997	1.49	0.368	4.0
120	0.997	1.36	0.372	3.6
130	0.998	1.24	0.375	3.3
140	0.998	1.14	0.378	3.0
150	0.999	1.04	0.381	2.7
160	1.000	0.97	0.384	2.5
170	1.001	0.90	0.386	2.3
180	1.002	0.84	0.389	2.2
190	1.003	0.79	0.390	2.1
200	1.004	0.74	0.392	1.9

Table 10.6A Properties of Water: Metric Units

t (°C)	C_p $\left(\dfrac{\text{cal}}{\text{g °C}}\right)$	μ $\left(\dfrac{\text{g}}{\text{cm sec}}\right)$	K $\left(\dfrac{\text{cal}}{\text{sec cm °C}}\right)$	$\dfrac{C_p \mu}{K}$ (dimensionless)
0	1.009	0.0179	0.00135	13.4
4.4	1.005	0.0155	0.00137	11.3
10.0	1.002	0.0131	0.00139	9.4
15.6	1.000	0.0112	0.00142	7.9
21.1	0.998	0.0098	0.00144	6.8
26.7	0.998	0.0086	0.00147	5.8
32.2	0.997	0.0076	0.00149	5.1
37.8	0.997	0.0068	0.00150	4.5
43.3	0.997	0.0061	0.00152	4.0
48.9	0.997	0.0056	0.00154	3.6
54.4	0.998	0.0051	0.00155	3.3
60.0	0.998	0.0047	0.00156	3.0
65.6	0.999	0.0043	0.00157	2.7
71.1	1.000	0.0040	0.00158	2.5
76.7	1.001	0.0037	0.00159	2.3
82.2	1.002	0.0035	0.00161	2.2
87.8	1.003	0.0033	0.00161	2.1
93.3	1.004	0.0031	0.00162	1.9

Conversions for English units and metric units are made with Tables 1.1 through 1.11. The heat is conducted through the aluminum plate, so that the conduction heat transfer equation is used.

$$\Delta t_2 = \frac{QL}{KA} \quad \text{(ref. Eq. 3.2)}$$

Given $\quad Q = 37.5$ watts $= 128$ Btu/hr $= 8.96$ cal/sec
$\quad\quad\quad L = 1.0$ in $= 0.0833$ ft $= 2.54$ cm
$\quad\quad\quad K = 100$ Btu/hr ft °F $= 0.413$ cal/sec cm °C

$$A = \frac{(2)(0.50)}{144} = 0.00694 \text{ ft}^2 = 6.45 \text{ cm}^2$$

In English units:

$$\Delta t_2 = \frac{(128)(0.0833 \text{ ft})}{(100)(0.00694)} = 15.4°F \qquad (10.11)$$

10.14 Sample Problem—Transistors on a Water Cooled Cold Plate

In metric units:

$$\Delta t_2 = \frac{(8.96)(2.54 \text{ cm})}{(0.413)(6.45)} = 8.5°C \quad (10.11a)$$

The forced convection coefficient across the liquid film in the coolant tube cannot be obtained until it has been determined if the liquid flow is laminar or turbulent. Therefore, the Reynolds number must be determined.

$$N_R = \frac{GD}{\mu} \quad \text{(ref. Eq. 6.11)}$$

Given $W = 1.0 \text{ gal/min} = 8.3 \text{ lb/min} = 62.8 \text{ g/sec}$
$A = (\pi/4)(0.312)^2 = 0.0764 \text{ in}^2 = 0.000531 \text{ ft}^2 = 0.493 \text{ cm}^2$

$$G = \frac{W}{A} = \frac{(8.3 \text{ lb/min})(60 \text{ min/hr})}{0.000531 \text{ ft}^2} = 9.38 \times 10^5 \frac{\text{lb}}{\text{hr ft}^2}$$

$$G = \frac{62.8 \text{ g/sec}}{0.493 \text{ cm}^2} = 127.38 \frac{\text{g}}{\text{sec cm}^2}$$

$D = 0.312 \text{ in} = 0.026 \text{ ft} = 0.792 \text{ cm}$
$\mu = 1.75 \text{ lb/ft hr} = 0.00722 \text{ g/cm sec}$ (Ref table 10.6) at 95°F (35°C)

In English units:

$$N_R = \frac{(9.38 \times 10^5)(0.026 \text{ ft})}{1.75} = 1.39 \times 10^4 \text{ (dimensionless)} \quad (10.12)$$

In metric units:

$$N_R = \frac{(127.38)(0.792 \text{ cm})}{0.00722} = 1.39 \times 10^4 \text{ (dimensionless)} \quad (10.12a)$$

Since the Reynolds number is well above 3000, the coolant flow is turbulent. The forced convection coefficient for the coolant is determined from Eq. 6.9. The Colburn J factor for the turbulent flow is determined from Eq. 10.2.

Given $h_c = JC_pG\left(\dfrac{C_p\mu}{K}\right)^{-2/3}$ (ref. Eq. 6.9)

$N_R = 1.39 \times 10^4$ (ref. Eq. 10.12)

$J = \dfrac{0.025}{(N_R)^{0.2}}$ (ref. Eq. 10.2)

$$J = \frac{0.025}{(1.39 \times 10^4)^{0.2}} = 0.00371 \text{ (dimensionless)}$$

$C_p = 0.997$ Btu/lb °F $= 0.997$ cal/g °C [ref. Table 10.6 at 95°F (35°C)]
$G = 9.38 \times 10^5$ lb/hr ft² $= 127.38$ g/sec cm²

$$\frac{C_p \mu}{K} = 4.8 \text{ [dimensionless; ref. Table 10.6 at 95°F (35°C)]}$$

Substitute into Eq. 6.9 for English units.

$$h_c = (0.00371)(0.997)(9.38 \times 10^5)(4.8)^{-0.666}$$

$$h_c = 1220 \frac{\text{Btu}}{\text{hr ft}^2 \text{ °F}} \qquad (10.13)$$

Substitute into Eq. 6.9 for metric units.

$$h_c = (0.00371)(0.997)(127.38)(4.8)^{-0.666}$$

$$h_c = 0.166 \frac{\text{cal}}{\text{sec cm}^2 \text{ °C}} \qquad (10.13a)$$

The temperature rise across the liquid coolant film in the tube is determined from Eq. 5.7.

$$\Delta t_3 = \frac{Q}{h_c A} \qquad \text{(ref. Eq. 5.7)}$$

Given $Q = 600$ watts $= 2047.8$ Btu/hr $= 143.4$ cal/sec
$h_c = 1220$ Btu/hr ft² °F $= 0.166$ cal/sec cm² °C
$A = \pi DL$ (inside surface area of 0.312 in diameter tube)
$A = \pi(0.312)(16$ in length$) = 15.7$ in²
$A = 0.109$ ft² $= 101.2$ cm²

Substitute into Eq. 5.7 for English units.

$$\Delta t_3 = \frac{2047.8 \text{ Btu/hr}}{(1220)(0.109 \text{ ft}^2)} = 15.4°\text{F} \qquad (10.14)$$

Substitute into Eq. 5.7 for metric units.

$$\Delta t_3 = \frac{143.4 \text{ cal/sec}}{(0.166)(101.2 \text{ cm}^2)} = 8.5°\text{C} \qquad (10.14a)$$

10.14 Sample Problem—Transistors on a Water Cooled Cold Plate

The temperature rise of the coolant as it flows through the cold plate is determined from Eq. 6.7.

$$\Delta t_4 = \frac{Q}{WC_p} \quad \text{(ref. Eq. 6.7)}$$

Given $Q = 600$ watts $= 2047.8$ Btu/hr $= 143.3$ cal/sec
$W = 1.0$ gal/min $= 8.3$ lb/min $= 498$ lb/hr $= 62.8$ g/sec
$C_p = 0.997$ Btu/lb °F $= 0.997$ cal/g °C (ref. Table 10.6)

Substitute into Eq. 6.7 for English units.

$$\Delta t_4 = \frac{2047.8 \text{ Btu/hr}}{(498)(0.997)} = 4.1°F \quad (10.15)$$

Substitute into Eq. 6.7 for metric units.

$$\Delta t_4 = \frac{143.4 \text{ cal/sec}}{(62.8)(0.997)} = 2.3°C \quad (10.15a)$$

The surface temperature of the transistor is determined by adding all of the temperature rises along the heat flow path to the inlet temperature of the coolant. In English units:

$$t_t = 95 + \Delta t_1 + \Delta t_2 + \Delta t_3 + \Delta t_4$$
$$t_t = 95 + 20.2 + 15.4 + 15.4 + 4.1 = 150.1°F \quad (10.16)$$

In metric units:

$$t_t = 35 + 11.2 + 8.5 + 8.5 + 2.3 = 65.5°C \quad (10.16a)$$

Since the transistor surface temperature is less than 160°F (71°C), the design is satisfactory.

SOLUTION—PRESSURE DROP

The pressure drop through the system is determined from Eq. 10.3. The flow through the cold plate is turbulent, as shown by Eq. 10.12. Therefore, the friction factor shown in Eq. 10.5 is used for the pressure drop calculations.

$$H_L = f\left(\frac{L}{D}\right)\left(\frac{V^2}{2g}\right) \quad \text{(ref. Eq. 10.3)}$$

Given $N_R = 1.39 \times 10^4$ (ref. Eq. 10.12) (dimensionless)

$$f = \frac{0.316}{(N_R)^{0.25}} \text{ (ref. Eq. 10.5) (friction factor)}$$

$$f = \frac{0.316}{(1.39 \times 10^4)^{0.25}} = 0.0291 \text{ (dimensionless)}$$

$D = 0.312$ in $= 0.026$ ft $= 0.792$ cm (diameter)
$g = 32.2$ ft/sec^2 $= 980$ cm/sec^2 (gravity)
$W = 8.3$ lb/min $= 0.138$ lb/sec $= 62.8$ g/sec (fluid flow)
$\rho = 62.4$ lb/ft^3 $= 1.0$ g/cm^3 (density of water)

$$A = \frac{\pi}{4}(0.312)^2 = 0.0764 \text{ in}^2 = 0.000531 \text{ ft}^2 = 0.493 \text{ cm}^2 \text{ (area)}$$

$$V = \frac{W}{\rho A} \text{ (velocity of water in pipe)}$$

In English units:

$$V = \frac{0.138 \text{ lb/sec}}{(62.4)(0.000531 \text{ ft}^2)} = 4.16 \frac{\text{ft}}{\text{sec}}$$

In metric units:

$$V = \frac{62.8 \text{ g/sec}}{(1.0)(0.493 \text{ cm}^2)} = 127.4 \frac{\text{cm}}{\text{sec}}$$

The length (L) of the coolant flow path is obtained from the length of the straight tubing plus the pipe fittings expressed in terms of equivalent pipe diameters, as shown in Table 10.3. The pipe diameter is 0.312 in (0.792 cm); see Table 10.7.

Substitute into Eq. 10.3 for English units.

$$H_L = (0.0291)\left(\frac{42.5 \text{ in}}{0.312 \text{ in}}\right)\frac{(4.16 \text{ ft/sec})^2}{(2)(32.2 \text{ ft/sec}^2)} = 1.06 \text{ ft H}_2\text{O} \quad (10.17)$$

Substitute into Eq. 10.3 for metric units.

$$H_L = (0.0291)\left(\frac{107.9 \text{ cm}}{0.792 \text{ cm}}\right)\frac{(127.4 \text{ cm/sec})^2}{(2)(980 \text{ cm/sec}^2)} = 32.8 \text{ cm H}_2\text{O} \quad (10.17a)$$

The head loss can also be expressed as a pressure loss by comparing the value with the standard atmosphere. This is where a standard atmosphere

Table 10.7 Diameter and Length of Various Fittings

Fitting Type	Number of Diameters	Length in	Length cm
180° bend, 6 diameters radius	10	3.1	7.9
90° elbow, 2 diameters radius	20	6.2	15.7
90° elbow, 2 diameters radius	20	6.2	15.7
Straight pipe, 8 + 8 + 2 + 1 + 2 + 6 in		27.0	68.6
Total equivalent pipe length		42.5	107.9

Figure 10.15 Tall double bay cabinet, with a blower at the bottom for cooling several racks of plug-in PCBs. (Courtesy Litton Systems, Inc.)

will support a column of water 34 ft high, so that 14.7 lb/in² equals 34 ft of water.

$$\Delta P = \frac{1.06}{34}(14.7) = 0.46 \frac{\text{lb}}{\text{in}^2} \qquad (10.18)$$

For metric units, a standard atmosphere is 76 cm of mercury, which represents about 1034 cm of water. This is then equivalent to a pressure of 1034 g/cm².

$$\Delta P = \frac{32.8}{1034}(1034) = 32.8 \frac{\text{g}}{\text{cm}^2} \qquad (10.18\text{a})$$

Figure 10.15 shows a tall double bay cabinet, with plug-in PCBs that are cooled by means of a blower in the base.

References

1. Dave S. Steinberg, *Vibration Analysis for Electronic Equipment,* John Wiley & Sons, 1973.
2. L. K. Boelter, R. C. Martinelli, and F. E. Rumie, An Investigation of Aircraft Heaters, Design Manual, NACA Report 5A06, August 1945.
3. U. S. Naval Air Development Center, Design Manual for Methods of Cooling Electronic Equipment, NAVWEPS 16-1-532, July 1977.
4. L. V. Berkner and H. Odishaw, *Science in Space,* McGraw-Hill, 1961.
5. R. Adler, L. Chu, and R. Fano, *Electromagnetic Energy Transmission and Radiation,* John Wiley & Sons, 1960.
6. Design Manual of Methods of Liquid Cooling Electronic Equipment, NAVSHIPS 900-195, Department of The Navy, Bureau of Ships 1960.
7. Design Manual of Natural Methods of Cooling Electronic Equipment, NAVSHIPS 900-192, Department of The Navy, Bureau of Ships, November 1956.
8. Design Manual of Cooling Methods for Electronic Equipment, NAVSHIPS 900-190, Department of The Navy, Bureau of Ships, March 1955.
9. Forest B. Golden, Analysis Can Take the Heat off Power Semiconductors, *Electronics Magazine,* December 6, 1973.
10. O. W. Eshbach, *Handbook of Engineering Fundamentals,* John Wiley & Sons, 1969.
11. To Determine the Effects of Ultrasonic Cleaning on Semiconductor Devices, Project 5960-1688, by Headquarters U.S. Army Command, Fort Monmouth, N.J., June 4, 1964.
12. Howard Dicken, Aluminum Bonding Wire Fatigue Induced by Ultrasonic Vibration, *Electronic Packaging and Production Magazine,* Kiver Publications, October 1978.
13. Arnold Wexler, *Humidity and Moisture,* Reinhold, 1965.
14. A. N. Nesmeyanov, *Vapor Pressure of the Elements,* Academic Press, 1963.
15. William H. McAdams, *Heat Transmission,* McGraw-Hill, 1954.
16. Warren H. Giedt, *Principles of Engineering Heat Transfer,* D. Van Nostrand, 1957.
17. A. I. Brown and S. M. Marco, *Introduction to Heat Transfer,* McGraw-Hill, 1958.
18. Marks, *Mechanical Engineers Handbook,* McGraw-Hill, 1951.
19. R. T. Kent, *Mechanical Engineers Handbook,* John Wiley & Sons, 1969.
20. S. L. Hoyt, *Metals and Alloys Data Book,* Reinhold, 1943.
21. MIL-HDBK-5B, Metalic Materials and Elements for Aerospace Vehicle Structures, Department of Defense, Washington, D.C. 1975.
22. *Materials Engineering Magazine,* Penton/IPC Reinhold, November 1976.
23. *Machine Design Magazine,* Materials Reference, Penton/IPC, March 1978.
24. Max Jacob, *Heat Transfer,* John Wiley & Sons, 1967.
25. E. Fried and F. Costello, Interface Thermal Contact Resistance Problem in Space Vehicles, *ARS Journal,* February 1962.
26. V. R. Stubstad, Measurements of Thermal Contact Conductance in Vacuum, ASME publication, November 17, 1963.

27. N. D. Weills and E. A. Ryder, Thermal Resistance Measurement of Joints Formed between Stationary Metal Surfaces, *Transactions of the ASME,* April 1949.
28. M. E. Barzelay, K. N. Tung, and G. F. Holloway, Effects of Pressure on Thermal Conductance of Contact Joints, National Advisory Committee for Aeronautics (NACA) Technical Note 3295, May 1955.
29. S. B. Marshall and R. F. Dewey, Plastic Power IC's Need Skillful Thermal Design, *Electronics Magazine,* November 8, 1973.
30. Carl J. Feldmanis, Network Analog Maps Heat Flow, *Electronics Magazine,* May 16, 1974.
31. R. P. Benedict, Two Dimensional Transient Heat Flow, *Electrotechnology,* May 1962.
32. Mary L. Rauhe, A Study of the Flat Pack Case, *Electronic Packaging and Production Magazine,* May 1966.
33. Fred C. Trumel, Six Ways to Cope with Thermal Expansion, *Machine Design Magazine,* February 10, 1977.
34. R. N. Wild, Some Fatigue Properties of Solders and Solder Joints, IBM Report No. 74Z00044S, July 1974.
35. S. S. Manson, A Designers Guide to Thermal Stress, *Machine Design Magazine,* November 23, 1961.
36. J. O. Hinze, *Turbulence,* McGraw-Hill, 1959.
37. C. F. Campen et al., *Handbook of Geophysics,* Macmillan, 1961.
38. U.S. Standard Atmosphere, National Aeronautics and Space Administration, USAF, 1966.
39. MIL-HDBK-217 C, Reliability Prediction of Electronic Equipment, Department of Defense, Washington, D.C., 1979.
40. John L. Alden, *Design of Industrial Exhaust Systems,* Industrial Press, 1948.
41. A. H. Shapiro, *The Dynamics and Thermodynamics of Compressible Fluid Flow,* Roland Press, 1953.
42. William C. Osborne, *Fans,* Pergamon Press, 1966.
43. W. M. Kays and A. L. London, *Compact Heat Exchangers,* McGraw-Hill, 1964.
44. Rotron Manufacturing Co., Fan Catalog, Woodstock, N.Y.
45. Gordon J. Van Wylen, *Thermodynamics,* John Wiley & Sons, 1959.
46. R. H. Sabersky, *Elements of Engineering Thermodynamics,* McGraw-Hill, 1957.
47. P. T. Landsberg, *Thermodynamics,* Interscience, 1961.
48. Lou Laermer, Air through Hollow Cards Cools High Power LSI, *Electronics Magazine,* June 13, 1974.
49. R. G. Hajec and H. L. Benjamin, Selecting Fans for High Altitude Cooling of Electronic Equipment, *Electronic Packaging and Production Magazine,* July 1966.
50. Gordon M. Taylor, Forced Air Cooling in High Density Systems, *Electronics Magazine,* January 24, 1974.
51. Shih-I Pai, *Viscous Flow Theory,* D. Van Nostrand, 1956.
52. W. J. Humphreys, *Physics of the Air,* McGraw-Hill, 1940.
53. Benjamin Shelpuk, Heat Exchangers Cool Hot Plug-In PC Boards, *Electronics Magazine,* June 27, 1974.
54. T. K. Tierney and E. Koczkur, Free Convection Heat Transfer from a Totally Enclosed Cabinet Containing Simulated Electronic Equipment, *IEEE Transactions on Parts, Hybrids and Packaging,* September 1971.
55. Lloyd M. Polentz, Add a component to Your Compressed Air System, *Hydraulics and Pneumatics Magazine,* September 1966.

References

56. Dave S. Steinberg, Quick way to Predict Temperature Rise in Electronic Circuits, *Machine Design Magazine,* January 20, 1977.
57. P. A. A. Laura and R. H. Gutierrez, Transient Temperature Distribution in Thermally Orthotropic Plates of Complicated Boundary Shape, Institute of Applied Mechanics, Naval Base, Puerto Belgrano, Argentina, 1977.
58. Francis J. Lavoie, Cooling with Heat Pipes, *Machine Design Magazine,* August 6, 1970.
59. A. Basiulis, Heat Pipes in Electronic Packaging, *Electronics Packaging and Production Magazine,* November 1978.
60. K. T. Feldman, Jr. and G. H. Whiting, Applications of the Heat Pipe, *Mechanical Engineering Magazine,* November 1968.
61. K. T. Feldman, Jr., and G. H. Whiting, The Heat Pipe, *Mechanical Engineering Magazine,* February 1967.
62. A. J. Streb, The Heat Pipe: Overcoming the Thermal Resistance Barrier, *Electronics Packaging and Production Magazine,* December 1971.
63. F. D. Yeaple, Intense Cooling Absorbs Heat of New Super Semiconductor, *Product Engineering Magazine,* February 1972.
64. W. E. Harbaugh and G. Y. Eastman, Applying Heat Pipes to Thermal Problems, *Heating Piping and Air Conditioning,* October 1970.
65. S. Katzoff, Heat Pipes and Vapor Chambers for Thermal Control of Spacecraft, NASA Langley Research Center, April 1967.
66. John A. Gardner, Jr., Liquid Cooling Safeguards High Power Semiconductors, *Electronics Magazine,* February 21, 1974.
67. 1979 Fluid Power Reference Issue, *Machine Design Magazine,* September 27, 1979.
68. Allan W. Scott, *Cooling of Electronic Equipment,* John Wiley & Sons, 1974.
69. Industrial Ventilation, American Conference of Governmental Industrial Hygienists, Lithoprinted by Edwards Brothers, Ann Arbor, Mich., 1952.
70. General Electric Heat Transfer Data Book, General Electric Co., Schenectady, N.Y., 1975.

Index

Absorptivity, solar: color effects, 134
  various materials, 136
Acoustic noise, 22, 175
Adhesives: conformal coating, 79, 80
  double-back tape, 79
  epoxies, 79
  RTV cement, 79, 233
AGREE test, 312
Air: density, 188, 199, 202, 264
  gap, 65, 81, 257, 282, 287
  molecule, 30
  pressure versus altitude, 200
  properties, 202
  seal, 29
Air-flow path, 9, 181, 261
Air-flow network, series and parallel, 294
Air-flow reversal, 166, 239
Albedo, 7, 147
Aluminum heat sink, *see* Heat sink
Amplifier: hot spot temperature, 324
  thermal capacitance, 327
  transient thermal, 324
Analog resistor network, 49, 53, 295
  natural convection, 121
ATR boxes, 6, 34
Auxiliary cooling, 5

Back side PCB, 3, 125, 179, 255, 277, 292
Black body, *see* Radiation
Board, printed circuit, *see* PCB
Bolted assembly: covers, 72, 244
  interface, 64
Bracket: cantilevered, 39
  component, 36, 39
  transistor, 88
British thermal unit (Btu), 14

Cabinet: chimney effect, 260
  electronic enclosure, 263, 279

flow losses, 261
flow restrictions, 267
flotation pressure, 262
impedance curve, 261
induced draft, 262
natural cooling, 260, 274, 279
PCB convection coefficient, 277
static loss factor, 268, 276
temperature rise, 267, 276, 279
warning note, 272
Calorie, 14
Capacitance, *see* Thermal capacitance
Card bucket, 274
Cassette tape, 11
Castings: die, 25
  dip-brazed, 22
  investment, 24
  large, 26
  plaster mold, 24
  sand, 25
Change of state, 5
Chassis: air-flow, 172
  die cast, 25, 243
  dip-brazed, 7, 22, 220
    plugged fins, 220
  impedance curve, 173, 189
  investment cast, 24
  sheet metal, 21, 243
  temperature cycle test, 312, 317
Circuit lamina, 47, 190, 231
Cleaning, cold plates: dip-brazed fins, 23
  salts, dip brazing, 23
Cold plate, 1, 6, 23, 82, 231
  cleaning, *see* Cleaning, cold plates
  fins, air-cooled, 219, 224
  liquid-cooled, 86, 346, 351
Communication shelter, 9
Components, electronic, 1, 39, 54, 78, 88, 99, 102, 133, 232, 256, 290, 305, 338, 353

365

case temperature, 159, 231, 252
heat density, max, 79
liquid cooling, 344
mounting, 87, 92, 95, 99, 101, 102
suspended mounting, 158
see also High power component
Conditioned cooling air, 207
air-flow curve, 210
Conductance: interface, 61, 65, 67
thermal, 54, 56, 62, 325
Conduction heat transfer, 3, 35, 53, 79, 354
component, 78
lead wire, 3, 255, 277, 284
two-dimensional, 54
Conductivity, thermal, various materials, 37, 38
Conformal coating: acrylic, 28
bridging strain relief, 28
epoxy, 28, 79
moisture resistance, 28
polyurethane, 28
silicone, 28, 79
Connectors: edge, printed circuit, 243
plug-in, 77
Console, *see* Cabinet
Convection: forced air, 176, 229, 251
induced draft, 260
liquid cooling, 348
natural air, 106, 109, 244, 260
table, geometric shapes, 205
Conversions, English and metric: area, 19
density, 17
heat-transfer coefficient, 15
length, 18
power, 14
pressure, 17
specific heat, 16
thermal conductivity, 15
viscosity, 16
volume, 20
weight, 18
Cooling air: conditioned, 207
flow reversal, 166, 239
forced convection, 176, 229, 251
natural convection, 106, 109, 244, 260, 268
temperature rise, 111, 175, 176, 252, 267, 276, 287
weight flow, 175, 208, 266, 296
Cooling cart, 5
Copper circuit, 44, 233, 290
Copper heat sink, 41, 78
Copper thickness, 44

Covers, sheet metal, 72, 244

Darcy equation, 221, 222
Density, *see* Air density
Desiccators, 32
Differential equation, 43, 315
Diode, 1, 39, 41, 77
DIP, 1, 77, 242
Dip-brazed chassis, *see* Chassis
Double-back tape, 79
Dry nitrogen, 31
Duct: air entry loss, 196
heat exchanger, 227
PCB, 185, 251, 285

Edge guide: altitude effects, 71
PCB, 70, 84
thermal resistance, 71
Effective length: pipe diameters, 342, 343, 359
pipe entrances, 343
pipe fittings, 342, 352, 359
Elbow losses: air flow, 197
liquid flow, 342, 359
Electronic box, 3, 22, 119, 126, 142, 154, 155, 173, 199, 208, 219, 317
Electronic enclosure, *see* Cabinet
Electronic equipment: airplanes, 6, 155
missiles, 6, 150
satellites, 6, 147, 150
ships, 8
spacecraft, 6, 147
submarines, 8
EMI, 8, 282
Emissivity: color effects, 134
combined, 137, 145
fin, effective, 161
geometric form, 137
metallic surfaces, 133
various materials, 135, 136
English units, 14
Entry losses: air flow, 196
liquid flow, 343
Epoxy: structural, 22
fiberglass, 48
Equivalent ambient, 157
Equivalent length, *see* Effective length
Extrusions, 26
porosity, 33

Failure rates, 157
Fan: air velocity, 170
axial flow, 175, 192
bleed air, 193

Index

blowing system, 165, 174
centrax point, 193, 194
cooling air direction, 165
entrance loss, 170, 182, 196
exhaust, 165, 183
flow impedance curve, 172, 181, 189
life, 243
minicomputer, 242, 249
muffler, 175
short circuit air flow, 166, 167
squirrel cage blower, 175
static head loss, 167, 170, 181
Fanning friction factor, 222, 276
Fiberglass, 48
Filters, air, 194
Fins:
    convection choking, 118
    convection film, 119, 220
    efficiency, 117, 120, 223, 229
    heat transfer, 121
    mathematical model, 122, 123
    natural convection, 117, 124
    radiation, 161
    spacing, 118
Flat pack, see Integrated circuit
Floppy disk, 11, 244
Flotation pressure, 262
Flow losses, air: chassis, 181
    contractions, 181, 198
    ducts, 182, 185, 196, 227, 251, 285
    exits, 165, 183
    expansions, 183, 198
    Fanning friction, 222, 276
    finned cold plates, 221, 237
    Hagen friction, 222
    inlets, 170, 182, 196
    sigma ratio, 210, 213
    sigma slope, 212, 216
    static, 170, 181, 187, 195, 209, 212
    through smooth pipes, 223
    tall cabinets, 261, 267, 273
    transitions, 184, 197, 198
    total, 169, 170, 219
    turns, 184, 197
    various altitudes, 196, 199, 209, 212
    velocity, 166, 168, 170, 186
Flow losses, liquid: elbows, 342, 359
    entry, 343
    Fanning friction, see Friction factor
    Hagen friction, 350
    pressure drop, 347, 357, 360
    tee's, 343
    valves, 342
Fluids, cooling, table, 345

Forced convection: cooling methods, 164, 173, 189, 192, 199, 205, 207, 225, 249
    convection film, 176, 202, 204, 229, 251
    fan performance, 165, 168, 171, 173, 181, 183, 189, 192, 194, 199
    liquid cooling, 348, 351
Free convection, see Natural convection
Friction factor: Darcy equation, 221, 222
    Fanning, 223, 273, 276
    geometric forms, 223
    Hagen Poiseulle, 222, 350

Gasket: flat rubber, 30
    cover, 31
    "O" ring, 29
    seal, 29, 31
Grashof equation, 107
Ground support system, 9

Hagen Poiseulle friction, see Friction factor
Heat balance, 42, 315
Heat exchanger: air-cooled, 226, 231
    hollow core PCB, 191, 226, 231
    water-cooled, 341, 347, 351
Heat flow: to back side of PCB, 3, 125, 179, 255, 277, 292
    balance, 42, 315
    radial, 73
    two-dimensional, 54
Heat pipe: applications, 336
    burn in, 334
    construction, 331
    contamination, 334
    degraded performance, 333
    performance, 334
    wicks, 332
Heat sink: aluminum, 46, 79, 81, 305, 317
    copper, 44, 54, 233
    finned, 219, 305
    internal, 94, 95
    warping, 47, 61
Heat sources, 1
Heat transmission, 2
High altitude: fan cooling, 196, 199
    interface conductance, 66
    natural convection, 129
    pressure loss, 209, 212
High power component: electrically isolating, 91
    insulating washer, 87, 91
    mica washer, 41, 93
    mounting, 87, 89, 346
    stud mount, 41

vibration and shock, 87
Hollow core PCB: air seal, 190
  with dip-brazed fins, 220
  forced convection, 191, 225, 227
  lap-soldered components, 190
  pressure drop, 236
  vibration, 191, 192
Humidity, 6, 26, 30
Hybrids, 1, 79, 225, 232
Hydraulic diameter: air cooling, 178, 251, 276
  liquid cooling, 348, 351
Hydraulic radius, 351

Impedance curve: fan cooling, 172, 181, 194
  induced draft cooling, 261
Induced draft, 260, 268
Insulators: beryllium oxide, 91
  epoxy fiberglass, 92
  hard anodized aluminum, 92
  silicone rubber, 91
  washers cracking, 87
Integrated circuit (IC), 1, 46, 80, 255
Integrating factor, 315
Interface, thermal: bolted, 64
  clamping force, 63
  conductance, 65, 68
  copper foil, 69
  high-altitude, 66, 73
  resistance, 41, 71, 319, 323
  riveted, 63
  spot-welded, 63
  surface finish, 65

Junction temperature, 88, 146, 207

Laminar flow, 111, 113, 177, 223, 230
Laminated PCB, 46, 78, 82, 231
Lead wire, strain relief, 28, 99
Lead-wire conduction, see Conduction heat transfer
Liquid cooling: cold plate, 82, 341, 346, 347, 351
  coolants, 341, 345
  direct, 340
  expansion tank, 343, 344
  heat-transfer coefficient, 340, 348, 356
  indirect, 340
  pipe fittings, 342, 343, 359
  pumps, 341, 342
Logarithmic mean area, 74
Louvers, air, 174, 243
LSI, 1, 77, 232, 242

Maintainability, 22
Mathematical model, 51, 53, 55, 97, 122, 123, 127, 245, 269, 283, 285, 295
Metric units, 14-20
Microcomputer, 10, 242, 252
Microprocessor, 10, 77, 242
  vibration, 104, 256
MIL handbooks: MIL-E-5400, 199
  MIL-HDBK-217B, 157
Minicomputer, 10, 242
Moisture, see Humidity; Water

Natural convection: characteristic length, 108
  development, 106
  enclosed air space, 125
  finned surface, 117, 119
  general equation, 107
  geometric constants, 108
  heat transfer, 110, 115, 125
  high-altitude effects, 129, 130
  horizontal plates, 109, 115
  laminar flow, 109
  PCB cooling, 123, 130, 245, 274
  with radiation, 154
  turbulent flow, 113
  vertical plates, 109, 111, 114
Node, 53
Nonuniform sections, 48, 50
Nusselt number, 107

Open-loop cooling, 8
Outgassing, 69

Parallel networks: air flow, 293, 294
  thermal resistors, 49, 51
PCB, 41, 44, 46, 71, 78, 123, 189, 244, 250, 253, 257, 274, 283, 319, 324
PCB vibration, see Vibration
Perforations, air, 243
Pipe fittings, liquid cooling, 342, 343, 359
Plate fins, 23, 79, 119, 221, 224
Plated through hole, 79, 81, 231, 233
Porous casting, 31
Potted module: aluminum filled, 94
  conformal coating, 95
  internal heat sink, 95
  machining, 94
  short circuits, 94
Power dissipation: components, 35, 39, 41, 46, 54, 65, 81, 89, 95, 173, 274, 283, 351
  density, 232
  specification, 12

Index

Prandtl equation, 107
Prandtl number: air, 202
  water, 353, 354
Prepreg cement, 48, 79, 231
Pressure drop, *see* Flow losses
Pressure relief valve, 29
Pump, liquid cooling, 342

Rack, *see* Cabinet
Radiation: black body, 133, 161
  between components, 142
  conformal coating effect, 143
  dual FET switch, 145
  electronic box, 150, 152, 244
  enclosed module, 283
  fins, 135, 161, 163
  grey surface, 134
  heat transfer, 134, 144, 152
  hybrid, 1, 78, 232
  minicomputer, 242
  with natural convection, 154, 245
  ratio α/e, 136, 148
  sealed enclosure, 283
  simplified equation, 151
  space, 147, 148
  temperature effects, 133
  various latitudes, 148
  various planets, 148
  view factor, 136, 153
Ram air, 7
Reliability, 157, 331
Resistance: component mounting, 78, 80, 83
  plated through hole, 38, 79, 231
  thermal, 48, 51, 231, 245, 283, 285
Resistor: component, 1, 51, 54, 77, 80, 159
  equivalent ambient, 157, 159
  network, parallel, 49, 51, 55, 245
  network, series, 49, 245, 283
Restrictions, air flow: contractions, 181, 183, 221, 236, 239, 261, 269
  electronic chassis, 166, 181
  tall cabinets, 261, 269
Reynolds number, 178, 201, 206, 228, 251, 276, 350
RFI, 8, 283
RTV cement, *see* Adhesives

Satellite electronics, 5, 6
Seals: covers, 29, 30
  electronic boxes, 8, 26, 29
  enclosures, 278, 281
  flat rubber gaskets, 30

"O" ring, 29, 31
  solder, 31
Series networks: air flow, 294
  thermal resistors, *see* Resistance, thermal
Shake and bake test, 312
Sheet metal chassis, *see* Chassis
Ships electronics, 8
Shock, 36, 104
Silicone grease, *see* Thermal grease
Simultaneous equations, 56
Solar absorptivity, *see* Absorptivity, solar
Solder: fatigue failures, 104, 256
  plated through holes, 98
  stress, maximum allowable, 98, 103
  wicking, 100
Solder seal, 31
Space vacuum, 7, 8, 67, 133, 147
Specific heat: air, 202
  metals, 329
  nonmetals, 329
Static head, 166
Static loss, 170, 181, 187, 213, 271
Static pressure, 166
Stefan Boltzmann constant, 134
Steady-state heat transfer, 4
Strain relief: air gap, 65, 81, 257
  component lead wire, 28, 99, 102
Submarine electronics, 8
Surface finish, 7, 62
Switches: cut off, 195
  thermal, 35, 196

Temperature: cabinet, 260, 267, 274
  case, 130, 159, 231, 236
  cycling, 2, 257, 313, 317, 323
  junction, 88, 146, 207
  liquid cooled systems, 347, 351
  parabolic distribution, 42, 45
  rise, 39, 45, 46, 64, 83, 91, 131, 174, 201, 232, 250, 267, 276, 281, 287, 305, 352
  transient, 312
  uniform distribution, 41
Thermal capacitance, 5, 7, 302, 327
Thermal conductivity: air, 202
  metals, 37
  nonmetals, 38
Thermal expansion, 47
Thermal grease, 7, 41, 66, 88, 351
Thermal inertia, *see* Thermal capacitance
Thermal resistance, *see* Resistance
Time constant, 303
Transient analysis:
  amplifier, 324

cooling cycle, 310
heating cycle, 304
insulated system, 300
temperature cycling, 312
thermal capacitance, 5, 7, 302, 327
time constant, 303
transformer heating, 301
transistor heat sink, 304, 310
Transistors: cooling cycle, 310
cooling TO-5, 205
heating cycle, 304
stud mount, 41, 88
Turbulent flow: air cooling, 113, 230
liquid cooling, 349, 355
Two-dimensional heat flow, *see* Heat flow

Vapor, water: bolt seals, 31
flat gasket protection, 30, 31
gas laws, 30
"O" ring seal, 29, 31
partial pressure, 30
Velocity head, *see* Velocity pressure

Velocity pressure, 166
Vibration: electronic components, 36, 104
microprocessors, 105, 256
View factor: cross string method, 137
line of sight, 142
parallel plates, 139
perpendicular plates, 140
Viscosity: air, 202
fluids, 345
water, 353, 354
VSI components, 1, 77, 232, 242

Warping heat sink, *see* Heat sink
Water: in cooling air, 189
drainage path, 27
properties, 353, 354
vapor, 30
Watt, heat conversion, 14
Wax, melting, 6
Weight velocity flow: air, 178, 201, 251
liquids, 349